T0183053

Brewing Science: A Multidisciplinary Approach

Michael Mosher • Kenneth Trantham

Brewing Science: A Multidisciplinary Approach

Second Edition

 Springer

Michael Mosher
Department of Chemistry and Biochemistry
University of Northern Colorado
Greeley, CO, USA

Kenneth Trantham
Department of Physics and Physical Science
University of Nebraska at Kearney
Kearney, NE, USA

ISBN 978-3-030-73421-3   ISBN 978-3-030-73419-0   (eBook)
https://doi.org/10.1007/978-3-030-73419-0

This Springer imprint is published by the registered company Springer Nature Switzerland AG
The registered company address is: Gewerbestrasse 11, 6330 Cham, Switzerland

Preface

To the Student

What do your professors do at the end of a day at college? Many, the authors included, will enjoy a craft brew and discuss the day's successes and opportunities for improvement. Thus began the discussion for the beginnings of this book. "Wouldn't it be awesome," we thought, "if we could highlight the science that goes into brewing? The students would just love a brewing science class!"

And that is what you'll find here. This text represents the topics that are taught in our courses in the Introduction to Brewing. These courses are very popular at our institutions. We've tried to write from your perspective and provide not only the processes that you'll encounter in the brewery, but also provide the reasons why those processes are completed and the science behind them.

Every few pages, you'll find CHECKPOINT boxes. These are designed to provide you with a chance to take a break and confirm that you've gathered the key topics of the discussion to that point. This is also how we've constructed the images that accompany the discussion. When they appear in the text, it is important to take a break from reading and examine the figures in detail. (Some professors, the authors included, find great quiz and test questions by looking at the figures.)

We sincerely hope that you enjoy your studies of this exciting topic. One thing you'll note from the start of your reading, brewing science requires an understanding of a wide range of topics from biology to chemistry to physics to history to almost every subject taught on campus. The purpose of the book is not to make you, the student, a physical chemist or a fluids engineer, but to give you a sense of what is possible in the brewery. And, it will provide you with an understanding behind why things are done the way they are in the brewery. Brewing science can be very technical, but our hope is that you find the subject just as fascinating as we do.

To the Instructor

The first incarnation of the course described by the topics in this text was directed at the general studies level. The science discussed in that type of course is more descriptive and general in nature. We found that the class attracted a wide range of majors with varying interest levels. This class is still taught at the University of Nebraska Kearney. However, we have found that the class also attracted those with more than a passing interest in brewing and wanted to dive deeper into the rich science that surrounds the craft brewing industry. So, we've included the detail that is appropriate for those courses that do this, such as the course taught at the University of Northern Colorado.

The text is written from a process-centric approach to uncover the principles behind brewing science. Instead of a discussion of brewing from the perspective of the four main ingredients (water, malt, hops, and yeast), this text is formatted and written from the perspective of the steps taken to manufacture beer (malting, milling, mashing, boiling, etc.). The topics are focused more on the technical aspects and design principles of brewing. As the students uncover the process of mashing, they explore the background chemistry needed to fully develop their understanding. As we explore wort chilling, we dive into the background in thermodynamics that explains this process. Thus, students learn what they need to understand as they need to know it.

We hope that this text will provide you, the instructor, with the greater detail needed behind each of the processes in the brewery and the insight into the interrelationships between the individual processes. We realize that there are parts of the book that may be mathematically challenging to a general audience. But, the language of science is mathematics – and with practice and motivation to be successful, the general audience can succeed.

Within each chapter are CHECKPOINT questions that provide key questions that students should be able to accomplish. At the end of each chapter are questions that expand upon these in-chapter questions. The summary section at the end of each chapter is also helpful in directing students as they move through the text.

Finally, each chapter contains at least one laboratory experiment that can help explain the material in the chapter. Both of the authors' courses in this subject have related laboratories that we've noted are extremely useful in developing student interest and motivation, and providing confirmation of topics in the course. Additional "laboratory experiments" can be obtained by modifying the laboratory analyses found in the American Society of Brewing Chemists Methods of Analysis resource.

It is our sincere hope that you, the instructor, find the information in this text to be helpful to you and your students irrespective of the level of your introductory course in brewing science. As the standalone text, or used in conjunction with handouts and additional readings, the material inside should be helpful to your students. Whether they are beginning their studies for a diploma in brewing from the Institute of Brewing and Distilling, satisfying a general studies requirement, or reading for interest, the student is sure to find interest in this topic.

Greeley, CO, USA   Michael Mosher
Kearney, NE, USA   Kenneth Trantham

Preface to the Second Edition

As our courses in brewing science have matured and expanded, we noted that the emphasis on certain topics within the field has changed. In addition, there were many additional topics that we wanted to add to the text. Hence, the construction of the second edition of this text was not only recommended by our students, but also needed by our students as they studied for exams.

There are a number of changes to the text from the first edition. These changes focus on the addition of a new chapter on clarification and filtration techniques and technologies, more information on food safety and best practices, and an enhanced section on the history of brewing beer.

Along the way, we've added more detailed images to show cutaway views of the equipment used in brewing. These images allow us to dive inside the different vessels and see where things are located and how the processes work. Our hope is that you, the reader, will benefit from these images and better understand how a modern brewery works.

The CHECKPOINT boxes remain throughout the text. These boxes ask key questions to confirm that the major points are finding their way into your toolbox as you read each chapter. Further checks include the questions at the end of the chapters. These can be used as homework questions or as jumping-off points to further explore the major topics of each chapter. To round off each chapter, you'll find laboratory experiments that help those who prefer a hands-on approach to learning about the subject.

As before, we hope that you find this text to be a good introduction to the field of brewing science. The subject is very interesting and with the sheer number of different topics (from biology to chemistry to physics to engineering) there is surely something here for everyone.

Greeley, CO, USA   Michael Mosher
Kearney, NE, USA   Kenneth Trantham

Contents

Introduction to Brewing Science

1

1.1 Science and the Brewer

Master brewers know a lot about the process and the product that they make. In most cases, they've spent a considerable number of years at their craft, experimenting with different malts and grains, different yeasts, and different processes until they've arrived at what they consider to be the perfect beer. Any courses or training that the master brewer takes are extremely rigorous, often involving multiple months or years of intensive study and potentially even an apprenticeship under an experienced master brewer. Those years of training to master the art of brewing beer require not only mastery of the processes and recipes, but also an understanding of how the science behind the process results in a particular flavor or product profile.

Many brewers would agree that knowing the science of brewing is important to the process of brewing beer. Not only does the science govern how hop oils protect beer from minor spoilage or how barley must be sprouted before it can be used to make beer, but the general principles used to practice science guide the brewer every day. What principle helps a brewer make everyday decisions?

1.1.1 The Scientific Method

The *scientific method* is a process of thinking about problems. The method is outlined in Fig. 1.1. Brewers, or anyone for that matter, that use this process methodically arrive at the answer to a problem, or develop a law or theory based on observations. This is the same way of thinking about the world around us that scientists use every day. An example helps guide us to understanding the scientific method.

A brewer starts with an observation. For example, suppose they notice that a recently brewed batch of beer has a buttered popcorn flavor. The brewer would then develop an educated guess, called a hypothesis, which would attempt to explain the origin or cause of the off-flavor. For example, the brewer may hypothesize that the

© The Author(s), under exclusive license to Springer Nature Switzerland AG 2021
M. Mosher, K. Trantham, *Brewing Science: A Multidisciplinary Approach*,
https://doi.org/10.1007/978-3-030-73419-0_1

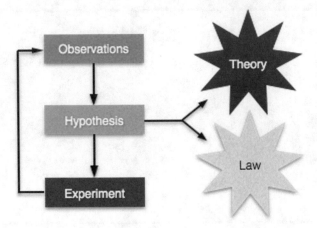

Fig. 1.1 The scientific method. A researcher makes observations, develops a hypothesis, and then tests the hypothesis with carefully designed experiments. This is a cyclical process that eventually results in a proven hypothesis that can either be called a theory or a law

temperature of the water used in the process was too hot. Then, the brewer would perform an experiment to try to eliminate that flavor, in this case by reducing the temperature in the next batch of beer. After making more observations (tasting the beer, running laboratory analysis of flavor components, etc.), the brewer would then compare those results to the original hypothesis. If the observations did not fit the original hypothesis, the brewer would modify the hypothesis and return to the brewery to perform another experiment. Then, they would modify or re-create the hypothesis and test it out in the brewery. This cyclical process would continue until the hypothesis did not need modification after repeated experiments. The brewer would have found the answer to the observation and know what to do if that situation ever arose again. When the hypothesis is confirmed in such a manner, the hypothesis is elevated to become a law or a theory.

A law is a tested and proven hypothesis that explains the initial problem. Laws do not explain why the problem occurs, but simply what happens. In our previous example, the brewer may eventually discover after multiple experiments that when the serving tap for the beer is cleaned immediately before use, the off-flavor disappears. The brewer's hypothesis would then become a law: Cleaning the beer tap removes the buttered popcorn flavor in the beer.

A theory, on the other hand, is a tested and proven hypothesis that explains why something happens. Theories are the most definitive statements that can be made. They are not just statements that identify the outcome of a particular step in the brewery. They predict the outcome by providing a detailed explanation of why that outcome occurs at any level of observation. To a brewer, and any scientist, a theory is the best statement that one can have about a process. Let's say, for example, the brewer does experimentation that shows bacteria in the tap line cause the buttery flavor in the beer. The hypothesis then becomes a theory: Because there are bacteria in the tap line, they cause the off-flavor in the beer; cleaning the line will remove the bacteria and remove the off-flavor.

In our daily practice as brewers, we use the scientific method as we explore issues surrounding brewing science. In this text, we'll learn some of the existing theories about the science, and use the scientific method to suggest additional directions that the brewer could go to advance their understanding of the process. But to get to that point, we need to take a brief look at where brewing started and how it became the process that it is today.

CHECKPOINT 1.1

In your own words, what is the difference between a law and a theory? Why would a brewer likely be satisfied with a law instead of a theory?

1.2   What Is Beer?

The word "beer" has many possible origins. Most likely this word derives from the Middle English word *bere* or from the Old English word *bēor*. The Old High German word *bior* may also be the precursor, as could the Middle Dutch word *bēr*. Just from looking at the origins of the word, we can easily observe that the word "beer" has roots in Europe from at least as early as the Middle Ages (fifth to fifteenth century). It turns out, the drink we call beer is much older than that. However, the words from the Middle Ages that gave us "beer" referred to a fermented drink made from malted cereals and flavored with a myriad of different ingredients. In some cases, roots or other starchy plant materials instead of cereal grains were used to make the drink.

Beer is vastly different from wine. In short, beer is brewed and wine isn't. Brewing, the process of converting starches into sugars that can be fermented, requires many more steps than just making wine. In other words, the starch found in grains and other materials must first be converted into sugars, and then those sugars can be fermented into an alcoholic beverage. Because fermentable sugars already exist in fruits, a fermented beverage made from fruits is not brewed at all, but still results in an alcoholic beverage. If grapes are selected as the fruit for the drink, we know the resulting fermented beverage as wine. In addition, mead (honey), cider (apple), cyser (a mix of apple and honey), and perry (pear) are other examples of beverages made by a process that begins with the existing sugars. The popularity of kombucha has increased dramatically in the last 5 years, but it isn't brewed. Instead, it is a mixed culture fermentation of a sweetened tea. These beverages do not require brewing. The sugars that will be consumed by yeast and other microbes are already present.

No one knows where or how the first beer was brewed. However, as beer moved across the world, different cultures came up with different ways to brew it. Each group used a recipe that, in some cases, vastly differed from others. Some made beer from barley; others used roots, wheat, oats, rice, and other grains to create the beverage. Flavoring agents, such as fruits, herbs and spices, that were added to the drink were equally as varied in those early days. In many cases they were completely absent from the finished product.

The process by which beers were initially made also differed from people to people. As we'll discover later in this chapter, one early culture used bread to start the process of brewing. Some cultures linked beer brewing closely with religion, with very strict rules on how it was to be made. Other cultures considered brewing as part of the daily ritual of food preparation with a process that was different from home to home. In many cases, the specific steps were dependent upon the types of equipment, jars, vessels, and mats that were available. And, where you lived in the world also dictated much about how you could brew your beer. Brewers in hot, dry climates used very different techniques than those found in cooler or wetter regions.

The multitude of factors led to tremendous number of beverages across the world. The flavor of beer was varied. In addition, the alcohol content in beer varied. Some early beers (and some beers that are still found today) had very little alcohol content, while others were quite potent. It all depended upon the process, the availability of ingredients, and the tastes of the peoples that made the drink.

CHECKPOINT 1.2

A brewer makes an alcoholic drink from yams. Is this a beer or a wine? Why? Why was beer not made with the same ingredients in different places of the world?

1.3   Some Common Conventions

In our studies of beer, we have to make sure that we can communicate with other brewers, malt and hop suppliers, yeast producers, customers, and tax enforcement officials correctly and accurately. To do so, a series of conventions have been established that help us in our conversations. These conventions ensure uniformity in reporting the values associated with volume, temperature, and weight measurements.

Unfortunately, this uniformity did not always exist. In the early years of brewing, descriptions of the amount of grain, the amount of water, and the amount of added fruits or herbs were akin to a trade secret. The brewer had a serious problem. How do you tell someone how to make beer but avoid using words such as kilogram, pound, or gallon? So recipes, if they existed, were developed using whatever measuring device was available. For example, if a certain amount of malt was to be added to the mashing vessel, the quantity might be described as equivalent to the weight of a person, or the same as would be used to make a dozen loafs of bread. It's easy to see that without a standard to compare, recipes were difficult to follow and very rarely was the same flavor of beer replicated.

The issue in reporting specific values lies within three main categories. An accurate recipe must be able to provide the brewer with ways to measure the volume, the temperature, and the mass of the ingredients. While the initial measurements might be done using reference to known things, once a standard was developed, the recipes were able to be accurately reproduced.

1.3.1   Volume

The volume of a liquid is the space that it occupies. That space in the United States is often referenced in terms of the number of fluid ounces, cups, gallons, quarts, etc. In the scientific world, volumes are referenced in terms of the dimensions of a cube that could hold the liquid. Reporting the volume in scientific terms means knowing the dimensions of the cube and describing it using the distances on each side. For example, a cube that is 2.0 inches by 2.0 inches by 2.0 inches is said to be 8.0 in^3 or 8.0 cubic inches:

$$2.0\,\text{in} \times 2.0\,\text{in} \times 2.0\,\text{in} = 8.0\,\text{in}^3$$

In the scientific world, a set of standards are used rather universally. Those standards are referred to as the Système international d'unités (in English, the International System of Units, or SI system). Distances in SI units are recorded using base units of meters. A meter is slightly more than the US yard, so the volume described by a cubic meter is a very large measurement. For example, in the United States, we still order concrete using the units of cubic yards (a volume of concrete that is 3 ft. by 3 ft. by 3 ft). To reduce this down to something more manageable, scientists often talk about volumes in terms of liters. A liter is the volume represented by a cube that is one tenth of a meter on each side (0.1 m by 0.1 m by 0.1 m), and is close to the same size as the US quart.

The use of SI units outside of the United States is very common. Brewers use liters to describe the volume of beer that they produce; or more accurately, brewers describe the volume of beer that they produce in units of hectoliters, where 1 hectoliter is equivalent to 100 liters. But in the United States, volumes are still reported using gallons and many of the older English measurements such as quarts, pints, and tuns. These were originally determined based on the different sizes of casks and other containers that were used to hold beer. Some of these units may seem foreign, but knowing them can be very helpful.

We can still measure the volume using a ruler and multiplying to get the number of cubic inches. All that is necessary at that point, then, is to convert the number of cubic inches to gallons or barrels or whatever other unit we wish to know. Table 1.1 lists some common volumes of beer and their relationship to the number of equivalent cubic inches and US gallons. The number of cubic inches is supplied in the table to aid in conversion from one unit to another, and to show just how large some of the measurements of volume can be.

It is important to note that the volume measurements outlined in Table 1.1 are specific to beer or ale. Believe it or not, if we talked about a different liquid or even

Table 1.1 Common conversions with beer volumes

UNIT	Number of Cubic Inches	Number of US gallons	Equivalence
1.0 liter	61.02	0.264	
1.0 hectoliter	6102	26.4	100 liters
1.0 gallon (US)	231.0	1	4 quarts
1.0 quart	57.75	0.250	2 pints, 64 oz (US)
1.0 pin *	1248	5.40	
1.0 firkin *	2497	10.8	2 pins
1.0 kilderkin *	4993	21.6	2 firkins
1.0 barrel (UK) *	9987	43.2	2 kilderkins
1.0 barrel (US)	7161	31.0	2 kegs (US)
1.0 puncheon *	19971	86.4	2 barrels (UK)
1.0 tun *	59912	259.2	3 puncheons or 2 butts
1.0 hogshead *	14978	64.8	6 firkins
1.0 butt *	29956	129.6	2 hogshead
1.0 keg (US)	3581	15.5	0.5 barrel (US)
1.0 quartel (US)	1790	7.75	1 pony keg
1.0 sixtel (US)	1194	5.17	1/6 barrel (US)

Note: The values marked with an asterisk reflect, where possible, the use of the most current usage of the imperial system. For example, the ale firkin until 1688 was equivalent to 2256 cubic inches, from 1688 to 1803 it was 2397 cubic inches, from 1803 to 1824 it was 2538 cubic inches. In 1824 it was defined in imperial gallons equivalent to 2497 cubic inches

a solid material like flour, our volume measurements would be completely different. What is very apparent is that there are a lot of different volume measurements for beer. And from the note to the table, volume measurements for beer were redefined a few times since they were introduced. Moreover, we can see that the current names for many of these units have been used for quite some time to relate quantities of beer.

There is more than one volume measurement known as a barrel. The two most common for beer are indicated in Table 1.1, but there are a lot of other definitions for the barrel (and as we noted above, the volume actually depends upon what you're measuring). The myriad of volume measurements is due to the long history of beer; different countries used different sized barrels to measure their liquids based on their own customs and how things were taxed. Standardization of some units, in fact, didn't occur until recently.

In the US brewing industry, the focus on beer volume measurement is on the use of the barrel (US) and smaller sizes. A small microbrewery in central Colorado may describe the amount of beer that they produce in each batch as 7 bbls. In the United Kingdom, this would be roughly equivalent to the brewery that manufactures 8.2 hectoliters. Smaller measurements in the United States are useful when the beer is sold to the consumer. These smaller volumes include the keg (1/2 of a barrel known also as a "full size keg"), the pony keg (1/4 of a barrel also known as a quartel), and the sixtel (1/6 of a barrel). The pony keg, typically a short wide container, is referred to as a slim quarter if the container itself is designed to be tall like a soda pop keg.

Many US homebrewers that keg their beer rather than bottle it prefer to use reconditioned or new soda pop containers known as Cornelius kegs after the company that originally manufactured these containers for the soda industry. "Corny" kegs come in many sizes and range from 2 to 10 US gallons (with the 5-gallon version being the most common). These kegs are rarely used in the brewing industry due to differences in the way beer is added and withdrawn from the containers. Cleaning and other issues with the soda kegs make them perfect for dispensing soft drinks, but not very suitable for the brewer.

While we often don't consider it as a dry material volume measurement, the bushel (bu) is just that. This unit is used quite heavily in the United States, and still has use elsewhere in the world. Unfortunately, even today, the bushel is a different measurement based upon what you are measuring and in which country you live. In the United States, a bushel of barley is 48 pounds, a bushel of malt is 34 pounds, and a bushel of corn is 56 pounds. In Canada, the bushel of oats (34 pounds) is a little heavier than in the United States (32 pounds). In the past, a bushel even was considered a different weight depending upon the state where you measured it. Iowa barley was 48 pounds to the bushel, yet, in Illinois the bushel was 44 pounds.

CHECKPOINT 1.3

How many pins are there in a kilderkin?

A startup brewery decides to make their beer in 5.3 barrel (US) batches. How many gallons will they make per batch? How many hectoliters is this?

1.3.2   Temperature

The temperature of the ingredients during the brewing process is very important. Temperature is a measure of the amount of heat, or thermal energy, contained by a substance. Things that are hot are said to have a large amount of this energy and things that are cold have less. When something has no thermal energy at all, it has the coldest temperature possible. This is known as absolute zero. At the other end of the scale, however, there is essentially no upper limit. We'll explore heat in much greater detail in Chap. 8.

The brewer prior to the invention of the thermometer definitely needed to know how much thermal energy was present during the brewing process. Recipes dating from the seventeenth century (prior to the use of the thermometer in brewing) often refer to an ingenious way to obtain a somewhat reproducible temperature for mashing. In that method, water was heated to boiling and then allowed to cool slowly. When the brewer could see his face in the liquid through the steam, it was deemed to be at the right temperature. This often resulted in a temperature very near to 150 °F – 160 °F, almost the perfect temperature needed for mashing malt.

In the early days of the development of science as a discipline, fascination with temperature was common among practitioners. Each was interested in devising an instrument to measure temperatures in the most accurate and practical method possible. In fact, there are at least 52 different thermometers that were invented. While most of these are no longer used today, a couple of them are used very heavily across the world.

Scientists, as we noted before, tend to prefer units and measurements that are outlined in the SI unit definitions. For temperature, this includes the use of the units in Celsius (°C), named after Anders Celsius (1701–1744), and Kelvin (K), named after William Thompson, Baron Kelvin (1824–1907). The Celsius scale places the temperature at which water freezes at 0 °C and the temperature at which water boils at 100 °C. The Kelvin scale simply adds 273 to these numbers so that the coldest temperature possible, absolute zero, is at 0 K, freezing water is at 273 K, and boiling water is at 373 K. Note that the Kelvin scale does not include the degrees symbol (°), and that reporting values in this unit one does not say the word "degrees." For example, if we report the value of 273 K, we say out loud "two-hundred seventy-three kelvin."

The brewer, especially in the United States, might use the Fahrenheit scale named after its inventor, Daniel Gabriel Fahrenheit (1686–1736). The definition of this scale is such that water freezes at 32 °F and water boils at 212 °F. Luckily, a very simple conversion between Celsius and Fahrenheit allows brewers to communicate with each other no matter which scale they prefer:

$$^\circ F = \left(1.8 \times {}^\circ C\right) + 32$$
$$K = {}^\circ C + 273$$

For the brewer, temperatures are often measured from just below the freezing point to the boiling point of water. The freezing point is simply the temperature at which a liquid becomes a solid. The boiling point, on the other hand, is the temperature where the vapor pressure of the liquid equals the pressure of the atmosphere.

As the temperature of any liquid increases, the amount of vapor above that liquid also increases. For example, when we heat a pot of water on the stove, the amount of steam above the liquid increases as the water gets hotter. When the pressure of that steam is equal to the pressure of the environment, the water is said to be boiling. Note that the boiling point is not the temperature where we see bubbles forming in the liquid. That temperature is close to, but not the same as the boiling point of the liquid. In fact, the bubbles we do see are actually steam (not bubbles of air) where locally, the water is hot enough to boil, but the entire pot of water hasn't reached the same boiling temperature yet.

As we increase our altitude where we perform our boiling water experiment, the temperature of the boiling water drops. At sea level, boiling water occurs when the temperature is 212 °F (100 °C). In Denver, Colorado, a mile higher into the atmosphere than at sea level, the boiling point of water occurs near 202 °F (94 °C). This occurs because the atmosphere is thinner in Denver; the pressure of the atmosphere in Denver is less than the pressure at sea level. Therefore, we don't have to heat the water as hot to get the pressure of the steam to equal the pressure of the atmosphere. It may save some energy to heat our water to boiling at a higher elevation, but the decreased temperature also results in a lower temperature of the boiling process in brewing beer. This decreased temperature means that we may have to boil the water longer in order to achieve certain chemical changes in the beer that we desire.

This is immediately apparent if we look at the back of a box of cake mix. There are often a special set of directions that help account for the decreased boiling points of water at higher altitudes. Usually, a small amount of additional water is added to account for the increased evaporation at the higher altitude. Table 1.2 provides the boiling point of water at a series of selected cities across the World.

Luckily the freezing point of water is independent of the pressure of the atmosphere (at pressures we would experience in the brewery). Therefore, the temperature of the freezing point of water in Miami, Florida, will be equivalent to the temperature for freezing water in Mexico City.

Other changes occur for water as the temperature changes, as we'll explore in a later chapter. Most notably, however, is the fact that as water gets hotter it swells in volume. This means that hot water occupies more space than cold water, a fact that we'll have to take into consideration as we're uncovering the science behind brewing.

CHECKPOINT 1.4

What is the temperature, in °C, of water that is reported to be 150 °F?

If a low-pressure weather system moves through town, what would you expect would happen to the boiling point of water?

If a brewer makes beer in Cusco, Peru, and must boil the wort for 60 minutes, what effect would the altitude have on his resulting beer?

Table 1.2 Boiling point of water versus elevation

Boiling point of water	Elevation	Example city
212 °F (100 °C)	0 ft., sea level	Miami, FL, USA
210 °F (98.9 °C)	1000 ft	Kansas City, MO, USA
207 °F (97 °C)	2500 ft	Tucson, AZ, USA
202 °F (94 °C)	5280 ft	Denver, CO, USA
198 °F (92 °C)	7250 ft	Mexico City, MEXICO
195 °F (90 °C)	9250 ft	Quito, ECUADOR
191 °F (88 °C)	11,000 ft	Cusco, PERU
187 °F (86 °C)	13,500 ft	Potosi, BOLIVIA

1.3.3   Weight

We often refer to objects by how much they weigh. For example, we might say "Add 2 pounds of Cascade hops to that kettle." The pound, a unit that measures the weight of an object, however, has a very specific scientific definition. Weight depends upon the force of gravity at the location where the object is measured. For example, a barrel of beer weighs six-times more on the earth than it does on the moon, because gravity is six times stronger here on the earth. Carried to the extreme, an object on the earth can have different weights based upon the minute differences in gravity where the object is placed on the scale. Because of this difference, scientists prefer a better term, mass, to define the quantity of a substance. The mass of a substance refers to the specific amount of material that makes up that substance. Mass is independent of the gravity and objects with a given mass on the earth have exactly the same mass on the moon.

While the definition of weight and mass are different, non-scientists often use the two words interchangeably. Just remember that a question such as "How much does that bag of malt weigh?" has a different answer than "What is the mass of the bag of malt?" Brewers know the difference between the words, but often follow the common lingo that is used where they live and sometimes interchange the words.

Mass, according to the SI system, is measured in kilograms, where a kilogram is approximately 2.2 pounds on the earth. Smaller masses can use other terms, such as the gram. Just as in the units for volume and temperature, there are a myriad of different terms and units used based on the area of the world where the unit was developed. Interestingly, the unit of weight known as the "grain" was originally devised based on the weight of a single grain of barley, one of the more important grains used in beer production. Table 1.3 lists many of these units.

Looking at Table 1.3, we notice that the ton isn't included. This is because the ton as a measurement of dry weight has many different definitions. In the United States, the short ton indicates 2000 lbs. and the long ton is 2240 lbs. In many other countries, the tonne (note the different spelling) is 1000 kg or 2204 lbs. While the term is used in the brewing and malting industry, the term is often qualified by calling it either the long ton, short ton, or metric tonne. Because each of these is rather a large amount, the pound, ounce, kilogram, and gram tend to be encountered more often.

Table 1.3   Common weights and mass units

Unit wanted	Abbreviation	Number of pounds	Simple conversions
1.0 stone	st	16	16 pounds
1.0 kilogram	kg	2.204	1000 grams
1.0 gram	g	0.002204	15.4 grains
1.0 pound	lb	1.0	16 ounces
1.0 ounce	oz	0.0625	16 drams
1.0 dram	dr or ʒ	0.0039	27.3 grains
1.0 scruple	sc or ℈	0.002857	20 grains
1.0 grain	gr	0.0001429	

Note: The dram, scruple, and grain were apothecary's units with many different definitions. While it is rare to encounter the dram and scruple, the grain finds use in the brewery because water hardness in the United States is sometimes reported in grains per gallon of water

CHECKPOINT 1.5
How many kilograms are 325 lbs. of malt?
How many grains are in a pound? How many stone does a 210-pound person weigh?

1.4   Yes Virginia, Beer Contains Alcohol

Part of the early allure to drinking beer was most likely the fact that it contained alcohol. While early beers, such as those discussed in Sect. 1.1, probably contained a very small percentage of alcohol, it was likely noted as an important part of the drink. That slight intoxication may have been recognized as divine euphoria. This could be the reason beer was involved in religious ceremonies in those early civilizations. To be sure, beer's intoxicating effects would have been thought of as a welcome release from the daily toils of the common citizen. Those effects were noted by many of the historians of the time, a hint that many new recipes for beer might have developed in search of ways to increase the euphoric feeling.

In addition to getting a little tipsy, the early brewers slowly began to realize that beer didn't spoil as quickly as other beverages (such as milk, juice, or even water). Milk, for example, spoiled in warm climates very quickly after it was obtained and had to be consumed soon or it would go sour. Beer, on the other hand, didn't appear to show a deterioration in its flavor as quickly and so it could be kept for a little while after it was made. In some cases, it likely tasted better when it was a little old.

The early brewers did not understand what gave drinkers the feeling of euphoria or allowed the beer to be kept for a while before consumption. Today we know that the small amount of ethyl alcohol (also known as ethanol, grain alcohol, or simply alcohol) produced in the beer during the brewing process is the reason for these effects. Some beers possess quantities of alcohol that are less than 1% of the total volume, others can be 10%, 20%, 30% or more alcohol by volume. While the preservative properties of alcohol are clearly recognized as beneficial to beer and were known long ago to be helpful during transportation of the beer to market, today we know that overconsumption of alcohol can have a serious detrimental effect on our health. Labeling on beer often reminds us of the hazards of alcohol (see Fig. 1.2).

GOVERNMENT WARNING: (1) According to the Surgeon General, women should not drink alcoholic beverages during pregnancy because of the risk of birth defects. (2) Consumption of alcoholic beverages impairs your ability to drive a car or operate machinery, and may cause health problems.

Fig. 1.2   The US federal government requires that all bottles containing alcohol for sale be labeled to indicate the hazards of alcohol consumption

Alcohol can impair one's abilities and capabilities. Thus, many of the states in the United States have limits on the amount of alcohol that can be in one's body if that person plans to be in public, drive a car, or operate any machinery. In the pre-1970s, laws that limited your consumption were rather relaxed. We've seen the old TV shows of the drunk driver receiving a ride home by the police after being stopped. That is definitely not the case today. Studies illustrating the increase in motor vehicle accidents and fatalities due to overindulgence and society's distaste for those that "couldn't hold their liquor" resulted in our current system of severe penalties for those that drink and drive. And those laws aren't there just to be admired. Today, enforcement of drunken-driving laws has become a very high priority across the United States.

Research has shown that impaired performance as a driver occurs with as little as 0.04 BAC (percent blood alcohol content). Most states in the United States (and throughout the rest of the world) have mandatory Driving While Intoxicated (DWI) laws that severely punish those with 0.08 BAC or higher. Penalties often include automatic loss of your driver's license, jail time, and hefty fines. In addition, violators often must attend rehabilitation or driver's education training. Of course, this assumes that the drunken driver didn't get in an accident; penalties for an accident caused if the driver was intoxicated can be extremely severe. Some states even have penalties associated with the lower level BAC (0.04 BAC) known as Driving While Impaired. And penalties can be nearly as harsh as the DWI. In short, it's just not worth the risk of spending the next couple of months in jail. If you've even been sipping an alcoholic beverage, don't even think about driving.

Table 1.4 indicates the typical BAC based on how many 12-ounce beers are consumed per hour. The table isn't 100% accurate as body mass and the strength of the beer factor heavily into the actual BAC that a person acquires. What is very evident

Table 1.4 Estimated BAC based on gender and body weight. The first number in the table is the BAC level for women, the second number is the BAC level for men. The shaded cells indicate BAC levels over 0.08

Standard Drinks per hour	Body Weight				
	100 lbs	140 lbs	160lbs	200lbs	220lbs
1	0.05 / 0.04	0.03 / 0.03	0.03 / 0.02	0.02 / 0.02	0.02 / 0.02
2	0.09 / 0.08	0.07 / 0.06	0.06 / 0.05	0.05 / 0.04	0.04 / 0.03
3	0.14 / 0.11	0.11 / 0.09	0.09 / 0.07	0.07 / 0.06	0.06 / 0.05
4	0.18 / 0.15	0.13 / 0.11	0.11 / 0.09	0.09 / 0.08	0.08 / 0.06
5	0.23 / 0.19	0.16 / 0.13	0.14 / 0.11	0.11 / 0.09	0.09 / 0.08
6	0.28 / /0.24	0.20 / 0.18	0.17 / 0.14	0.14 / 0.11	0.11 / 0.09

from the data is that it doesn't take much to be intoxicated by the rules of law. And it even takes less to be impaired. For example, one beer an hour might keep some people below the 0.08 BAC, but severely impair the abilities in others.

We've all heard phrases at the parties: "one more won't hurt" or "let's have one more for the road." Peer pressure and ignorance have led to many things that people take as truth about drinking. Unfortunately, most of the sayings we've heard about drinking are in fact misconceptions and untruths. For instance, many people believe that drinking coffee will reduce their level of intoxication and their BAC levels. This is simply not true. Research has shown that drinking coffee, water, or other beverages does very little to the BAC level. The caffeine in coffee may make a person feel like they're less intoxicated, but they aren't. Others believe that vigorous exercise will reduce a BAC level, and still others think that eating certain foods before drinking will stop them from being intoxicated. These beliefs, as well, are not true. While eating certain foods may reduce the rate at which alcohol is absorbed by the body, the effect is minimal and the alcohol will still eventually be absorbed into your bloodstream. In every case, a BAC level drops ONLY by waiting. Every hour that passes, on average, can reduce your BAC by about 0.01%. Note that BAC levels rise very quickly when you drink but decrease very slowly over time.

In addition to laws against drinking and driving, some US states and local governments have public intoxication laws. These laws possess penalties that are less harsh than Driving While Intoxicated laws, but they can include fines, stays in jail overnight, and community service. Drinking in public, being drunk in public, and abusing alcohol are things you shouldn't do, and that society does not appreciate.

Abusing alcohol not only comes with legal penalties but doing so can be damaging to your health. The US National Institutes of Health defines binge drinking as four drinks for women or five drinks for men over a two-hour period. Repeated binge drinking episodes can lead to heart problems such as stroke or heart attack, severe liver damage such as hepatitis or cirrhosis, and even an increased risk of cancer. Consuming more than one to two drinks a day is definitely not worth the risk.

Most importantly, intoxication is contrary to the reason we should be drinking beer. Beer is about enjoying flavors and aromas and about enjoying and appreciating the hard work of the brewer. Drinking beer should not be about getting drunk. Put another way: it's hard to appreciate a fine barley wine if you're prostrate on the floor. Brewers know that the consumer wants to savor and appreciate the work they've done. So, they spend time to develop finely crafted beverages that produce pure satisfaction to the entire palate. We should do our part to savor the efforts of their labor and drink responsibly. In short, we should drink for the pleasure of the beverage.

CHECKPOINT 1.6

Investigate the laws where you live and determine the penalties for driving
   intoxicated. Use Table 1.4 to determine how many drinks you could have
   to reach this level of intoxication.
Use the Internet and explore what chronic binge drinking does to the kidneys.

1.5   A Short History of Beer in the World

In order to really understand how beer and brewing work to produce what we argu-
ably think about as the world's greatest beverage, we have to examine its origins.
This means diving into the history of the beverage to see how it evolved over time,
to really explore the nature of the process. Only then can we explore the modern
processes and understand why they work the way they do. In addition, we can keep
our eyes out for details, facts, and interesting recipes that might be needed while
we're working in the industry. For example, many of the styles that are currently
enjoyed today were, at one time, lost and forgotten. Only by looking back were they
remembered, explored, and brewed again. So, where do we start? Let's start at the
beginning…

1.5.1   The Very Early Years

Plenty of debate exists as to when the first beer was made. Some scholars have suggested that early humans were frugivores (fruit eaters) that instinctively sought out
the ripest fruits to eat. Metabolically, the ripest fruits offered a significant caloric
intake that was beneficial to opportunistic scavengers. In some cases, these overly
ripe fruits were starting to rot when they were eaten. When that happened, some of
the sugars in the fruit had started to be consumed by natural yeast that lived on the
skins of the fruit. The flavor and effects drove early humans to continue the trend in
seeking ripe fruits. Others have suggested that the collection of grains by groups of
early humans went hand-in-hand with the discovery of beer. Whatever its origins,
the beverage was part of civilization across the globe.

It is likely that cultivation of grains and other plants began as early as 12–14
thousand years ago as cultures of people began settling into long-term camps.
However, records of this do not exist. Neither are there records from this time that
describe beer or brewing. However, archaeologists have recently found evidence of
a beer-making operation by the Natufian culture located in what is modern-day
Israel. The site was dated to the same timeframe as the likely cultivation of cereals.

As early as 9000 years ago, beer was being made in China. The evidence takes
the form of residues found on pottery excavated from ancient sites. The rice used in
this process was most likely germinated before brewing or chewed by the brewer to
promote the release of starches and their conversion into fermentable sugars. While
no text has yet been found that describes the process or provides a recipe that was
used, it is clear that the ancient peoples of China enjoyed beer. By 5000 BC, barley
became part of some Chinese beer recipes. Researchers postulate that the introduction of barley in China was to aid the brewing process rather than as a food source.

Written records of beer and brewing did eventually become made. For example,
clay tablets dating from 2500 BC confirm that the Sumerians were producing beer
in cities across their domain. A specific tablet was found that appears to be a receipt
for an order of beer from the city of Ebla. Evidence exists that beer was likely a
staple long before these even these writings were made. In the same region of the

Fig. 1.3 Egyptian wooden model of beer making in ancient Egypt, located at the Rosicrucian Egyptian Museum in San Jose, California. (Photograph by E. Michael Smith)

world, tombs and figurines from ancient Egypt illustrated beer making and drinking (Fig. 1.3). These reliefs and figurines date to about the same time as the Sumerians were enjoying the fruits of their labors. Interestingly, the Sumerians and Egyptians made their beer almost the same as it was described in the Hymn of Ninkasi.

As written records became more prominent in the culture, references to beer and brewing become more prevalent. And, everywhere people lived, beer was ingrained in their society. We know this because of numerous written references to beer beyond stories and tales. In particular, mention of beer parlors were noted in the Code of Hammurabi. This list of laws was written around 1772 BC by the sixth Babylonian king, Hammurabi. The code explained the laws for the operation of their civilization; the phrase we know as "an eye for an eye" comes from these laws. Beer was noted in many places within the laws as well, hinting that beer predated this time enough to have pervaded society by 1772 BC.

Specifically written in the Code were severe penalties for beer parlor owners who overcharged for their product. They were to be drowned as punishment. This is harsh punishment to be sure, but most likely considered equivalent to restricting citizens from their beer. The laws even noted that high priestesses were to be executed by fire if they were caught in establishments that served beer. While priestesses often drank and used beer in religious ceremonies, they were not allowed to do so in a bar. This indicates that a separation between the religious uses of beer and the common consumption of beer was desired in their society.

Other evidence of beer is also found in this region of the world. Ninkasi, a deity worshipped by the Sumerians, was known as the goddess of beer. Archaeologists uncovered a tablet written in cuneiform dating from about 1800 BC that outlines just how important Ninkasi was to her worshippers. The Hymn to Ninkasi, either chanted or sung to a tune that wasn't written down, details how to make beer. Most likely the hymn was created as a way for the average citizen to remember the recipe, because only highly skilled people could read and write during those times. It is clear that the beer made by the recipe was used not only for religious rituals but was also considered one of the staples made by the women of the household. The second verse of the hymn explains this use:

May Ninkasi live together with you! Let her pour for you beer [and] wine,
While I feel wonderful, I feel wonderful, drinking beer, in a blissful mood

The recipe described in the Hymn refers to the preparation of a bread made from malted barley. The bread was to then be soaked in water, mixed with grapes and honey, and allowed to ferment. The resulting mixture, with sort of a soft oatmeal-soup-like consistency, was then poured into vessels to be consumed. This differs greatly from today's brewing process, but still results in a drinkable beer if you don't mind drinking soggy bread-beer.

Early Greek historians indicate that there were a multitude of different types of drinks that we would call beer today. And when you consider peoples outside of Africa and Mesopotamia, the variability in what would be called beer would be staggering. The different recipes used to make beer resulted in a similar variability in the number of words that refer to the beverage. For example, the Sumerians called it "ka." The Egyptians called it "bouza." Even within a culture there were multiple names for the beverage, most likely because of many different recipes or reflecting different uses of the beer. For example, the Egyptians referred to seventeen different types of beer, including one known as "joy bringer" and another known as "heavenly." The ancient Babylonians called the beverage "sikra."

These early beers differ greatly from the beers we find in the stores today. We currently consider beer to be made solely from malted grains (typically barley), hops, and water and then fermented with yeast to provide an alcoholic drink with varying degrees of bitter flavor. The ancient beers, from what evidence can be found, were likely very malty, bready, and in some cases sweet or dry depending upon the addition of fruits and their type. It is very evident that there was little, if any, bitterness to them. The development of beer from the early fermented bread-water mixture to that of today is a long a storied history that we will explore later in this chapter.

The initial development of beer as a religious and spiritual tool in Mesopotamia and Egypt gave rise to a large number of different recipes and names for the alcoholic beverage. Most beer was made in the home as part of the normal preparation of food. However, there were some attempts at mass-production via government-run or commercial enterprise. Some breweries uncovered by archaeologists in ancient Syria could produce up to 100 gallons per batch. Beer, likely the reason for civilization itself, was being consumed regularly.

As we noted, beer in its early stages was a soupy mix of semi-solids and liquid. People either drank their beer like a thin slush or sucked the liquid through a clay or reed straw. Most beer drinkers preferred the straw method as the vessels used to hold beer were much too heavy to lift comfortably. In fact, consumption of beer was a group event; groups of people gathered around the beer jar and passed the straw as they consumed the drink. Hieroglyphs, cylinder seals, and wall images from pre-2000 BC in the Middle East depict many scenes of people sitting around a container of beer drinking it through straws. The use of a straw to drink beer is still done today in regions of the world where the beer is made as an unfiltered soupy mixture in large vessels. Some evidence suggests that the straw might have been used to filter the solids from the beer, but this is likely not the only reason for the use of the straw.

Just as in the Hymn to Ninkasi, beer at that time was prepared by essentially making bread from barley, mixing it with water, heating the soup to cook and distribute the ingredients evenly, and then cooling it. Wild yeasts and bacteria in the air or living in the pottery or wooden spoons used to stir the mixture would grow quickly in the cool mixture and fermentation would result. After few days, the fermentation would have slowed dramatically, and the beer would have been consumed. The mixture was most likely served as it was at this point, or only partially clarified by decanting or crudely filtering the liquid away from some of the solids that settled to the bottom of the vessel.

1.5.2   Beer in Europe Before 1500

The brewing process changed as time passed. One notable change was the way in which the ingredients were mixed and "cooked" before the fermentation started. Malted grains, spices, and other additives were poured in a vessel and heated slowly to boiling. The slow increase in the temperature converted the starches to sugars and boiling helped to thin the mixture and clarify it. The early brewer probably also learned that stepwise heating should be part of the process. At these early stages, however, none of the grains were removed until after the boil was complete. After it was cooled, the liquid was poured off into large vessels and the mixture allowed to ferment. From those fermentation vessels, the beer was transferred into mugs or cups for each consumer to drink individually. With the smaller mug and clearer beer, the consumer could consume the beverage without the use of a straw.

By the time of the Roman Empire, beer and wine were an integral part of society as sustenance and for use in religious services. In other words, having an alcoholic beverage was part of what citizens did during their daily lives. And as the Empire grew, the Romans took their preferred recipes with them. Prior to the invention of the wooden casks, beer was made, stored and transported in clay pots or amphora (Fig. 1.4). These containers often weighed more than their contents and were difficult to transport and use. By 21 AD, the Romans had obtained wooden casks from the Celts and put them to good use holding, transporting, and serving beer into smaller mugs. The much lighter and easier to transport casks meant that beer did not

Fig. 1.4 Amphora as they might be stacked for transport in a ship. (Image courtesy of Ad Meskens https://creativecommons.org/licenses/by-sa/3.0/deed.en)

have to be made in the same town it was produced. While some transportation was possible, these early beers did not survive the rigorous handling that accompanied being moved very well. Most beer was still consumed close to where it was made.

There is no evidence that the Celts, the makers of the wooden barrel, provided the recipes to make beer to the Romans. But similarities in the brewing processes of the two peoples reflected at least more than a casual interaction. For example, *celea* and *cerusia* were Celtic words used by the Romans for beer made directly from grain rather than the ancient Sumerian method of brewing from bread. Posidonius, in 100 BC, wrote that beer was the drink of the common Celt and wine was the drink of the aristocrats. Your favorite recipe for making beer or wine, part of the staples of life, moved with you as you moved across the Empire.

As Posidonius noted, not everyone preferred beer as the drink of choice. Most Romans preferred to drink wine or *posca* (a sour wine diluted with water and herbs). In fact, some did not even find the beverage palatable. For example, Xenophon (430–354 BC), a Greek historian, noted in his writings that beer was consumed in many regions of Europe, but to him, it was an "acquired taste." Tacitus (56 AD – 117 AD), a Roman historian, also found the drink distasteful and preferred wine to beer. Emperor Julian, in 361 AD, agreed completely and reported that while wine smelled like nectar, beer smelled like a goat. Nonetheless, the soldiers and many Roman citizens developed a taste for the beverage. This was the case in many of the outposts of the Roman Empire. Whatever the tastes of those living at that time, beer was greatly enjoyed by many across Europe.

Beer was not the only alcoholic beverage that was consumed. Mead, a drink made from fermented honey and sometimes spiced with herbs and fruit, was preferred by some in Northern European society. While honey itself was more expensive than grains such as barley, diluting a batch of honey prior to fermentation could result in significant reduction in the cost of the resulting drink. Thus, even the poorer classes could afford to drink diluted mead. Members of the upper classes preferred the more expensive, more flavorful, and more alcoholic versions of mead. Braggot, mead that was flavored by adding beer or brewed with grains in the same manner as beer, was also popular.

Wine, the drink we typically associate with Roman culture, was not consumed much by the average citizen. Why was this so? Wine was relatively expensive; grapes cost more than grain and they were less plentiful as you moved to less temperate climates. In addition to the much longer fermentation times required for making wine, the higher alcohol content in wine meant that it was not something to consume during the workday (especially if your productivity would suffer). In fact, most wine served at the time was diluted with water and herbs. In addition to making the drink more palatable, this increased the amount of drink that you had. However, the average citizen and those living at the edges of the Empire still preferred beer to wine.

The use of beer as a staple meant that the average citizen consumed the beverage throughout the day. A glass in the morning would accompany the start of the day and quench the thirst of those who could afford it (or made it at home) during the workday. At night, it filled your glass during your evening meal and was the last thing you consumed before you went to bed. By the start of the Middle Ages, nearly everyone in Europe consumed significant quantities of beer daily. Overuse of the beverage was not typical because the alcohol content was very low for these beers. After all, this was a drink consumed as part of the daily activities – beer was considered food. It wasn't free (purchasing or growing the ingredients cost money) and stretching the amount of grain used to make the beer was the norm. This meant that the beer was alcoholic, but only slightly so.

Recipes and the overall process improved as time passed. Instead of making a beer that was often much less than 3% or 4% alcohol by volume, these improvements increased the alcohol content significantly. As the alcohol content increased, overuse of the beverage in some places appeared to be a problem. Some brewers would mix their grains with beer instead of water in order to make the drink stronger. This "double beer" had the effect they sought and became extremely popular. In England, King Edgar (959–975) saw the overindulgence as a threat to the way of life in his kingdom. Thinking that the intoxicated people he saw were the result of too many breweries, he instituted a rule outlawing more than one brewery per town and placed limits on how much could be consumed during a visit. These laws did little to stop the production of more alcoholic beers, let alone the number of citizens that overindulged. Instead, it encouraged the production of beer in the home and the production of even stronger beers.

It was King Edgar who promoted the use of pins or pegs attached to drinking vessels. A typical drinking vessel held 2 quarts of beer and had 8 pegs along the side. Patrons were then only allowed to drink from one peg to the next (a half-pint). This actually encouraged challenges to drink your entire measure before passing the vessel along – and it even became a sport. This is also where the saying "knocking someone down a peg" comes from.

Beer was so ingrained in culture and society that wars, invasions, restrictions, laws and culture changes did not reduce its production and consumption. Beer was a pastime, a staple in the diet, and used in religious ceremonies. People made beer at home if they couldn't get it elsewhere, often this was relegated to the duties of the housewife or, if you were wealthy, to the kitchen staff. Large breweries serving groups of people, such as communes and monasteries, became important in the early seventh and eighth century (see Fig. 1.5). Those breweries were typically included in the plans when a new monastery was built. And the plans kept a keen eye on making sure the space in the brewery was enough to supply the monks and their visitors with an endless supply of beer.

In Northern France in 822 AD, hops were recorded as one of the flavoring agents being used in beer. Gaius Plinius Secundus (a Roman author), also known as Pliny the Elder, had written almost 800 years earlier about the hop plant. And even though mention was made of the cultivation of hops by 736 AD, beer was typically made without hops. In addition to imparting a very desirable flavor to the finished product, beer flavored with hops tended to last longer. Therefore, it didn't have to be consumed as quickly as un-hopped beer. However, hops were difficult to obtain everywhere in Europe, and the quantity of hops needed for a batch of beer was hard to determine because of the variability of the amount of bittering compounds in the wild plant. This meant that as people traveled, the beer they consumed had vastly different flavors resulting from the availability of what ingredients were found near the breweries. We can imagine that those regional flavors might not be palatable to every traveler.

Some governments, realizing that beer was an integral part of the society in which they reigned, set out to standardize the flavors of beer across their kingdoms. Doing so would ensure that beer in each of the cities and towns was of equal quality. One of the laws that were instituted in present-day southern Germany required beer to be made with gruit (pronounced groot or groo-it).

Today, we don't really know what gruit was. While there are some records of what was found in gruit, its exact nature really isn't known. Some historians believe gruit was likely a mixture of spices and herbs that likely had a variable recipe across a region. The actual mixture of herbs and spices was likely only passed from person to person by word of mouth. Our best guess today is that gruit included dried crushed leaves from the bog myrtle plant, wild rosemary, and other common herbs. Bog myrtle is a small bush that grows in wetlands and is relatively common across northern Europe. The leaves have a bitter flavor that, while an acquired taste, would be recognizable and probably acceptable as a flavor in beer.

More recent hypotheses about the nature of gruit seem equally as plausible. It has been suggested that gruit was a mixture of spices, herbs, and fermenting malt water. In other words, that it was a "starter" for a batch of beer. Anyone making beer would then obtain the gruit and add it while they are brewing. The result would instantly begin to ferment and make beer that contained the prescribed spices and herbs.

Fig. 1.5 This drawing from a book was completed about 1426. The book listed the members of a fellowship or commune in Nuremburg. The text states: "The 46th brother, who died here, was named Herttel Pyrprew." Pyrprew was the word for brewer. (Image in Public Domain: CC-PD-US)

Governments dictated who could make beer by controlling the gruit. In some cases, a gruitrecht, or right to produce gruit, was granted to certain citizens under their rule. Clergy, towns, and even prominent citizens could be granted a gruitrecht. This was done as a favor to the recipient, or to repay a debt that the government owed. Anyone interested in making beer, as most families were, had to visit the gruithaus (the brewery that made the gruit) owned by the holder of the gruitrecht. There, they would purchase enough to make their batch of beer. If a family had the financial ability to do so, they could purchase a daily supply of beer from a monastery or a municipal brewery that had already acquired the gruit. Monasteries and communes even supplied free beer, often the lowest quality beer, to travelers or the poor.

Money charged for the gruit was essentially a tax that was collected by the holder of the gruitrecht. Governments used these funds to support the kingdom. While brewing was a domestic practice, it was clear that towns and monasteries were operating more and more breweries in an effort to assist families with supplying their daily intake of the nourishing beverage. For example, the Weihenstephan Abbey obtained its gruitrecht in 1040 AD. It continues to brew beer to this day almost 1000 years later.

Beer consumption by the average citizen grew as the years progressed. By the Middle Ages, beer was more than a mealtime supplement. Everyone from infant to the convalescent senior drank beer. In addition to a glass at every meal, if you were thirsty, you drank a beer. Moreover, those that could afford it or had the supplies to make it at home preferred beer to water. This was often because water smelled bad or tasted bad. And the beer did not have those off-putting characteristics. Beer was even prescribed as nutritional supplement for nursing mothers. Beer was food and drink.

Prior to the modern age in Europe, cities were not the cleanest places to live and work. In fact, they were incredibly dirty. People dumped their garbage and raw sewage into the street outside their house. Rivers, streams, and lakes also served as garbage dumps. Children played in the garbage and people walked through filth. With no control over waste, citizens of those cities had no choice but to drink the same water that they polluted. At the very least, the appearance and flavor of drinking water was not pleasant. At worst, drinking that polluted water resulted in constant outbreaks of diseases such as cholera. Did the average citizen know that drinking polluted water was bad because it was contaminated and could cause disease? Unfortunately, they probably did not. They were, however, definitely aware that contaminated water tasted and smelled bad.

In fact, the unsanitary nature of the water sources in the cities meant that disease was commonplace. And once a disease started in a city, it spread very quickly to everyone, and even to neighboring towns. Drinking beer instead of water, which was boiled as part of the brewing process, provided a much more pleasant experience and offered some minor protection from water-borne illnesses. It was only minor protection as washing with contaminated water still spread many diseases.

As an example, let's examine what London looked like for the average citizen to see how beer was so important. About 1200 AD, there were roughly 350 ale houses (taverns) for the approximately 80,000 London residents. Brewing supply stores that sold malted barley and other ingredients were even more common. Records from the time indicate one brewing supply store existed for every 60 residents. Everyone was making and drinking beer. In fact, the citizens of London had many different choices when it came to beer, from making it at home to buying it in an ale house. And, it was not uncommon for many city dwellers to send their children to the ale house to buy the family's beer for the day. By 1309, it is estimated that there was one vendor for every 12 persons living in London.

Unfortunately, the makeup and quality of beer across Europe varied greatly. Depending upon where you lived, you could be drinking beers flavored with hops, gruit, nothing at all, or a whole host of other spices or herbs (some were even toxic). You could drink beer made from malted barley, wheat, oats or almost any other type of starchy vegetable. Some were high in alcohol; some very low. If you were traveling, it was likely that the beers you'd encounter each day were not even close to those "like mom made." Some likely didn't even come close to being enjoyable or familiar to what the average traveler was used to consuming.

The argument for maintaining quality eventually became important enough that the Duke of Barvaria, William IV, in 1516 issued a law known as the Reinheitsgebot (the Barvarian Purity Law). This law was constructed not only to assure some uniformity in beer production, but also to ensure that enough grain was available to make bread for people to consume. The Reinheitsgebot dictated that beer had to be made from barley, hops, and water. (Yeast was not known in the sixteenth century as a separate entity, and thus was not added to the list of acceptable ingredients.) This law was officially still on the books until 1987. Barley malt was thus restricted to the production of beer, a task for which barley is very well suited. Grains, such as rye and wheat, which were much better for use in bread making, were not to be used to make beer. The Reinheitsgebot also dictated the price that a brewer was allowed to charge for a liter of beer.

The Industrial Revolution swept across Europe in the mid- to late 1800s, resulting in a dramatic impact on the production of beer. Engines that could operate machines, the invention of the thermometer and hydrometer to monitor beer production, the development of coal-fired ovens used to kiln malt, and small but very useful improvements to packaging beer for sale in bottles, served to increase the output of municipal and commercial breweries. In fact, that increase in production signaled the beginning of a shift from brewing beer in the home to brewing beer commercially. The economies of increased scale meant that beer became cheaper and easier to buy than it was to make in the home.

CHECKPOINT 1.7
Why was beer so important to societies in early Mesopotamia?
What was gruit and what was the purpose of gruit?

1.5.3 Colonization and the New World

Travel on ships could be a perilous adventure. Early transportation and exploration by ship typically meant sailing close to shore. Captains did this for many reasons, including being close enough to land in case the provisions ran low or running into a cove to get out of bad weather. As ships began exploring farther from land, and trips on those ships began to take longer and longer, food and drink became more and more important as a stable supply on board.

Because refrigeration was not available, food was stored in wooden barrels, bags, or crates for the long ocean voyages. The typical ship brought along large amounts of food that didn't need refrigeration, such as hard bread (known as "hard-tack"), salted beef, and dried beans or peas (see Table 1.5). Some ships were also outfitted with fishing nets and tackle so that fish could be caught to supplement the diet of salted or dried meats. Other items might also be included depending upon the generosity of the captain or the nationality of the ship. Nevertheless, the vast amount of storage space required for the food would limit just how long the ship could be away from a port to re-provision itself.

Meat, which was often stored in very salty water known as brine, could still spoil or take on very unpleasant flavors. Cooking it with spices for a long time would make it somewhat palatable. Bread in the form of hard tack or flatbread was also included as a large part of the diet. The harder breads were chosen as they tended not to spoil as quickly. However, spoilage still occurred, and insects such as weevils were a constant problem. It was said that most sailors were advised to eat quickly and not look at the food while they ate in order to avoid the unpleasant nature of the food, and the occasional weevil.

Vegetables and fruits were not included in the typical sailor's diet because they didn't last long before spoiling. This posed a serious problem on particularly long voyages. For example, scurvy, a disease resulting from vitamin C deficiency, crippled those that didn't eat properly over long periods of time. The English Navy eventually learned that citrus fruits had to be included regularly in the diet to combat the incidence of scurvy. Limes were often chosen as a daily supplement the diets on the ships, and English sailors became known as "limeys" because of it. This was a small price to pay in order to avoid the pain of scurvy.

Drink on board a long voyage included grog and beer. Fresh water was included but not often consumed by itself, especially after the first week of a voyage. Water was the most abundant of the provisions accounting for approximately 45% of the

Table 1.5 A week's rations for a British sailor in the late 1700s. It's likely that passengers ate better or worse than this depending upon what they paid for the passage or brought onboard

Item	Weight	Item	Weight
Salted beef	4 lbs	Salted pork	2 lbs
Hard bread	7 lbs	Butter	6 oz
Oatmeal	3 pints	Cheese	12 oz
Beer	7 gal	Dried peas	2 pints

total weight. Unfortunately, in wooden barrels, water quickly turned green from algae growth and became less drinkable each day. So as to not waste anything, the water was made palatable and relatively safe by adding some rum and other spices or flavorings. The result was known as grog. Grog became common on ships that provisioned rum. Typically, sailors enjoyed a pint of grog in the morning and another in the evening. The rum itself was reserved for the officers of the ship or destined to be sold for profit when the ship reached port.

Luckily, beer was very plentiful on board a ship. Not including the water provisions, beer accounted for 50 to 75% of the remaining weight of the provisions. Rations included a gallon of beer a day for each sailor depending on the ship. Some reports indicate that many sailors could not even drink that much in a day. The result was that the beer tended to last a little longer than predicted.

Needless to say, stores of food and drink would eventually run out unless the ship could be re-provisioned. This was the case for the sailors on the Mayflower. John Carver, the leader of one of the Pilgrim migrations to the New World, booked two ships to carry the colonists. They had negotiated to start a settlement in the Virginia Colony near the mouth of the Hudson River (at that time, Virginia stretched as far north as the Hudson). The Mayflower left London in mid-July 1620 and joined up with the Speedwell in Southampton. On August fifth, 1620, the two ships left for the New World, but the Speedwell started leaking shortly after leaving port. The two ships put into harbor at Dartmouth where the Speedwell was repaired. By August 23rd, the ship was repaired, and the voyage attempted again. Unfortunately, the Speedwell started leaking again and the two ships returned to England. Eventually, sensing the urgency of beginning their new life as colonists and the fact that their provisions were being used, the Pilgrims crowded onto the Mayflower and left port without the Speedwell on September 6th. This left the Mayflower heavily overloaded with 102 passengers and about 30 crew for the voyage across the Atlantic. Due to rough weather along the way, the trip took much longer than expected (66 days). The crew sighted the shores of Cape Cod on November ninth as their provisions started to look rather bare.

Attempts to round Cape Cod and head south to the Hudson River were met without success due to storms. The captain of the Mayflower, Christopher Jones, became even more concerned about the remaining supplies on board, especially the supplies of beer and grog. He needed to ensure that he had the provisions for the return voyage to England and he would not entertain trying to round the cape again. Most likely it was he who convinced John Carver to identify a suitable site for their settlement as quickly as possible.

The settlers took many trips inland and along the coast trying to find the best location for a town. Unfortunately, the winter weather hampered their efforts. As one of the settlers wrote in their diary during an exploration:

> We could not now take time for further search of consideration, our victuals being much spent, especially our beere.

It wasn't until late December 1620 that the settlers finally decided where to set up the colony. Because it was too late in the season for the Mayflower to return to England, the ship stayed at anchor until the Spring. The settlers of the Plymouth Colony spent the brutal winter of 1620–1621 on dry land while the crew of the Mayflower stayed on the ship. The winter conditions, however, were not much improved over those on the ship and the construction of houses was hampered by the winter. While they did have some provisions remaining from the voyage and were able to find some caches of corn and beans left by the native peoples, there obviously wasn't enough to support everyone well. There are many accounts of Pilgrims begging for supplies from the ship's stores that were only, albeit begrudgingly, supplied to those colonists who were ill.

While the Mayflower voyage was a perilous one, many settlements across the east coast of the New World fared much better. Most left early enough in the year to have a much shorter voyage, arriving in August or September in time to build a village, gather additional food, and prepare for the coming winter. Ensuring that the brewing of beer would be able to replace what was used on the voyage was often of paramount importance. Some of the colonies were visited regularly by ships and could import their beer from Europe to supplement what they were able to make. The first of such imports occurred in 1607 in the Virginia colonies. This was an expensive alternative though.

The majority of the colonies set about building a brewery as one of the first buildings in their new villages. Then, the brewer in the colony could start making beer to satisfy the entire colony. In fact, the earliest help wanted advertisement in the New World occurred in 1609 for a brewer. The supplies brought over to start the brewery in the new settlements often contained both barley seed for growing a new crop and bags of hops, malt, and raw barley for use in making beer immediately. To be sure, the stockpile wouldn't last long. New barley would have to be grown and malted quickly.

In many settlements along the coast, the European barley strains would not grow well. Supplements and even replacements for the barley were needed. Wheat, rye, oats, and other grains were chosen for the purpose. Those immigrants that befriended native tribes learned to use corn as a replacement for barley. The first recorded beer made from corn by settlers to the New World in 1587. In fact, the natives in the New World had been making their own version of beer using corn and other ingredients for centuries. Their knowledge coupled with the immigrant's skill in making beer resulted in a beverage that was said to be better than an English ale. This could have been self-deception, but it was clear that the production and consumption of beer was so important that the colonists would try anything.

While beer was undoubtedly brewed as part of colonial activities much earlier, the first formal brewery in the New World was established in 1612 in New Amsterdam (later to become New York City), roughly 8 years before the Pilgrims began their voyage. Had the Mayflower found its way to its planned destination in the Virginia colony near the mouth of the Hudson, the Pilgrims and crew of the ship would have been close enough to access this brewery and get the help of those colonists immediately. As it was, the Pilgrims followed suit with many of the other settlements and built their own brewery.

Evidence of the importance of brewing in the New World is clear. For example, in 1640, a law was passed in the Massachusetts Colony stating that no one should be allowed to brew beer unless he is a good brewer. Other laws were enacted to ensure the quality of beer. In 1655, in the New York Colony, a law required brewers to have sufficient skill and knowledge in the art and "mystery" of brewing if they planned to sell their beer. Either there were people making very bad beer or being very wasteful with the precious ingredients.

As the settlements grew, brewing followed. The task was still accomplished in the home for those that had the ingredients, time, or could afford to make the beer the way they liked it. In settlements without a communal or commercial brew house, families grew or bartered for the ingredients to make beer. As towns popped up, trading between settlements for supplies became the norm. And the place to meet to conduct the trade was the local brewery or tavern. In many cases, the two were the same place.

Taverns often had a communal room to serve meals and beer and bedrooms or bunkrooms for those that stayed the night. The first recorded tavern was licensed to Samuel Cole in 1634 in present-day Boston. These places served as restaurants, hotels, and gathering places where the locals would obtain the day's news (Fig. 1.6) or meet others to conduct business. When the colonies started to stretch their legs and proclaim independence, many heated discussions were held in taverns. For example, some delegates to the 1787 Constitutional Convention roomed at the Indian Queen Tavern in Philadelphia. Many discussions in the dining room of that tavern likely helped form the beginnings of a country.

Fig. 1.6 Scenes such as this were common in the colonial tavern. Taverns were more than just places to drink beer. Patrons visited with each other, learned the news of the day, conducted business, and enjoyed a relaxing time. "Sea captains Carousing in Surinam" by John Greenwood. (Image in Public Domain: CC-PD-US)

Taverns served a major role in the American Revolution. When a State would "call up" a militia, the commander of the militia would usually buy a barrel or two of beer at a local tavern and give away the beer to anyone that showed up. Able-bodied men from all around would appear for their free beer or two (or three) and join the militia. George Washington, a skilled brewer at home, quickly learned the power of the tavern. As a young lieutenant in the Continental Army, he found out that supplying beer to the polling places dramatically increased one's chances of being elected to office. Later, in his role as General, he insisted that his troops received a quart of beer each day. During that fateful winter at Valley Forge, he decreed that the troops received their daily ration before the officers.

As America entered the nineteenth century, small breweries became common-place. Brewing was still done in the home, but the time-consuming nature of the process meant that there was a place for the commercial brewery. These breweries produced their beer primarily for local use, storing the beer in casks until it was consumed. By 1810, there were 140 commercial breweries in the United States. Production wasn't very large for these breweries and most of the business was local, but those breweries that were successful did quite well. Vassar College, for example, was founded from the profits earned by brewer Matthew Vassar in 1860.

While some growth in the industry happened in the early 1800s, large growth in the industry was not really observed until the decades after the end of the American Civil War. The first Federal tax on beer production was imposed in 1862 to help offset the costs of the war. Unfortunately, at the end of the war, the incidences of heavy drinking had increased. The war had been very traumatic on everyone in the country, and for many, solace was found in over-indulgence of alcohol. Advocates for less alcoholic alternatives to "hard" liquor became more and more vocal.

In addition, as immigrants from Germany began coming to the United States, their taste for lagers over ales helped fuel the increase in beer production and the development of more and more breweries that specialized in this style of beer. With the invention of artificial refrigeration, the mass production of beer and transportation by railroad to different parts of the country became commonplace. Annual production of beer in the US rose from 3.7 million barrels in 1865 to 39.5 million barrels by 1900. The dramatic increase in production meant that beer was available year-round in nearly every part of the country. This resulted in a similar increase in the per capita consumption to nearly 16 gallons per person per year.

CHECKPOINT 1.8
Why did ships carry barrels of beer when they left port?
What was the purpose of a tavern in Colonial America?
What events occurred that resulted in National Prohibition in the United States?

1.5.4   Beer in Post-1700 Europe

While consuming beer continued to be a daily routine in much of Europe, a new craze was born in England in the early 1700s. Gin production was greatly encouraged by the government as a way to use excess grains and increase taxes. This encouragement worked and distillers flooded the market with inexpensive gin. In 1714, production levels were just over 2 million gallons. By 1733, London alone was producing over 11 million gallons. To combat the crisis of intoxication that was sweeping the nation, Parliament prohibited the sale of small quantities of gin and increased the taxes dramatically. It wasn't until the late 1740s that gin consumption started to decline. This was primarily due to the fact that drunkenness was starting to be looked down upon, workers needed to be sober at their jobs in the industrial revolution, and quality beer started to become cheaper and tastier.

The Industrial Revolution had a dramatic impact on the production process in the brewery. In 1765, the steam engine was invented. This was followed shortly by the thermometer and the hydrometer. They had been invented much earlier, but due to some trepidation by brewers who were used to their "tried-and-true" methods, were not implemented in the brewery until the 1770s. Each of these devices meant that the brewery could be more efficient and produce better quality beer. This also had the effect of shifting consumption from distilled spirits to beer across Europe.

Beer was often produced locally and consumed from taps at the local pub or tavern. Children would often stop by the brewery or tavern to collect beer for the family using a pail if the family could not stop by to get their own. And some families made beer at home as part of the daily routine. In fact, most families made beer two or three times a week as part of the normal household chores.

The introduction of exotic alcoholic beverages into Germany in the 1700s and 1800s led to a reduction in the number of breweries in that region. The drinks from the New World were just in high demand. Unfortunately, governments in the region continued to impose taxes and regulations on the brewing industry. Coupled with the fluctuating costs for the ingredients, brewing began to decline in this timeframe. Only the largest and most wealthy of the breweries could remain open. The same situation is happening again in this region today. Currently, the number of imported beers into Germany is starting to reduce local production. The decline in production and domestic consumption actually started in the late 1980s, just as the United States started to allow home brewing and saw large growth in the craft beer market. In spite of the decline, Germany itself has seen almost 100 new breweries join the market in the 2009–2015 timeframe. In 2015, there were over 650 microbreweries in the country.

1.5.5   Beer in the Far East

About 7000 BC, the same time as beer was being made in Mesopotamia, the Chinese began making their own version of the beverage. While their methods of production were quite similar to those in the Middle East, whether they discovered the drink on

Table 1.6 Top 5 Countries
for beer production in 2018

Country	Production (million hL)	Rank
China	381.2	1
United States	214.6	2
Brazil	141.4	3
Mexico	119.8	4
Germany	93.7	5

their own or learned about it through trade is unknown. Beers tended to focus on the use of rice as the main starchy component of the beer. And just as in the early beers of Europe, they were flavored by the addition of honey, fruits and other ingredients. China has grown tremendously in terms of beer production in recent history. In 2018, China produced 381 million hectoliters (325 million barrels) of beer. The United States ranked second in beer production (see Table 1.6).

In 2014, researchers examined the yeast strains that are capable of making the lager style of beer. They found that the genetics of the modern strains pointed back to one location as the source, Tibet. Given that the eastern spice trading routes began around 2000 years ago, it seems likely that the yeast that inhabited the Tibetan plateau caught a ride to central Europe. Over time, it co-mingled with the ale yeast that existed. And through the process of brewing that made lagers, it became selected as the main yeast for the process.

Evidence of yeast transfer to Asia from Europe exists. As European countries grew and set up colonies in the region, they also constructed breweries using the brewing practices with which they were familiar. In the late 1820s in northern India, Edward Dyer set up an English brewery to supply the troops in the area. The Dyer brewery produced an IPA known as Lion as the first European brand produced in Asia (the brand was reformulated in the 1960s to a lager). This brand can still be found in India but is not to be confused with the lagers produced by the Lion Brewery in Sri Lanka. The Lion Brewery was initially established in Sri Lanka in 1849 as the Ceylon Brewery by the English explorer Sir Samuel Baker. It wasn't until the 1940s that the brewery began making its famous Lion Lager, and in 1996 changed its name to the Lion Brewery.

In Japan, Dutch trade with the country resulted in the introduction of European beer to the country in the 1700s. By the mid-1800s, the Kirin Brewery Company and the Sapporo Brewery were started to supply the country with beer using European brewing techniques. After the Liquor Tax Laws were relaxed in the mid-1990s, the craft brewery scene opened in Japan. By the mid-2010s, there were over 200 such breweries in operation.

1.6   Beer in the United States

Beer has a very long history in the United States. Reports from early settlers and traders to the New World indicated that the native population prepared, and consumed, beer made from corn and other ingredients. In many cases, it was the native peoples that showed how corn and other starches could be used to prepare beer when malted barley was scarce. The fact that beer was made by cultures across the Americas provides further evidence that beer was likely a discovery that predated the cultivation of grains. Could the knowledge of beer have been brought with people as they migrated onto the continent about 15,000 years ago?

1.6.1   Beer Unites the Nation

Whatever the initial source of beer in the New World, it is clear that the European settlers consumed beer in much greater quantities than the native peoples. Beer was used as a food supplement and as a beverage. Beer was known, trusted, and preferred over other beverages. Even though there was plentiful fresh water in the New World, the colonists focused on drinking beer.

Everywhere that the colonists set up camp, a brewery was eventually started. Initially, this was done in the towns. But as the settlers moved inland to better farming, they set up breweries in their homesteads. Then as more settlers gathered in the area, operating a commercial enterprise became a better option.

As the nation started to form, many of the councilmen and legislators gathered in the town's meeting house. In some cases, this was an actual building constructed for town meetings, business, and judicial system. In other cases, the meeting house was also the local tavern or brewery. For example, the Mosby Tavern in Virginia was built in 1740 and served for a time as the county courthouse and jail. The White Horse Tavern, in Newport, Rhode Island is another example of a tavern that was not just a hotel and bar. Used as a tavern starting in 1673, it also took on roles of the community courthouse, assembly meeting house, and the city hall.

And when the courthouse, city hall, and meeting house were located in a different building, the local tavern often served as a gathering place after hours. Some council members who traveled in to conduct business at a meeting house, often spent the night in the local tavern. Local residents knew that this was the case and would congregate at the tavern to hear the latest news of the day. Debating issues, conducting trades, and drinking beer went hand in hand at the tavern.

Bringing the peoples of the town and area together at the tavern was the best way to pass news along. And the success of the tavern as a gathering place on days when the news was a little old was based in the presence of beer.

1.6.2   Expansion Across the West

As the nation grew and expanded into uncharted territory, beer went along for the ride. Trappers and explorers of the American West often set up trading posts to supply settlers that came their way. Because beer did not have a long shelf life in the cask and pasteurization of bottled beer wasn't invented until much later, the early settlers of the continent had to make their own beer when they wanted it.

Many brewers realized that the people moving west would want beer. Some moved west with the wagon trains and set up shop just for this purpose. For example, Charles Barrett, and English immigrant, opened the Portland Brewery and General Grocery Establishment in 1854. By 1859, breweries were located in many of the new towns that sprang up at the end of the Oregon Trail. In San Francisco, California, the Adam Schuppert Brewery opened at the height of the gold rush in 1849. This brewery supplied beer to everyone in the region. By 1852, there were over 350 bars and taverns in the growing city. This was definitely a great place to start a business.

This was repeated everywhere across the new nation. As people moved into an area, they wanted beer. But since it didn't last long, and shipping was not much of an option, they had to make it where they lived. In most cases, it was very difficult (and expensive) to obtain malt and other ingredients to make your own beer at home. So, purchasing it from the local brewery was the best option. Towns like Chicago, Milwaukee, St. Louis, and Denver became supported by large breweries that supplied thousands with beer. The advent of a railroad that stretched across the country by 1869 meant that some places could receive beer shipped from these larger breweries. However, in many cases it was much more economical to make the beer on site.

The tavern or saloon often sprang to the rescue. If a brewery wasn't available, obtaining your beer was as easy as stepping inside the saloon. For example, by the 1880s, Silverton, Colorado's saloons supplied thirsty miners with the beer and entertainment they demanded. In fact, the saloons outnumbered other businesses in Silverton three to one. Unfortunately, public intoxication was often an issue in the rough-and-tumble western towns of the 1870s and 1880s. Many fights, brawls, and even murders were attributed to overindulgence. The attitudes to drinking and associating with alcohol started to change.

1.6.3   Temperance and Prohibition

During the growth of the United States, there became an awareness that the consumption of alcohol was not always beneficial to health. This was most apparent where settlers and Native Americans traded. Often, alcohol was used as part, or all, of the trade. In fact, some tribes began campaigning against the consumption or trade in alcohol as early as the 1730s.

In 1791, the US Federal government instituted a tax on whiskey production to pay for the expense of the Revolution. The tax was not implemented on beer due to the fact that it was consumed locally and usually lasted only a short time before it spoiled. This tax was extremely unpopular as whiskey had become a drink of choice at the time. Following a failed revolt in Pennsylvania in 1794, sentiment for the tax waned and by 1802 it was removed from the books. The opposition was more to the requirement to pay a tax than it was on the taxation of whiskey. However, consumption of distilled spirits did decline slightly as a result of the tax.

The temperance sentiment took a much firmer hold in the United States in the early 1800s as organizations were formed to promote abstinence and moderation as a way of life. The most noteworthy of these was the American Temperance Society (ATS) that was founded in Massachusetts in 1826. The ATS promoted abstinence from wine and distilled spirits but allowed moderation in the consumption of beer. ATS quickly grew to as many as 1,250,000 members by 1833. The organization spread easily by setting up local chapters that could manage the affairs of the local members. The success also spawned the formation of a large number of additional temperance and abstinence groups. While it was suggested that beer was better (less intoxicating) than liquor, every alcoholic drink quickly became included in the temperance movement. The organizations grew as they considered temperance and abstinence to be morally correct. The fact that many church leaders were heavily involved in the temperance movement increased the growth in the number of people that joined. By the early 1850s, it was said that every community in the United States had at least one temperance society chapter.

The organizations took on abstinence as forcefully as they could. In 1851, Maine was the first State to outlaw the manufacture and sale of alcoholic beverages. By 1855, twelve states had adopted similar "Maine Laws" making it illegal to sell beer. In that same year, a riot in Maine over perceived access to beer broke out – the law was extremely unpopular with the working class and immigrant families in Maine. By 1856, the law was repealed in Maine.

Temperance movements were stalled across the country during and immediately following the American Civil War. In addition to the huge demand for beer and other alcoholic beverages to dull the pain of the war, there was a profound recognition of the need to tackle more important issues such as emancipation and women's suffrage. The federal government also needed to generate funds to pay for the war, and taxing alcohol seemed the obvious answer. The first excise tax on beer was only $1 per barrel of beer in 1862. By the war's end in 1865, the tax generated $3.7 million in revenue.

This tax, however, wasn't repealed at the end of the war. In fact, as consumption increased across the country during its westward expansion, the funds grew exponentially. The tax eventually became $6 per barrel and by 1919, it provided $117.5 million in revenue for the country. The funds were used for many different purposes and it seemed that everyone was used to having them around.

1.6.4 Prohibition in the United States

In the years after the American Civil War, many temperance societies were formed and tried to gain a footing with the people as they had done in the early 1800s. While some were moderately successful, the formation of the Women's Christian Temperance Union (WCTU) in Ohio in 1873 did make big strides in curbing consumption of alcohol. Their goals of abstinence, purity, and evangelical Christianity appealed to many. The WCTU promoted social reforms on labor, prostitution, health, and women's suffrage. In fact, the organization eventually included many international chapters and gathered the support of many politicians across the globe. While its membership is only 10% of what it was in the 1930s, the organization continues to this day to work within communities to complete its mission to:

> ...educate all people, with the help of God, to total abstinence from alcohol, illegal drugs, and tobacco as a way of life.

As many other temperance groups had done, the WCTU recommended that its individual chapters construct drinking fountains to quench the thirst of people who might be tempted to have a beer instead. These temperance fountains, most constructed in the 1870s and 1880s, can still be found in city parks across the country.

The Anti-Saloon League (founded by Howard Hyde Russell) was a highly organized and powerful abstinence group started in 1893 in Ohio. Supported by the WCTU, the anti-saloon league used pressure tactics to advance their cause. They assisted in electing "dry" politicians to the US Congress. By the 1910s, the stage was set to have the Federal legislature enact a mandate against alcohol.

During the late 1800s, a large influx of immigrants from Ireland, Germany, and England flooded the United States to escape famine, crop failures, and rising taxes. This was welcomed as they filled a vital need for more workers across the country. However, growing anti-German sentiment in the United States during and after World War I empowered the temperance movement. Because families of German descent operated many of the larger breweries in the country, the link to abstinence was fueled.

In 1917, the US Congress passed the 18th Amendment to the US Constitution. Thirty-six of the 48 states had ratified the amendment by the end of 1919. The law providing enforcement of the Amendment, known as the Volstead Act, went into effect on January 17th, 1920. It provided penalties of up to $1000 in fines and up to 1 year in prison for "manufacture, sale, and transportation of alcoholic beverages in the United States and its possessions." Overnight, the future of alcoholic beverages in the United States changed.

Many of the small breweries closed shop and sold off their equipment to anyone that was interested. Some just closed their doors and waited, hoping that Prohibition wouldn't last. Others retooled themselves and produced non-alcoholic beer, soda, root beer, and even ice cream. For example, the Anheuser-Busch brewery manufactured and sold almost 5 million cases per year of Bevo, a non-alcoholic beverage. While this helped the bottom line, it was not a way to stay profitable for many of the breweries.

Prohibition didn't stop Americans from consuming alcohol. Overnight, speakeasys popped up across the country. People learned how to make alcohol at home – so-called bathtub gin. Everywhere you looked, there was alcohol to be had. Unfortunately, the price of a drink was much greater than it was before Prohibition. Coupled with a growing crime rate due to bootlegging and rum-running, the citizens became extremely dissatisfied with the law. Many felt that the law was only for the poor who could not afford to pay the prices for a beer at a speakeasy. Others felt that lost taxes and jobs from the thousands of workers in the brewing industry had a damaging effect on the economy. The start of the Great Depression in 1929 fueled the discontent. Many saw the repeal of Prohibition as the only way out of the economic disaster.

Prohibition was finally repealed in 1933 as the United States started to come out of the Great Depression. While the repeal did not remove the depression, it did a lot to help. Over $1.3 billion in excise taxes due to alcohol manufacture and sales flooded into the federal registers and helped pay for President Roosevelt's New Deal. The regulation of the manufacture, sale, and consumption of alcohol was returned to the states. Many states delegated the regulation to the specific counties and parishes. In 1937, Kansas was the last state to set regulations for alcohol. However, even today there are some counties in the United States that still outlaw the sale of alcohol.

1.6.5   Post-prohibition

Shortly after the repeal of Prohibition, beer production began in earnest. Those breweries that were able to do so, retooled, hired employees, and opened for business. Initially, only beer containing 3.2% alcohol or less by mass (4.2% by volume) was allowed for sale. This worked well for the breweries in that less malt was required to make such a beer. In addition, the initial shortage of malt as the breweries came back online meant that other adjuncts would need to be used. Thus, rice and corn use in the manufacture of beer in the United States entered the scene en force. The demand for wheat and oats for bread and other foods, coupled with their propensity to create a hazy product, limited their use in beer production. In addition, breweries wanted to make beer that everyone could purchase, and due to the limited employment in the years of the Great Depression, using the cheaper rice or corn as an adjunct surely helped keep costs down.

The breweries that weathered the years from 1919 to 1933 were easily able to retool and produce vast quantities of adjunct-rich lager style beer for the thirsty American public. Immediately, the number of breweries increased as there was a seemingly endless market for beer. By 1934, there were over 700 breweries producing beer in the United States. Unfortunately, the initial rush to produce beer in quantity was immediately followed by World War II. Rationing of malt and other staples during the 1940s meant that the adjuncts initially used to reduce the cost of the beer were to become an integral part of the standard recipe. The continued production of the American Lager meant that the average US citizen learned to love and accept this as the only possible flavor of beer. After all, many beer drinkers in the 1940s and beyond had only this style available their entire life. After World War II, the production of the American Lager continued with vigor. And until the 1980s, there was very little else in the way of domestic beer styles to choose from.

The invention of a can to hold beer in 1935 by the American Can Company meant a shift in how Americans drank their beer. Prior to Prohibition, most beer was consumed in the bar, saloon, or tavern from a cask or keg. By the end of World War II, 62% of all beer was sold in cans or bottles (by 1980, nearly 88% of all beer is sold this way). This meant that beer could be purchased from any brewery at a store and consumed at home. Thus, those breweries who distributed beyond their local region did much better than those who could not. Those breweries with good shipping and distribution started to purchase the smaller breweries. By 1974, the five largest breweries in the United States produced 64% of all the beer sold to Americans. This led to the loss of most local and regional breweries in the late 1900s, with only 89 breweries remaining in 1978.

1.6.6   Returning to the Home

Also in 1978, a bill came to President Carter's desk to sign. This bill, HR 1337, included an amendment by Sen. Cranston and Rep. Steiger that removed the taxes from the production of beer in the home. While it cannot be sold, up to 200 gallons of beer or wine per year per household can be made for personal or family use. This law had an immediate impact across the country as people started to make their own beer. Since reproduction of the beer found in the stores was difficult, many home brewers ended up exploring other styles of beer. Some they liked a lot. And a demand for more was made.

In one example, Bert Grant opened doors to a brewpub in Yakima, Washington in 1982. Scottish by birth, he was genuinely interested in making good beer and pushing the envelope. His brewpub, Yakima Brewing and Malting, was the first brewpub in the United States post-Prohibition. And not to be outdone by that accomplishment, he was the first to produce an Imperial Russian Stout in the United States, and the first to make a Scottish Ale and an IPA since Prohibition. These beers were extremely popular and helped drive his business. The brewpub closed in 2005, but not after it had spread the desire to push the envelope and create amazing beers and different beer styles.

By 1985, there were 110 breweries across the country. Many sprang up in response to the demand from consumers for beers that were not in the mainstream. There were over 8000 breweries in the United States in 2019. Most of these breweries are relatively small and operate on business models that require sales at their taproom to survive. While some do package beer for sale in cans and bottles of all sizes, the social aspect of gathering at a brewery taproom, visiting with friends and the brewery staff, and tasting the latest styles of beer are the driving forces for these companies.

Along with the huge growth in the local brewery and appreciation for different styles of beer is the huge growth in homebrewing. Homebrewing clubs, magazines, and help from the local brewery staff continue to build the hobby. People can get involved by purchasing kits from local stores, building their own systems with help from local homebrew stores, or explore on their own using tools from online video and websites. Across the country, there are many homebrew competitions that remind one of the state-fair atmosphere. And bragging rights to an award-winning beer can now even include adoption of the recipe by a local or regional brewery.

While some households approach their maximum annual production of beer and wine each year, most do not. Even so, the impact on the brewing industry has been minimal. As one homebrewer has remarked, making beer at home is fun, a hobby done to relieve stress, and a great way to explore.

1.7   The Current Market for Beer

By the 1970s there were less than 100 breweries remaining in the United States. The variety of beer styles that were available in the local supermarket or liquor store was limited primarily to the American Lager style with very few imported beers that did not sell well. Public perception of beer was that "beer is beer.". Very little variety existed and if you didn't care for the flavor of an American Lager there was not much you could do. The American Lager was only supplemented in the market by the American Light Lager that catered to those interested in limiting their caloric intake.

Table 1.7 lists the top 10 breweries in the country in 1980. The market had changed dramatically from the pre-Prohibition era. Thirty years earlier, the top ten

Table 1.7   Top 10 Breweries in 1980

Brewery	Brand names included	Rank in 1980	Fate
Anheuser-Busch Inc.	Budweiser, Busch	1	
Miller Brewing Company	Miller	2	Merged with Molson Coors in 2007
Pabst Brewing Company	Pabst Blue Ribbon	3	
Joseph Schlitz Brewing Co.	Schlitz	4	Bought by Heileman in 1981 then Stroh in 1982
Adolph Coors Company	Coors, Coors Light	5	Became Coors in 1989; merged with Molson in 2005; merged with Miller in 2007
G. Heileman Brewing Co.	Heileman's Old Style	6	Acquired by Stroh's in 1996
Stroh Brewery Company	Stroh's	7	Acquired Schlitz in 1982; acquired Heileman in 1996; acquired by Miller in 2000
Olympia Brewing Co.	Olympia, Hamms	8	Acquired by Pabst in 1983
Falstaff Brewing Company	Falstaff	9	Acquired by Pabst in 1990; discontinued label in 2005
C. Schmidt & Sons	Schmidt's	10	Acquired by Heileman in 1987

companies produced nearly equal volumes of American Lager, but by 1980 mergers and acquisitions were the game of the brewery. If a company wanted more customers, it would just buy one of the other companies. Large companies were more successful than the small ones through large marketing campaigns, economy of scale, and consistency of their product. Between the 1980s and the present day, the top three companies, Anheuser-Busch, MillerCoors, and Pabst, remained on top by acquisition of the competition in the marketplace.

President Jimmy Carter, in 1979, was instrumental in the development of the current beer market. (His brother, Billy Carter, is well known for the Falls City Brewing Company's Billy Beer brand.) The President helped enact a Federal law that allowed beer production in the home. Homebrewing, or the production of beer in the home for personal use, was still illegal (both formally and informally) in some states until 2014. Currently, the majority of the states allow the annual production of 100 gallons of beer per adult in a household, up to a maximum of 200 gallons, for personal use. The specific rules that homebrewers must follow are highly dependent upon the state in which they reside. These specifics may include obtaining a permit, not transporting the beer outside of the home, making less than the Federally approved 100 gallons, or adhering to other restrictions. Federal law and most state laws require that production beyond the 100-gallon limit involves filing documentation and taxes as a beer producer. And for those that don't follow the rules there are severe penalties including fines and jail time.

But the freedom to produce beer in the home again meant Americans could make whatever style they wanted to try. The average person's tastes for styles other than the American Lager grew and continue to do so as more and more people explore the different styles of beer. This exploration of styles opened the brewing industry for the craft-brewing and brewpub industries. By 1988, there were over 200 breweries in the United States (Fig. 1.7). As the American consumer began to appreciate different beer

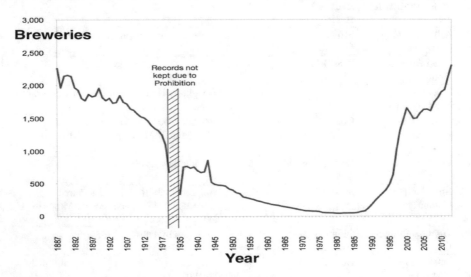

Fig. 1.7 Breweries in the United States by year. (Data are from the Beer Institute beerinstitute.org)

Table 1.8 Top 10 US breweries in 2011

Brewery	State	Brand names included	Rank in 2011
Anheuser-Busch Inc.	MO	Budweiser, Bass, Beck's, Busch, Goose Island, Landshark, Michelob, Rolling Rock, Shock Top and Wild Blue	1
MillerCoors	IL	Coors, A.C. Golden, Batch 19, Blue Moon, Colorado Native, Herman Joseph, Keystone, Killian's and Leinenkugel's	2
Pabst Brewing Co.	IL	Pabst Blue Ribbon, and Schlitz	3
D. G. Yuengling and Son Inc.	PA	Yuengling	4
Boston Beer Co.	MA	Sam Adams	5
North American Breweries	NY	Dundee, Genesee, Labatt, Magic Hat and Pyramid	6
Sierra Nevada Brewing Co.	CA	Sierra Nevada	7
New Belgium Brewing Co.	CO	New Belgium	8
Craft Brew Alliance, Inc.	OR	Kona, Red Hook, Widmer Brothers, Omission	9
The Gambrinus Company	TX	BridgePort, Shiner and Trumer	10

styles, the number of breweries grew even faster. That number continues to grow today, with over 2000 breweries producing nearly every beer style possible – and pushing the envelope on many of those styles. Note in Table 1.8 that the top 10 breweries in 2011 include three craft-brewers. Also note that many of the major breweries that specialize in the American Lager have expanded their style offerings as well. In some cases, this has occurred by the purchase of smaller craft-breweries, in other cases by the development of styles on their own.

Mergers and acquisitions continue to emphasize the changing face of the brewing market. The purchase of SABMiller by AB InBev in 2015 further mixes up the field of the top 10 beer producers in the world. So, what does the ever-changing face of the brewing industry mean for the consumer? In the end, not much. Most, if not all, of the brands that consumers are used to buying will still be on the shelves. The cost of those brands, the quality of the product, and everything else one can think about as a consumer likely won't change either. The biggest change will be in how this large merger may require divestiture of brands in the United States and China due to anti-competition rules. This may mean that some smaller companies may return to the top 10 lists as individual producers. Only time will tell.

Nonetheless, beer production in today's market contributes significantly to the economy. Table 1.9 illustrates the initial growth in beer production after the Civil War and the impact of Prohibition on the industry. Per capita consumption (how many gallons of beer are consumed per person in the US) of commercial beer rose dramatically after the economic downturn of the late 1970s, but has leveled off in

Table 1.9 Beer production in the United States

Year	Production over previous 10-year period (in million gallons)	Per capita consumption (in gallons)	Federal tax on beer, imports and domestic (in millions of US dollars)
1880	118	3.8	–
1910	1844	20.0	–
1930	114	0.9	–
1960	2931	16.3	800.9
1980	5840	25.8	1547.6
2010	6049	19.6	3651.0

recent years. In fact, consumption has slightly declined each year since the high of 25.9 gallons per person in 1983. Despite the reduction in per capita consumption, alcoholic beverages amount to about 12% of a US family's annual food and beverage budget. Even that number is down from the average over the period since the end of World War II. One reason for this decline is likely due to consumption of homebrewed beer.

Tax revenue from beer consumption is a major factor in state and federal budgets. In fact, the Federal government's first tax on beer production has been in place since it was instituted. The tax in 1862, initially set at $1.00 per barrel of beer to help the United States pay for its debts during the Civil War, has increased over the years to the current maximum of $18.00 per barrel (for large producers and for more than 60,000 barrels for small producers; small producers only pay $7.00 per barrel for the first 60,000 barrels). According to the Beer Institute 2015 report (beerinstitute.org), the 2014 US federal tax on beer production generated over $5.4 billion and over $5.2 billion in state taxes. When taken into consideration with the more than $37 billion in business and personal taxes for beer production, this represents a very large influx of money destined for the operation of federal and state governments. Recent legislation in the US Congress has been proposed to modify this tax to encourage small brewery growth.

Currently, beer production injects more than $250 billion (2014 data) into the US economy. It results in numerous jobs directly involved in the brewing process, but also indirectly impacts many other industries. Advertising companies benefit from logo design, TV ads, and other marketing products. Secondary sellers of t-shirts, glasses, coasters, and other paraphernalia benefit. Farmers supply barley, wheat, hops, and other ingredients for manufacture of beer. Trucks, trains, and other shipping move the raw materials and finished products. The cattle industry benefits from the purchase and use of spent grains after mashing. And let's not forget the immediate sale of the end product in bars, restaurants, and other retailers. Each of these industries benefits from brewing.

CHECKPOINT 1.9
What events occurred that resulted in the ability of US citizens to homebrew? What are the reasons for the large increase in the number of breweries in the United States post-1980?

Chapter Summary

Section 1.1
The scientific method is used to evaluate a problem and arrive at a solution that can result in a law or a theory.

Section 1.2

Beer as a beverage is first recorded in Mesopotamia and Egypt about 4000 BCE, although it is probably used much earlier.

The Hymn to Ninkasi outlines the first known recipe used to make beer.

Section 1.3

The volume of a liquid is the space that the liquid occupies. Conversions from one volume to another involve unit-conversion mathematics.

The boiling point of water decreases as the atmospheric pressure increases.

The mass of a substance is a measure of the quantity of that substance. The weight of a substance is a measure of the force of gravity on the substance. While not interchangeable, the two words (mass and weight) are often used interchangeably.

Section 1.4

Blood alcohol content increases dramatically as alcohol is consumed, but decreases very slowly only with time.

Penalties for operating motor vehicles, or being in public, when one is intoxicated can be very severe.

Section 1.5

Beer spread quickly through Europe via the Romans and other peoples.

The Reinheitsgebot and other laws were created to govern the production of beer.

Section 1.6

Breweries and brewing were introduced to the United States as peoples from Europe migrated to the New World (although beer already existed in the New World).

Commercial breweries became numerous and spread across the United States between the Civil War and World War I.

Temperance societies led the push to enact national prohibition. This severely impacted the brewing industry.

The American Lager style became the major style of beer produced in the United States after Prohibition.

Section 1.7

Relaxing the laws for beer production in the 1970s resulted in people trying new beer styles. This encouraged the growth of the craft brewery in the United States.

Currently, beer is a multi-billion dollar industry supporting hundreds of other industries.

Questions to Consider

1. A brewer tests a hypothesis and comes to the conclusion that "cooling wort quickly reduces the off-flavors in the finished beer." Is this a law or a theory? Explain.
2. Another brewer comes to the conclusion after much experimentation that "calcium in the strike water binds to tannins and reduces the tea-like flavors in beer." Is this a law or a theory? Explain.

3. How many quarts are in a barrel (US)? ... in a puncheon? ... in a liter?
4. A cooking pot holds 14 quarts, how many gallons of beer will fit in the pot?
5. How many tuns are in the cooking pot described in question #4?
6. How many firkins are there in the volume described by the pot in question #4?
7. What is the temperature of water in °C if the water is found to be 32 °F?
8. Determine the boiling point of water in the city in which you live. You may need to use Table 1.2 to estimate the value.
9. How many scruples does a 1.0-ounce sample of hops weigh?
10. If a recipe calls for 0.25 stone of malt, how many pounds is this? What is that mass in ounces? ... in kilograms?
11. A brewer in Denver, Colorado, wishes to heat their water to 150 °F. What is that temperature in °C? Does the altitude have an effect on this process?
12. If a mash tun contains 250 pounds of mash and the materials are 7.5 inches deep, how deep would the bed be if the mash tun contained 400 pounds of material?
13. If a person drinks one 12-ounce beer and their BAC increases to 0.05, what would be their BAC if they consumed a 22-ounce beer in the same amount of time?
14. Does the percent of alcohol in a beer have an effect on the BAC for the consumer of that beer? What is the "standard" alcohol found in a beer? What BAC would you expect for a 140-pound person that drinks a 12-ounce glass of beer containing 8% alcohol?
15. Describe the impact on the Pilgrims if the Speedwell had been able to make the voyage across the Atlantic Ocean.
16. A cooking vessel is 12″ in circumference. How many inches deep will 1 gallon of wort be in that vessel? (Use the internet to determine the volume of a cylinder.)
17. Why did the US consumer of the 1950s prefer the American Lager style?
18. What is the oldest continuously operating brewery in the United States? What style of beer are they known to produce?
19. If the 102 passengers and 30 crew aboard the Mayflower consumed 1 gallon of beer per day, how many barrels (US) of beer did the crew have to bring to survive a 66-day voyage?
20. Use the internet to find the text of the Hymn to Ninkasi; then construct a recipe using modern ingredients and methods to make the beer.
21. Use the internet to find the text of the Reinheitsgebot. What were the penalties for violation of the law? How would being punished that way be a severe penalty?
22. Visit the website for the state where you live. What are the penalties for driving with a BAC greater than 0.08?
23. In your own words, describe why colonists to the New World were so concerned about setting up a brewery quickly. Be sure to factor into your description whether the colony was interested in trading with Europe or not.
24. It was stated in the Chapter that the native peoples of the New World were making many different styles of beer prior to the arrival of the Europeans. For

example, use the internet to look up "chicha" and discover how it is made. Based on that information, speculate how these peoples developed similar styles of beer.

25. Visit the website for a brewery that is closest to your home. What styles of beer do they produce? What volume of beer do they make each year?
26. What is the reason for the decline in the number of breweries from the 1880s to the start of Prohibition, and from 1945 to 1980?
27. Two brewers decide to make a batch of beer while they are on trips in different parts of the world. One, in Denver, Colorado, heats some water to 100 °F. The other, in Cuzco, Peru, repeats the exact process. Which brewer will notice more steam coming from their water?

Laboratory Exercises

Familiarization with Laboratory Measurements.

This "experiment" is designed to familiarize you with the standard types of laboratory equipment used in the analysis of beer and its components. This is very useful if you are not familiar with the SI system of units, and serves as very good review if you are.

Equipment Needed
Laboratory scale to 2 decimal places
100 mL graduated cylinder
Pint-sized mason jar
Thermometer
Approx. 200 g of malt, corn, or other grains

Experiment

Obtain a graduated cylinder, beaker, and thermometer. Also obtain a zip-lock bag of malt from your instructor.

Mass First, estimate the mass of a single grain of malt. Use the balance to weigh 10 kernels of malt and record the mass in grams to 2 decimal places. From this mass, determine the average mass of a single kernel of malt. Then, obtain the mass, in grams, of the entire bag of malt (be sure to not include the mass of the bag.) Estimate the number of kernels of malt in the bag, then calculate the number of kernels using the average mass of a single kernel and the mass of the entire bag.

Volume fill the beaker approx. ¼ full with water. Estimate the volume of the water in milliliters. Then, tare the graduated cylinder and pour the water from the mason jar into the graduated cylinder to determine the volume of the water.

Density note the mass of the graduated cylinder containing the water. Subtract the mass of the empty graduated cylinder to determine the mass of the water. Then divide the mass of the water by the volume of the water to determine the density of water in grams/milliliter. Repeat the measurement of density for the malt by filling the graduated cylinder with 20 mL of water and adding a known mass of malt to the graduated cylinder. Swirl until all of the malt is below the surface of the water and determine the new volume of the material in the cylinder. Use the mass of the malt, and the difference in the volume in the cylinder to determine the density of the malt. Does your value agree with your prediction for the density of the malt?

Temperature

fill the mason jar approximately ¼ full with crushed ice and a very small amount of water until you have a slushy-like mixture. Place the thermometer in the ice and record the temperature. Does this agree with your estimate of what the temperature should be?

Exploring the Internet

This 'experiment' involves using the Internet to further explore some of the history of beer and brewing.

Experiment

Choose one of the topics from the list below and write a one-page summary of the information you find about this topic using the Internet, the library, or any other source available to you. Present your findings to the rest of the class.

Topics (others may be presented by your instructor):

Pulque	Kvass
Chicha	Carrie Nation
Taverns in the colonies	Temperance Fountains
Taxes on Beer in the 1800s	States that ratified the 18th Amendment
US counties that remain "dry"	The problems with Prohibition
Anton Dreher	George Washington's Beer Recipe
Small Beer	Alcohol Belt in Europe

Breweries in the New World before 1700
How archaeologists know that a vessel contained beer

Tavern signage	When lagering was "invented"
Beer bottles before 1940	The glass tax in the 1700s
The avoirdupois system	The apothecary system

Why temperance failed to be a national US movement in the 1800s

Adolph Coors	Women's Christian Temperance Union
Louis Pasteur	The invention of the crown cap
Billy Beer	Breweries in Philadelphia in 1885

Beer Styles

2

2.1 Judging Beer

As a brewer, there are many different beer competitions. Some of the competitions are local (such as a competition within a homebrew club for the best 'summer' ale), regional (such as competitions between homebrew clubs or for the best brew in the mountains), or even national (such as the Great American Beer Festival™). The contestants take these competitions very seriously. Much like the BBQ cook-offs outlined on TV, care is taken by the brewers to make sure that they have done the best job possible. Not only can cash prizes be awarded, any medal itself is highly coveted and proudly displayed by the brewer.

Each of these competitions requires that the brewer submit a couple of bottles of beer, properly labeled, and including the particular category to judge the beer under. Because there are only so many spots open in each competition, failure to submit the samples properly is often grounds for elimination from the judging. Once the samples have been accepted, the brewer just has to wait as the judges meet, taste, and evaluate each of the beers.

The results, when they finally arrive, outline the beer in terms of its appearance, aroma, and taste. This information can be very helpful to the brewer. It can be used to provide feedback on the quality of the beer as it conforms to the judge's perception of it. For example, it may have a buttery flavor and knowing that the flavor exists may be of assistance to the brewer in tuning the flavors for future batches. The beer may be unusually bubbly, indicating to the brewer some information about the packaging process.

Judges for a brewing competition can have a wide variety of experience. In some competitions, no experience is required. In others, at least one, if not all, of the judges on each tasting panel is required to have some formal training in judging beer flavors. Programs that can provide that training exist and include the Beer Judge Certification Program (BJCP; bjcp.org). This international volunteer-run organization provides testing and verification of judges.

© The Author(s), under exclusive license to Springer Nature Switzerland AG 2021
M. Mosher, K. Trantham, *Brewing Science: A Multidisciplinary Approach*,
https://doi.org/10.1007/978-3-030-73419-0_2

Training for certification as a judge can also be useful, as judges need to not only know about the possible flavors that can exist in beer, but they must be able to taste them in beer. This means training your palate to identify the banana and clove components in a beer, to recognize and evaluate the carbonation that may or may not be present, and to have experience in recognizing how these components come together in a particular beer.

What areas do beer judges evaluate? This depends upon the competition. But most competitions judge a beer based on its appearance, aroma, taste, and feel in the mouth. The beer is carefully evaluated in each of these categories by the judges and a score generated. Then, beers can be compared based on the scores. With multiple judges, the beer with the greatest score is selected as the winner.

2.1.1   Beer Styles

If all of the beer in the world was made from an identical recipe, judging would be solely based on the process of the brewer, i.e., did the brewer follow the recipe correctly? But this isn't the case. Beer in the world is not the same from region to region and even from brewery to brewery. We noted this in Chap. 1.

Beer was historically made from local ingredients. In today's world, this still holds true, in a way. If a brewer can't get the ingredient, they do not make beer with it. And to carry this a step farther, barley, hops, water, and yeast have many different varieties or cultivars. With the added additions of adjuncts, variability in recipes, and even variability in the brewing process, beer is different everywhere you go.

However there are beer recipes and processes that are relatively similar to each other. Many different attempts have been made to group these similar beers together into a common classification, known as a *style*. It should be noted and stressed that beers are classified into styles based on their similarities. Beers begat styles, rather than styles giving rise to beers.

For example, the beer style known as the American Lager became solidified after the US Prohibition years. This style began when brewers in the United States noted that to increase production with limited barley supplies, the heavy use of adjunct cereals (such as corn and rice) was needed. Recipes were formulated and explored. The consumers knew what they liked and over time they preferred the taste of the beer that they could get. The result was the style of the American Lager.

2.1.2   Conforming to a Style

Conforming to a style means that the beer being examined has characteristics that match those expected for the style. This doesn't mean that the beer is the best example of the style, nor does it mean that it is a beer that everyone would enjoy drinking. All this means is that the beer has flavor and other characteristics that would be expected.

As we just noted above, everyone doesn't enjoy every style. To be successful, a brewer in today's market should make beers that customers will drink. This means that a brewer might make a series of styles and test their sales. Those that sell quickly might find a place in the brewery lineup. Those styles that don't sell or sell sluggishly might either be retired from the brewery, or be relegated to seasonal production.

The brewer needn't construct recipes that fit within a particular style. Using the "customer is always right" theory, or during the exploration of ingredients, the brewer might put together a beer that falls in-between styles, combines the characteristics of two or more styles, or just doesn't conform to any styles. For example, the interest of the beer consumer for the hoppier flavors found in the India Pale Ale style moved the brewery industry in some parts of the United States toward these ales. The success of the IPA style led to breweries trying to push the boundaries of what customers would buy. And the rate at which the new hoppier IPAs were gobbled up led to even more pushes on the boundary. Today, the original IPA style is only a distant memory of what it once was.

Today, many IPA-style beers sold on the west cost of the United States are very "hop-forward." In these brews, hops are used for bittering, for flavor, for more flavor, and for aroma. Then, more hops are added after the aroma hops just to make sure enough hop aroma is present. Some brewers even add another batch of hops just to really be sure. And customers line up outside the brewery when the new batch is finally released. Names of the beers hint at the sheer masses of hops used in their creation; Hoptimus Prime, Hopasaurus Rex, Modus Hoperandi, Hop Stoopid, Hoptopus, Hop Slam, Hops of Wrath, …and the list goes on. The new ultra-hoppy IPA caught on and is now ingrained in the typical brewer's portfolio. The pervasiveness of the new version of the IPA has led to beer-style classifiers declaring a new style: the American IPA.

Jokingly, but in most cases accurately, the authors of this textbook have often declared that the US version of any style is "more" than what the style guidelines for the non-US version dictate. Many of the American versions of the styles are more alcoholic than their European cousins. Many are more malty, more hoppy, or more sugary. The deviations are so "out-of-style" from the versions that they are attempting to emulate, that classifiers have granted new styles for these versions of the beers.

In short, conforming to a style doesn't have to happen. A brewer can spend the time to do so, but there are no rules that require the brewer to follow the existing guidelines. A brewer can formulate any recipe that they wish to create.

CHECKPOINT 2.1
Why do we say that a beer doesn't have to conform to a style?
What is a beer style?

2.2   Parameters That Classify a Beer Style

There are many characteristics of a beer that can be evaluated or measured. In this section, we'll explore those characteristics and outline the basic parameters that dictate the flavor characteristics for beer styles. To ensure that the current batch of beer is the same as those previously made, the brewer can use these parameters. Alternatively, the consumer can use these values to judge the flavor or other properties of the beer prior to tasting. For example, a consumer may choose a beer with a certain alcohol content, or one with a heavier bitterness to go with the food they're eating.

2.2.1   Physical Parameters

IBU   The bitterness of the beer is often reported using a numerical scale known as the International Bitterness Units (IBU). In the United States, this is the scale of preference. In Europe, the scale used is known as the European Bitterness Units (EBU). In theory, both IBU and EBU should give the same values. However in practice, the EBU value is slightly lower due to the differences in how the measurements are made.

The bitterness measurement is actually a quantitative measure of the amount of iso-alpha acids that are present in the beer or wort. It is not a measurement of the perceived bitterness when someone takes a drink of the beer; very malty or alcoholic beers have a perceived bitterness that is lower than the measured IBU value would suggest.

Brewing scientists measure the bitterness in beer using a spectrophotometer. The measurement is performed by extracting the acids into a solvent and then determining the amount of light that is absorbed at 275 nm (in the ultra-violet region of the spectrum). The test in the lab is fairly straightforward, but does require some time as the beer and solvent must be shaken vigorously for 30 min.

ABV   Alcohol content is often one of the first physical parameters that are evaluated by the consumer during sampling of the beer and the brewer during the construction of the recipe. The brewer measures this parameter by many different methods. First, the alcohol content can be estimated by measuring the amount of sugar that has been converted to alcohol. And, if the brewer has access to a laboratory, the alcohol content can be determined by experiment.

In fact, the American Society of Brewing Chemists (ASBC) lists seven different methods for the experimental determination of alcohol content in beer. The variety of methods includes distilling a small sample of beer to move the alcohol and some water away from the rest of the beer sample, then measuring the density of the distillate (Fig. 2.1 and Table 2.1). Other methods include use of chemical instruments such as a refractometer, a gas chromatograph, or an instrument known as an

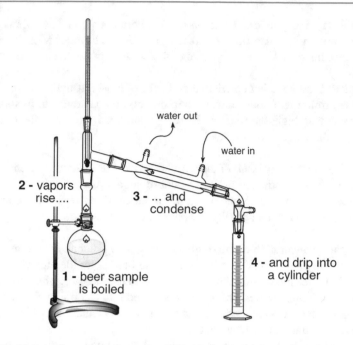

Fig. 2.1 Distillation of sample to determine ABV. A sample of beer is distilled and the alcohol is collected in a separate vessel. When diluted back to the original volume with pure water, the density or refractive index of the sample can be used to determine the amount of alcohol in the original beer sample

Table 2.1 Density of alcohol–water mixtures. The concentration of alcohol in water can be determined by accurately measuring the density of the water at 20 °C

ABV (%)	Density (20 °C)	ABV (%)	Density (20 °C)
0	1.000	50	0.9318
5	0.9927	55	0.9216
10	0.9864	60	0.8927
15	0.9808	65	0.8810
20	0.9753	70	0.8692
25	0.9698	75	0.8572
30	0.9638	80	0.8449
35	0.9572	85	0.8324
40	0.9496	90	0.8194
45	0.9412	95	0.8057

alcoholyzer. For concentrations below 0.5%, the ASBC recommends use of an enzyme (alcohol dehydrogenase) that is purchased separately. The enzyme converts ethanol into acetaldehyde and then the acetaldehyde is oxidized to acetic acid. An ultraviolet-visible spectrometer (UV-vis) then measures the concentration of the alcohol.

Alcohol is typically reported in terms of the amount of alcohol by volume (ABV). Concentrations in beer can run from 0.1% to 12% and beyond. Most beers in the United States have an ABV in the range of 4–7%. In Europe, the average ABV is a little lower.

ABV does have an effect on the overall flavor of a beer. If the amount of alcohol is high, the drinker will notice a warmth in the mouth and stomach. In some cases, if the ABV is very high, the beer will have a pronounced warming effect that might detract from the other flavors of the beer.

OG The original gravity (OG) of the wort before it is fermented is often measured. This parameter is an indicator of the amount of sugars in the liquid. A higher wort original gravity dictates that there are a lot of sugars (both fermentable and unfermentable). A lower original gravity says there are fewer sugars.

The sugar concentration of the original wort doesn't dictate the amount of alcohol that will be produced. While it can give the brewer an idea about the alcohol concentration, the amount of unfermentable sugars isn't known. The unfermentable sugars, as their name implies, will not be converted into alcohol by the yeast during fermentation. Instead, these sugars contribute to the sweetness and thickness of the beer after fermentation has taken place.

OG can easily be measured using a glass device known as a hydrometer (see Fig. 2.2). This instrument is placed into a sample of wort and allowed to float. The higher it floats in the sample, the higher the original gravity. The lower it sinks into

Fig. 2.2 The hydrometer. This simple device is inexpensive and extremely versatile in the brewing laboratory. It is placed into a sample of liquid and the density of the liquid read from the scale at the top

the sample, the lower the original gravity. The gravity of the liquid can also be measured easily by simply obtaining the weight of an accurate volume of the liquid. For example, the brewer could determine the gravity by obtaining the mass of a quart of liquid.

Gravity is just another name for the density of the sample. Density, or the mass of the sample per unit volume, can be represented by the brewer in the units of grams/milliliter (g/mL) or kilograms/cubic meter (kg/m³). And if the gravity (or density) was determined by measuring the mass of a quart of liquid, the brewer would divide the mass of the liquid by the volume of a quart. Typically, though, this is done using SI units such as grams and milliliters. For example, a standard pale ale might have an original gravity of 1.048 g/mL or 1048 kg/m³. Sometimes a brewer may refer to a wort's density by just reporting the last two numbers of the density. This type of reporting has been called the gravity units (GU). Therefore, an OG of 1.048 may be referred to as 48 gravity units (or 48 GU).

The gravity can also be determined using other scales that reflect the amount of sugar directly. In the brewing industry, the scale created in 1900 and named after the German Fritz Plato is used. The scale is known as the Plato scale and represents numbers in degrees Plato (°P). This scale was based on one developed in 1843 by Karl Balling (i.e., the Balling scale), but improves upon that scale by using percent sucrose (by weight). Dissolving a known mass of table sugar in a known mass of water provides the scale. For example, 10.0 grams of sugar in 90.0 grams of water gives rise to a 10% solution (10.0 g sugar/90.0 g water +10.0 g sugar) that is reported as 10 °P.

Degrees Plato and density are not identical in values nor do they have a linear relationship between themselves. The relationship is very close to linear, but isn't, especially as the amount of sugar increases in the water. However, a quick way to convert the gravity to an approximate °P value is to divide the decimal portion of the density by 4. Alternatively, the formula below will provide the conversion. In other words, a gravity measurement of 1.048 g/mL would be approximately equal to a solution of 12 °P. Some hydrometers have both the density and the Plato scales written on them.

$$\left(Gravity - 1\right) \times 250 \approx° P \tag{2.1}$$

FG   Final gravity (FG) is the density of the beer after fermentation has been completed. Just like OG, this is measured in units like grams/milliliter (g/mL) or kilograms/cubic meter (kg/m³). This is a measure of the amount of sugars that remain in the beer and that weren't fermented. In other words, this is a measure of the amount of unfermented sugars.

In some cases, it is also a measure of the amount of fermentable sugars that were not fermented by the yeast. This would occur if the yeast quit working on the production of alcohol due to some reason (such as going dormant when the concentration of alcohol got too high for the yeast to remain living in solution.)

FG is also measured easily using a hydrometer. Alternatively, the final gravity can be determined by weighing a known volume of the beer. To measure the density, though, factoring in the density of the liquid alcohol must be taken into account. Luckily, the density of alcohol–water mixtures is well known (see Table 2.1). And, the density of the alcohol water mixture doesn't deviate much from that of water alone in the ABV range of most styles of beer. However, as the concentration of alcohol climbs above 10% ABV the amount of error in the FG reading increases.

SRM The standard reference method (SRM), used in the United States, refers to the measurement of the color of the beer sample. The value of the color can also be reported as the EBC (European Brewery Convention) value. SRM and EBC differ by about a factor of 2. Measuring the amount of light absorbed by the beer sample at 430 nm and multiplying the value by 12.7 determines the value of SRM. The EBC method multiplies the value by 25. The ASBC methods of analysis require that the sample of beer is degassed (so that bubbles do not interfere) and that the sample is entirely free of turbidity (or cloudiness). In some cases this means filtering the solids or centrifuging the sample. For example, measuring the color of a hefeweizen (an often-cloudy German wheat-style beer) dictates that the sample be filtered. Alternatively, the SRM values can be determined by comparison of the beer color with a chart or colored disks that represent the colors along the scale from 1 to 40.

Colors for beers can range from a very light straw or yellow color (SRM 1–2) to a very dark or almost black beer (SRM > 40) as shown in Fig. 2.3. A spectrum representing all of the wavelengths of light that typical beer samples absorb illustrates that there is a continuous set of colors that beer can absorb. Coupled with the fact that multiple samples with fairly similar SRM values can have very different appearances, beer researchers have been working on identifying a different way to quantify the color of a beer. For example, it is entirely possible that an Irish Red Ale could have the exact same SRM as a Belgium Dubbel even though they have a very different appearance.

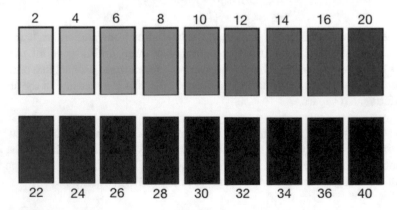

Fig. 2.3 SRM Colors for beer and wort analysis

One such way to report the color of the beer involves the use of tristimulus color theory. This theory suggests that all colors can be accurately reported using the combination of three primary colors in differing amounts. When working on a computer, we sometimes create colors by inputting values for the amount of red, the amount of green, and the amount of blue to mix to make the color. This is known as the RGB value for the color, one example of a tristimulus method for reporting the color.

For beer, however, the tristimulus color is reported in L*a*b* values, where L* is a number representing the "lightness" of the color (0 = black; 100 = white), a* is a number representing the position of the color between green and red (<0 = green; >0 = red), and b* is a number for the position of the color between blue and yellow (<0 = blue; >0 = yellow). The three numbers are required in order to represent the actual color of the sample.

The ASBC has developed a method for determining the tristimulus color of a beer sample. The method requires the measurement of the absorbance of light at many different wavelengths at a given angle from the source of the light. This is important, because the color of a beer sample is highly dependent upon the angle at which the sample is viewed in relationship to the source of light. In other words, if you view the beer sample at 90° from the light source, you would get a different set of L*a*b* values than if you view the sample from a 10° angle. The values for the measurement are also very dependent upon the level of carbonation and the degree of turbidity in the sample.

Turbidity, or cloudiness, in a beer sample has a significant amount of impact on many of the measurements made by the beer scientist. While some beer samples are naturally going to be turbid, such as a hefeweizen, the effect causes a lightening of the sample. Someone viewing the sample gets the impression that the beer is lighter in color than it actually is. Measuring the color using the tristimulus color method, then, requires that the beer be filtered.

CHECKPOINT 2.2

It is said that the amount of alcohol, as ABV, in a beer can be calculated using the eq. (OG – FG) * 131. What is the ABV if the OG is 1.048 and the FG is 1.008? If the ABV is 6.5% and the OG was 18 °P, what is the FG?

2.3   Common Beer Styles

There are many different style definitions for beer, as we noted above. These include the Beer Judge Certification Program's 2015 guidelines that are periodically updated and/or retooled to account for new style classifications. They also include the Brewer's Association's guidelines that are annually updated to reflect changes in the classification details. While both are available online, they represent some significant details about beer styles that are beyond the scope of this textbook. However, with that said, we will explore some of the basic classification details that allow us to make some statements about beer styles. To aid this process, we will break the styles down into two main categories and each category into regional styles. A hybrid style breakdown will capture the majority of the rest of the styles.

2.3.1 Lagers

The lager style of beer was likely first started in the 1500 or 1600s in Germany. By 1860, the style had grown significantly in popularity. It is this "new" beer production process that caught on in the United States as German immigrants moved to the United States.

Lagers are characterized by the use of bottom-fermenting yeast (*Saccharomyces pastoranus*). This yeast strain requires the use of low temperatures during the fermentation step in brewing, and followed by maturation or conditioning of the beer under very cool conditions. During the initial stages of the development of this major style classification, beer was fermented in caves or cellars where the temperature did not get very high. In fact, lagering may be part of the reason for statements in the Reinsheitgebot about why different prices for beer were acceptable at different times of the year. During winter months, beer was less expensive because it was fermented and aged when the temperatures were warm. Beer fermented and aged at these higher temperatures often had many more flavors added from the yeast. The warmer temperatures ended up producing a beer that was not as crisp or clean. During the summer, the beer that was sold was fermented and aged over the previous winter (when the temperatures were colder.) The colder temperatures produced fewer off-flavors and gave rise to a crisper, cleaner beer. But this took longer to ferment and age, hence the slightly higher price:

> From Michaelmas (September 29) to Georgi (April 23), the price for one Mass or one Kopf, is not to exceed one Pfennig Munich value, and From Georgi to Michaelmas, the Mass shall not be sold for more than two Pfennig of the same value, the Kopf not more than three Heller.

The advent of refrigeration in the early 1800s meant that lagers were possible in locations that did not have access to year-round cool places. Because of the low temperatures during fermentation and conditioning, the yeast slowly converted sugars in the wort into beer. The process produced very few off-flavors. Therefore, lagers, in general, have a crisp clean flavor. They represent almost every color possible with other styles, but typically the head on the beer is white or off-white. Many lagers are very carbonated and bubbly, and many are relatively still.

2.3.1.1 European Lagers

European lagers are quite varied. While some accounts suggest that the lager style began as early as the mid-fifteenth century in Munich and its surroundings, the repeated brewing season over the winter months likely selected the lager strain of yeast. In the colder temperatures, the yeast that fell to the bottom of the fermenter had a better chance of surviving the cold. And thus the style was born.

Many of the European lagers are represented in the traditional beers of Germany, France, and the Czech Republic. These lagers differ in the ingredients that traditionally are grown in the specific areas, but all have a fairly similar set of characteristics.

Pilsner   The beer produced and consumed near the town of Plzen, Czech Republic, best represents the Pilsner style of lager. This style, first brewed in the early 1840s, is characterized by its golden color and slightly malty flavor. Noble hops tend to complement the flavor. Boiling portions of the mash as the beer is made (in a process known as decoction mashing) helps to darken and flavor the beer. It is relatively lower in alcohol (~ 4% ABV) and is very drinkable and refreshing.

Amber and Dark Lagers   A slightly darker style of lagers characterize these styles. Anton Dreher (Vienna Lager) and Gabriel Sedlmayr (Oktoberfest) created two of the common members of this style after the lager yeast was identified and isolated. The style is characterized by a malt-forward profile with a slightly higher alcohol content than the pilsner style (~5% ABV). The roasty malty flavor of the beer is coupled with crisp finish. Many of the examples of these styles are found served at festivals in October in Germany.

Bock   This style of beer, which includes the dopplebock and eisbock, was historically made by brewers in the German town of Einbeck. When consumers in Munich first tasted this style of beer, it was an instant hit. They referred to the style as the beer made in "einbeck." Over time, the style became known as "ein bock." Ein bock in German literally means "one goat." The name stuck and is the reason beers of this style often have a goat on the label.

The main style of this category tends to be fairly malty with a relatively high alcohol content (~6.5% ABV). The dopplebock (or double-bock) has typically the same amount of hops as the bock, but a little more alcohol (ranging from 7% to 10% ABV). Partially freezing a bock and then removing the ice results in the eisbock version (eisbock literally translates to "ice bock"). The ice that is removed contains mostly water, but also contains a small amount of the off-flavors that are made during fermentation of the bock. Thus, the style becomes a little more malty with a greater alcohol content (ranging from 8% ABV on up). Repeated partial freezing and removal of the ice can continue to "purify" the flavor and increase the alcohol content. Some commercial examples of the eisbock have greater than 30% ABV.

2.3.1.2 English Lagers
Lagers in England are not the style that was originally preferred by consumers. However, in modern times, the lager style has become one of the best selling products. It is a bottom-fermented beer that is very crisp and clean in its flavors. Adjuncts, such as rice, are used to lighten the flavor of the beer, but most are made solely with

lighter kilned malts. The lighter color and flavor of these beers compliments the lighter alcohol content (ranging from 3.5% to 4.5% ABV). The low alcohol content, light color, and crisp finish are likely the reason for their appeal.

2.3.1.3 American Lagers

The American Lager style is probably the best selling beer style worldwide (based on the amount of consumption of this style in North America alone). It is typically made with a large amount of adjunct cereals (such as rice and corn). Both of these adjuncts ferment very cleanly and lend themselves well to the crisp, clean flavor of the beer. And in some cases, the beers mimic the flavors of the pilsners made in the Czech Republic. This particular style ranges on the lighter end of the ABV spectrum (from about 3.5 ABV to 7.0 ABV) and on the lighter end of the SRM scale (from about SRM 2 to 6). The drinkability of the style also arises from the high levels of carbonation. Many, the authors included, have noted that there is nothing better to drink on a hot summer day after mowing the yard.

While some classifications make a distinction between the American Lager and the lower calorie versions of this style, we consider that they are in the same major classification. Similarly, rather than make a distinction between those made exclusively in the United States versus those made in Canada, Mexico, or other countries, the major style classification is somewhat identical across the entire set of examples from these countries.

2.3.1.4 Other Lagers

The lager style is made in many other regions of the world where it has taken on a completely different flavor profile that suits the consumer. The lager style, in fact, is the predominant style consumed in Asia. Rice plays a predominant role in the recipe for these lagers (in some cases amounting to more than 60% of the total grain bill. For those cases, it seems out of place to call rice an adjunct cereal when it is the main component.

The flavor profile due to rice is extremely light and crisp. The alcohol concentration in the beer tends to be in the 3–5% ABV range. The result is a drink that is smooth, refreshing, and able to be paired with a lot of subtly flavored dishes. Just like many of the beers in the lager family, these beers are often highly carbonated, free of any haziness, and crisp without the hop bitterness or bite.

CHECKPOINT 2.3

What is the relationship between *Saccharomyces pastoranus* and *Saccharomyces bayanus*? (You may need to look this up on the internet.)

Using the information in this section, propose a reason why the lager styles are more prevalent and diverse in Europe than they are in the United Kingdom.

2.3.2 Ales

The traditional method for the preparation of beer used yeast that fermented on the top of the wort. The top-fermenting yeast is *Saccharomyces cerevisiae*. This species of yeast prefers a slightly warmer temperature during fermentation, and with this warmer temperature, the yeast produce additional chemical compounds that give flavors to the beer. The flavors of ales tend not to be as crisp and clean as the lager, but contain some lingering flavors and mouthfeel that truly influence the experience of drinking. Because of this, a wider variety of ales exist.

Homebrewers often start their exploration of the craft by creating ales. This is an easier style of beer to make and is a great one in which to practice the craft of brewing. The style doesn't require the colder temperatures of fermentation and conditioning that the lager styles require. And, the ability to create malty, hoppy, sour, and other flavors easily is very accessible.

The microbrewer and startup brewery also tend to focus on the preparation of this style for the same reasons. Ales are easier to produce and the variability of the ingredients and yeast varieties alone gives rise to a wide variety of flavors. Furthermore, the need for expensive refrigeration, decoction, and/or longer time from production to sale makes the ale the obvious choice.

As we noted, there are a very wide variety of ales. Many are simple modifications of the original. Basic groupings of these into three main style classifications are discussed here.

2.3.2.1 European Ales

Beers from Belgium, France, and to some extent Germany predominate this classification. The German version of this class of styles tends to be made with wheat or rye. In many cases, the wheat must occupy at least 50% of the total amount of grains used to make the beer. Because of this, many of the styles are cloudy due to the high levels of wheat proteins. They tend not to be very bitter but, instead, showcase the flavors of clove and banana. Some of the German ales mimic the bock style, but are made using the top-fermenting yeast rather than being lagered.

The Belgian and French ales showcase rich, deep, and complex flavors. They can be divided into two main categories based on this.

Sour Ales These range from slightly tart or barely perceivable sourness to the very sour. The base of the beers comes from the use of pale malt (and sometimes from the use of large amounts of wheat). As a class of beer, the malty flavors are complemented with complex fruity and/or spicy notes. The level of alcohol ranges dramatically based upon the specific style or region in which the beer is made. The most predominant flavor is that of the sourness due to the use of wild yeast, Lactobacillus (which produces lactic acid – a sour flavored compound), or Brettanomyces (a yeast strain that produces many off-flavors along with a mild sourness).

Sour ales in this category include styles known as Saison, Oud Bruin, Witbier, Berliner Weisse, Lambic, and Geuze (not to be confused with Gose – a beer made from salty water with added coriander spices.) Many of the beer styles in this category originate from what is referred to as a "farmhouse" style. This refers to the early production of beers by residents making their own beers at their home. With limited equipment, the beers often were inoculated using spontaneous fermentation. The results of allowing spontaneous fermentation are the inclusion of some wild strains of yeast and bacteria. These strains become stable in the overall beer-making process as they inhabit the casks that are used. Mild and wild flavors (sourness, complex fruity, spicy) often result from this method.

Belgian Ales Ales made in Belgium are typically malt-forward with an alcohol content that is greater than the sour ales (ranging from 6% to 12% ABV). These beers also have a very complex flavor and darker colors. Aromas and flavors of prunes, raisins, and other dried fruits can be found in the styles. While hops are used, these strong fruity flavors and malty background gives the drinker the impression of wine. The higher levels of alcohol in the beers give rise to their names: dubbel, tripel, quadrupel, etc.…

Particularly noteworthy examples of this style are those made by Trappist Monks. The Trappist order originated in France, and throughout the Middle Ages, this order produced beer for the local communities. Over the ages, the beers and style that were produced became quite desirable by those fortunate enough to sample them. Today, 11 Trappist breweries exist and are the only ones licensed to brew beer and call it a Trappist Ale (the oldest brewery with this designation is the Basserie du Rochefort that began in 1595). For a brewery to be allowed to use the Trappist designation, its beers must be made inside a Trappist monastery, not as the primary work of the monastery, and the proceeds from the sale of the beers must only go to cover the living expenses of those in the monastery. Any remaining profits are donated to charities.

Related examples of the Trappist Ales are known as Abbey Ales. When breweries are unable to be classified as Trappist, they can still produce beers that are similar in style, although they are not held to the same rigorous style guidelines. However, brewers are not just allowed to use the Trappist name to describe the flavors that they have created. So, they use the name "Abbey" to refer to their beers.

2.3.2.2 English, Scottish, and Irish Ales

The classification of the ales from England, Scotland, and Ireland includes a wide variety of relatively similar products. The similarities are not very obvious, but exist in the type of malt and hops that are used in the creation of these beers. While grouping the myriad of different styles into this one category eliminates many of the special styles that deserve individual mention, the general trends in the differences of the styles are the intent.

English Ales This class of styles are quite varied. With the exception of the barley-wine, imperial stout, and robust porter styles, the entire class of ales typically has an alcohol content at the lower end of the spectrum (ranging from 3% to 6% ABV, with many of the examples falling in the 4–5% ABV range.) Malt tends to predominate the main flavors with a noble hop aroma and flavor. Most are very well balanced between malty and hoppy.

As the demand for these ales outside of the country became high, the brewers began adding additional hops or increasing the alcohol levels in order to improve the shelf life of the beers. This was necessary as many of the ales had to be shipped overseas on long voyages. When the lighter colored pale ales were stored in kegs with additional hops added and then shipped to colonies in India, the beer style known as the India Pale Ale was born. When the darker more malty beers were brewed to contain higher levels of alcohol and shipped to Russia, the Imperial Stout and Robust Porter styles were produced. But for the exported beers, the majority of the styles in England were very drinkable throughout the day with lower alcohol contents.

One of the more interesting styles in this category is the Porter. This beer was initially produced in the 1700s as a drink for the workers in the shipyards (i.e., the porters who loaded and unloaded ships). It was first made around 1721, though historical records are somewhat confusing as to which brewery is likely the producer. Brewers noted that they could sell beer that was flavorful, but had a lower ABV, to these customers at lunch, dinner, or after a shift, if it were inexpensive to purchase and not inebriating. And the beers sold quite well as a thirst quencher. At the time, most of the inexpensive malt was kilned over wood fires and had a smoky aroma and flavor and a brown color. The beers, thus, also had these flavors and colors. The smoky flavors likely were reduced by relatively longer maturation times in wooden casks or kegs. As the Industrial Revolution brought about cheaper ways to make malt (without the smoky flavor or darker color), the brewers modified the recipes by adding chocolate malts and roasted barley to give back the darker colors to the finished product. The porter style almost died out entirely when customer preference for lighter ales became the trend, but has since found a resurgence in the craft beer drinker.

The porter style is a deep dark-colored beer that is very malt forward. It typically has a lower alcohol content (ranging from 4% to 6% ABV) with complex caramel, coffee, and/or chocolate notes. It has a light colored, almost white, head that appears contrary to the dark color of the beer. And, it is very drinkable.

The porter is not to be confused with the stout, but is currently a very close cousin to the stout. The stout began as simply a stronger version of beer; a customer might ask for a "stout porter", meaning a stronger version of the porter. In fact, the first mention of a stout in 1677 likely predates the production of a beer known as a porter. The stout, as a current style, has a dark, almost black, appearance and a pronounced bitterness that complements the roasty, toasty, somewhat "darker" flavors of caramel. The head can be white, but also can be tan to brown in color. Alcohol concentrations range from 4% to 6% typically, but can be much higher in the imperial or extra categories of stout.

Scottish Ales The more malty flavors of the Scottish Ales are obvious in this class. The limited use of hops is apparent with some earthy or estery flavors. The result is a malt-forward beer that has very little hoppiness or bitterness. While this is the case, the result is not an overly cloying or sweet flavor. The alcohol content tends to range from about 2% to 5% ABV, and is often denoted in the number of shillings associated with the name of the beer. For example, a 90-shilling Scottish Ale would have a higher ABV than a 60-shilling version. The color of the ale also can be related to the number of shillings and ranges from a golden to copper color.

Irish Ale The most commonly thought of Irish Ale is likely the stout. This style of beer is characterized by a very dark, almost black, color and a hop bitterness that is accentuated by the use of roasted and darker malts and barleys. The darker roasty flavors and astringency from the use of these grains, is characteristic of the style. The creamy tan-colored head on the beer is a necessity. Increased alcohol contents are found in the export versions of the Irish Stout. Guinness makes one of the more recognized versions of the Irish Stout. Initially, the beer we know as Guinness Extra Stout was known as Extra Superior Porter, reflecting the appropriation of the word "stout" as a stronger version of the porter style.

Also in this class is the Irish Red. This beer style is malty but has a strong bitterness from a hop addition early in the production of the beer. Often the grains used are such that the color of the beer is deep amber to red or copper. Very little, if any, hops are noted in the aroma or flavor of the beer. And the alcohol content tends to be in the 4% to 6% ABV range. The result is a very drinkable beer with a dry finish that invites another sip.

2.3.2.3 American Ales

As expected with this classification, the sky is the limit for any of the ales produced. Many are mimics of the styles found elsewhere on the globe. However, in mimicking the style, many of the brewers exaggerate or downplay a characteristic or two (or more) for the original style. This does not mean that these are poor versions of the original style, but instead that they are examples of an American version of the style.

For example, many different breweries in the United States have produced mimics of the India Pale Ale style from England. The more successful IPAs, in terms of consumer preference, are extremely hop-forward. As we noted in Sect. 2.1.2, this particular style has been taken to the far extreme.

Another example lies in the American version of the Porter style. This style is often very flavor-forward with emphasis on the use of chocolate or coffee flavored malts. In some cases, the American Porter is flavored with the addition of cold coffee extract or chocolate nibs (cocoa beans) to emphasize those flavors in the finished beer.

2.3.3 Hybrids

While the majority of beers can be assigned to styles that clearly define the lager or ale categories, there are some beers that span across the two main categories. For example, a hybrid style is one in which the yeast strain used during fermentation is mismatched with the process used for the other yeast strain. Such is the case in the California Common style of beer where a lager yeast is used to ferment the beer at the higher temperature typically used for ale yeasts. The higher temperatures result in the production of a lot of esters and other flavors that would normally be considered out-of-place in a lager. Other styles that fit into this category are the Blonde Ale, the Cream Ale, and the Kölsch.

CHECKPOINT 2.4

How is it possible that the same flavor of a farmhouse style ale from Belgium could be obtained from spontaneous fermentation?

The Gose style is mentioned in this section. Compare and contrast it to the Geuze style. (Use the internet to find information on these two styles.)

2.4 Historical Beer Styles

Many beer styles have been lost to time, but were described or written about in historical documents. These styles either fell out of favor with the consumers, their ingredients became difficult to obtain, or laws and rules eliminated the options available to make these beers. The rapid growth in interest for craft beers has resulted in these styles being reintroduced.

For example, George Washington, one of the founders of the United States, was not only a statesman, but a wealthy plantation owner in the eighteenth century. As we uncovered in Chap. 1, almost every household brewed their own beer for consumption. In Washington's case, this meant the production of fairly large quantities of beer to supply everyone that lived at Mount Vernon. One such recipe for the construction of a table-strength or small beer (low alcohol for consumption any time of the day) survives from George Washington's personal notes:

> Take a large Sifter full of Bran Hops to your Taste -- Boil these 3 hours. Then strain out 30 Gall. into a Cooler put in 3 Gallons Molasses while the Beer is scalding hot or rather drain the molasses into the Cooler. Strain the Beer on it while boiling hot let this stand til it is little more than Blood warm. Then put in a quart of Yeast if the weather is very cold cover it over with a Blanket. Let it work in the Cooler 24 hours then put it into the Cask. leave the Bung open til it is almost done working -- Bottle it that day Week it was Brewed. – George Washington, personal papers, 1757

Attempts to reconstruct this beer have been made by both homebrewers and commercial operations. Similarly, many other beer recipes have been identified in the literature and have been converted into beers. Some of these have even become very well-known commercial products. For example, the kottbusser style was originally descriptive of the ale made from wheat, oats, malt, honey, and molasses. A couple of commercial breweries in the US have revived the style.

Some US brewers take specific steps to research and then produce a style of beer that hasn't been tried in the modern ages or is fairly rare on the US market. This includes the Kottbusser ale, the Kentucky Common ale (pre-prohibition ale common in Louisville), the Gose ale (made with salt water, coriander, and sour-producing bacteria), and Grodziskie (a Polish smoked-wheat ale). Additional styles and recipes continue to be located in the literature nearly every day. It's likely the better tasting ones will find their way to the commercial market.

One particular example of a little known style in North America is quite common in South America: chicha. Chicha is a drink that pre-dates the European migration to the New World. This beer style is entirely corn based (with flavorings added both pre- and post-fermentation). Since the corn isn't malted prior to fermentation, masticating (chewing) the corn provided the enzymes needed to mash the corn. Spontaneous fermentation from wild yeast produced the finished cloudy beer. Yes, at least one commercial brewery produces chicha in the United States. And yes, the brewers take turns chewing on the corn and then spitting it into the mash tun. This definitely sounds like a beer everyone should try.

A search of old manuscripts, letters, and documents from the library (or digital library) may reveal a new beer recipe that has fallen into obscurity or even extinction. A little research and you may become the brewer that resurrects the style.

CHECKPOINT 2.5

The Reinheitsgebot outlawed styles such as the Kottbusser. Why?

Write down the recipe from George Washington and then "translate" it into masses and volumes that would be used today.

2.5   How to Sample and Taste Beer

Professional beer tasters have a specific set of steps that they follow to drink a beer. Those specific steps allow them to examine every characteristic of the beer, from the flavor to the appearance and even beyond into the after-effects of the beer on the palate. Their method gives them the best chance to determine flavors that shouldn't belong, evaluate the level of carbonation, or recognize the mouthfeel of the beer as it is consumed.

Every beer taster follows the steps in a slightly different way. Their individual modifications of the steps play to their strengths in analyzing beer. Some tasters have a very good eye for bubbles and gaze deeply into the beer. Others have a weaker sense of taste for certain off-flavors and spend considerable time slurping, swishing, and breathing while consuming a small sample. But all take their time to make sure that their evaluation is the best it can be.

For the non-professional, there are a few steps that can be taken to ensure that we are evaluating the beer in a way that allows us to appreciate every part of the brewer's invention. Of course, one could always rely on the use of laboratory analyses to give the impression of the beer, but this isn't the best method. When it comes to determining if the beer tastes good, we have to rely on our senses. Thus, we have to taste it.

2.5.1   Beer Glasses

The best way to analyze the beer is to sit down and pour it into a glass. That doesn't mean "drink the beer from a glass bottle." That means we should open the beer container and pour it into a glass. The packaging is simply just that, packaging. While many of us grab a cold one and drink it right from the bottle or can, this isn't how the brewer likely intended their hard work to be consumed. To honor the brewer's work, the glass should be used to truly enjoy the beer.

In fact, specific types of glasses go with specific styles of beers. It is the beer drinker's responsibility to make sure that the beer they are examining is consumed from the glass that works best. That is the glass the brewer envisioned as they constructed the recipe, and it is the glass that best accentuates the style of the beer. And with the myriad of styles that exist, there are an accompanying wide variety of glasses to choose from.

The basic styles of glasses are shown in Fig. 2.4. They include glasses that are tall and short, fat and skinny, and round and straight. The Pilsner glass is intended for light clear bubbly beers, such as the pilsner style. The Mug or Tankard is for lower alcohol beers. Typically, when made of glass, this is the vessel chosen to serve light colored beers, such as those of the Oktoberfest style, but even other low alcohol beers can also be used. Related to the Mug or Tankard is the beer Stein. The Stein is often a ceramic vessel with a pewter lid (originally put there to keep out flies), but wooden, leather, and glass versions do exist.

The Weizen glass is a tall glass with a protrusion in the wall that gives the top of the glass a globe-like shape. This glass is often used to serve wheat beers; the wheat beers tend to have fairly large heads and the large area at the top of the glass allows it to hold the foam. In addition, the wide area at the top enables the drinker to really understand the aroma of the beer as they consume it. The Pint, Nonic and the Stang glasses are related to the Weizen by shape, but have very different purposes. The Pint glass is the standard ale glass used in pubs, taprooms, and bars everywhere to provide patrons with a pint of beer and a large area at the top to smell the aromas. It has straight sides. The Nonic differs in that it has a bulge near the top that is wider than the glass is at the top. This bulge stops the glasses from getting chipped on their lips when they are stacked or clinked together. The Stang is a straight-walled glass used for many German beer styles such as the Kölsh, Dunkel, and Doppel Bock.

The Chalice, Snifter, and Goblet glasses are related in their shapes. They tend to be used for beers where the consumer can sink their nose into the glass and smell the aromas. The Chalice is squat and wide at the top, the Snifter is round and narrow

Fig. 2.4 Glasses used in drinking beer. From left to right: pewter mug, chalice, flared top stang, goblet, pilsner, pub glass (cross between a weizen and a pint), stein, and pint. (From the author's own collection)

at the top, and the Goblet has the shape of a wine glass and is narrow at the top. These also tend to be used for higher alcohol more flavorful beers.

Many breweries have made their own versions of different glasses, and some have even created new glass styles. For example, some breweries serve their beers in Mason jars, others in glasses that mimic the aluminum can, and still others in the shape of a boot, ball, or other object. In many cases, these are the glasses the brewer intends you to consume their hard work. Yet, some are simply unique glasses used as advertising for the brewery.

2.5.2 Serving Temperature

Serving temperature for a beer is more important than the glass that is used. Beer that is consumed at a temperature that is too high will accentuate the off-flavors in the beer. While that might be good for trying to determine those flavors, it is not appropriate for trying to get the best flavor out of the beer. Moreover, when the temperature of the beer gets too high, the sensations of carbonation and the bitterness from the hops decrease. This is not good for an IPA at all. Conversely, beer that is consumed at too low of a temperature will de-accentuate all of the flavors in the beer. In other words, too cold and you cannot really taste anything other than the carbonation and bitterness.

Table 2.2 Suggested serving temperatures

Style	Temperature (°C)	Style	Temperature (°C)
Amer. Lagers	2 °C	Wiezen	7 °C
Pilsner	4 °C	IPA	9 °C
Blonde Ale	4 °C	Porter	10 °C
Belgian Ales	7 °C	Stouts	10 °C
Sour Ales	7 °C	Trappist/Abbey	12 °C

This leads to a common mistake in beer drinking, the "frosty mug." This should never be done. In addition to freezing the beer as you pour it into the glass, it adds water to the brewer's creation and dilutes the flavor. The beer becomes very cold and can eliminate the flavors that are intended in the product.

The best temperature for a particular beer style seems to mimic the SRM value or "richness of flavor" for the beer. Lighter beers, such as the American Lager and Pilsner styles, tend to favor the lowest temperatures. The darker beers, such as a barleywine or Belgian Tripel, prefer a much warmer temperature to allow the consumer to taste everything in the beer, as shown in Table 2.2.

Beers that are served directly from a cask (i.e., cask ales) are lightly carbonated and the flavors of the beer should be the focus of the beer. This style of ales is seeing a strong resurgence in the United Kingdom and elsewhere in the world. Serving temperatures for these beers (no matter the specific style) should be around 12 °C. Contrary to popular opinion, they are not served at room temperature. In fact, no beer should be served that warm.

CHECKPOINT 2.6

Suggest a glass and serving temperature for a Märzen style beer. (Use the internet to learn about this style.)

Assume a new glass has been invented that constantly forces bubbles to form at the bottom of the glass. Which styles of beer would benefit from the use of this glass?

2.5.3   Sampling and Tasting

There are four specific qualities in beer that need to be analyzed by the senses. They rely on our eyes, nose, and mouth to do the measurement, and require that the beer be served at the appropriate temperature, and in the appropriate glass, in order to get the most from the sampling. Some judging takes place by pouring the beer samples into small taster glasses. While the judges understand how to adjust their analysis to account for the small glass, this is not even close to the appropriate method for maximum enjoyment of the brewer's project.

The key features of the sensory evaluation that are typically explored are:

- Appearance
- Aroma
- Taste
- Mouthfeel

The appearance of the beer is determined by evaluating the beer in two steps. First, as the beer is poured, special attention is paid to the quality of the "pour." Is the beer thick and syrupy, or is it thin and watery as it is poured. Second, the beer is examined in the glass like a fine gemstone. It is held up to the light to judge its color, which also gives the judge the opportunity to examine the bubbles and the head on the beer itself.

The amount or thickness of the head often results in the perceived quality of the beer. Is the head thick and rich with small beer-covered bubbles, or it is small and thin? Are the bubbles uniform in size and mostly small? Is the head similar throughout, or are there regions where the bubbles seem to clump into large "icebergs"?

Then, the next step in analyzing and enjoying the beer is to hold the glass close to the nose and allow the aroma to be sampled. The aroma can be based on the hops that are added near the end of the boil, but can also come from other sources. Does the malt shine through into the aroma? Can you smell the off-flavors of diacetyl (butter), apple (acetaldehyde), vinegar (acetic acid), or creamed corn (dimethyl sulfide)? While some of the off-flavor aromas are needed and required as part of the style, many aren't and shouldn't be found in beer.

Multiple ways to pull a sample of the aroma of the beer into the nose exist. One such way is to rapidly pull small amounts of the beer into the nose with miniature exhales. In this method, approximately 4–5 small puffs of the aroma are brought into the nose within as many seconds. The beer should then be removed from under the nose immediately after the puffs are sampled. In another method, the beer is placed near the nose and the sampler then pulls a long deep draw into the lungs. Again, the beer is removed away from the nose as quickly as possible after the sample. If the glass is small enough, such as the sampling glasses found at a judging competition, the hands can be cupped around the mouth of the glass and then around the nose. In this way, only the aromas from the beer are pulled into the nose.

Of course, with any of these methods, it is imperative that the sampler not be wearing heavy cologne or perfume. The use of hand creams should be avoided (even the unscented ones can have an impact on the aroma of the beer). Other distractions, such as loud music, talking, and overly bright lights, can have a negative impact on the sampler's ability to judge the aromas of the beer.

The beer taster then puts a small sample in their mouth (about 10–15 mL or 0.5 ounces) and swishes it around. There are many ways to do this. One of the easier ways is to pretend that you are eating a bit of food and chewing the beer. In other words, the consumer moves their jaw up and down while sampling the beer. The effect is to add oxygen into the beer and allowing the vapors from the beer to mix as they go to the back of the throat. Those vapors are the aroma of the beer and must hit the palate fully oxygenated in order to give the appropriate result.

Sometimes the beer is spit out of the mouth. For a judging process or for those beer aficionados visiting a beer festival, this is not a bad thing. With the shear volume of samples, it isn't required that the beer be swallowed in order to determine the appearance, aroma, flavor, and mouthfeel. As Anton Ego, a food critic in the movie *Ratatouille* said:

I don't like food – I love it! And if I don't love it, I don't swallow.

There is nothing wrong with that approach to tasting and drinking beer.

The final sensory experience that assists in determining the overall characteristics of a beer is known as the "mouthfeel." To use a tautology, mouthfeel is how the beer feels when it's in your mouth. The sensations that the beer causes when it is in the mouth (not the flavor or taste, although there is some overlap) are referred to as the mouthfeel. This characteristic covers three main areas of sensations: after effects, carbonation, thickness. Some descriptors (words that describe the sensation) that are related to the mouthfeel of a beer include warming (from alcohol), astringent (tea-like dryness), flat (no bubbles), gassy (opposite of flat), creamy (coats the mouth like milk), and thin (like water).

Judges often record their notes in each of the four categories to assist in their grading of the beer sample. In some cases, this is very helpful to the brewer (whether they win first place or not), because the notes can provide information about another person's perception of the beer. When the information is detailed, the data can be used to adjust a recipe or a process.

But that isn't the only place information about a beer should be obtained. One of the major and most important analyses to perform is the sensory analysis. In fact, even if a microbrewer or nanobrewer lacks a laboratory to calculate the key aspects of a brew, they should still have access to a panel of tasters. By giving samples to the tasters, the brewer can learn about off-flavors, consistency, and other aspects of the beers that they make. These results should be obtained constantly to determine the quality and customer satisfaction with the beers the brewer makes. We'll uncover this in much greater detail in Chap. 12.

Chapter Summary

Section 2.1

Beer styles represent a classification system that groups beers with similar characteristics together.

There is no requirement that a beer must be made to fall within a style category.

Section 2.2

Physical characteristics that can be measured for a beer include OG, FG, IBU, ABV, and SRM.

Section 2.3

Ales and lagers are the main categories of beer, classified based on whether the yeast is top-fermenting or bottom-fermenting.

Most styles can be classified into European, English, or American categories.

Section 2.4

Historical styles are on the rise as brewers and consumers look for new tastes in beer.

Section 2.5

Beer glasses come in many different shapes and are based on the style of beer that should occupy the glass.

Beer should always be served at the appropriate temperature to ensure the best experience.

Tasting and evaluating a beer is very important in determining the flavor characteristics.

Questions to Consider

1. Describe the key differences between an American Pale Ale and a European Pale Ale.
2. Use a website that describes beer style information. Use that site to compare and contrast the ABV and SRM for the bock and doppelbock styles.
3. Repeat question #2 to determine the flavor characteristics of the helles and kölsch styles.
4. Describe how the alcohol content could be determined for an American Lager style beer. Would your description change if the beer style was an Irish Red ale?
5. Use the internet to look up and report the style characteristics for the California Common style. What is unique about this beer style?
6. What would you say to a brewer that wanted to make a beer that had some of the flavor characteristics of a weizen and some of the characteristics of a porter?
7. Why would the German ales likely not conform to the Reinheitsgebot?
8. Describe the best glass to use to drink a Porter.
9. Use the internet to provide the history of the porter style. Be sure to include a discussion of "three-threads."
10. Given the information in this chapter, how would you arrange a sampling room to obtain the best results of a sensory analysis of a single beer?
11. Is there a relationship between OG and FG?
12. Is there a relationship between OG and SRM?
13. The alcohol content of a beer can be reported in ABV and ABM (alcohol by mass). Given that the density of pure alcohol is 0.789 g/mL and water is 1.000 g/mL, which measurement (ABV or ABM) would provide the larger number?
14. Using the information in question #13, calculate the ABM for a beer that is 4.0% ABV.
15. Why is the SRM difficult to report above 40?
16. If a brewer wished to report the SRM of a beer that was a 60, how could this be done using only the beer and water?
17. Use the internet to look up information on how chicha, umqombothi, and chibuku are made.

18. Look up a flavor wheel on the internet and describe the flavors for at least two different key descriptors.
19. What is the density of a 10 °P wort? ...16 °P? What is the °P for a wort with a density of 1.069 g/mL? ...1036 kg/m^3? ...52 GU?
20. What is the ABV of a beer that had an OG of 14 °P and an FG of 1 °P?
21. Explain how two nearby towns could be responsible for two completely different beer styles.
22. Throughout South America, the native population produced chicha long before the arrival of the European settlers. Explain why this style was found across this wide region.
23. Use the internet to determine how a refractometer works.

Laboratory Exercises

Density Measurements

This "experiment" is designed to provide the student with a clear understanding of the term "density" and when and where it makes sense to use it. When coupled with use of refractometry, the information in this experiment gives excellent background on determining alcohol levels in beer.

Equipment Needed

Laboratory scale to 2 decimal places
Graduated cylinder, 10 mL and 100 mL
Erlenmeyer flasks (3), 125 mL or 250 mL
Beaker, 50 mL and 250 mL.Hydrometer and hydrometer tube
Refractometer, handheld or Abbé style
Bag of cane sugar
Ethanol, 100%, denatured or punctilious

Experiment

Prepare three solutions. The first solution should be 10 grams of sugar dissolved in enough water to make 100 mL. The second solution should be 10 mL of ethanol and 90 mL of water. The third solution should be 5 grams of sugar and 5 mL of ethanol dissolved in enough water to make 100 mL.

Measure the density of the solutions using the hydrometer by pouring the solution into the hydrometer tube and using the hydrometer. Then, the density of the solution is measured using the 10 mL graduated cylinder. To do so, mass the cylinder empty and then add at least 5 mL of the solution to the cylinder. The density is recorded by dividing the mass of the solution in the cylinder by the exact volume that is in the cylinder. Compare the hydrometer and the density numbers to determine the accuracy of the two methods.

The refractive index for each solution is then determined by placing a drop of each solution on the refractometer and reading to at least 4 decimal places.

The density and refractive index for pure water and pure ethanol should be determined as well.

Finally, a plot of the density as a function of the percent sugar in the solution should be made. Is this plot linear or is there some other relationship?

A plot of the refractive index as a function of the percent sugar should also be made. Also make a plot of the refractive index as a function of the percent alcohol. Is either of these two plots linear or is there some other relationship?

What conclusions about the use of density and refractive index can be made?

SRM Determination

This "experiment" is designed to provide a rapid evaluation of the color of a beer using a spectrometer to determine the actual value for the SRM.

Equipment Needed

Visible spectrometer capable of reading 430 nm and 700 nm
Graduated cylinder, 10 mL
Test Tubes (6), 20 mL and a test tube rack
One bottle of a clear "dark" beer such as a Porter
One bottle of a clear "medium" beer such as an Amber Ale
One bottle of a clear "light" beer such as an American Lager
One bottle of a cloudy "light" beer such as a hefeweizen

Experiment

Obtaining a 10 mL sample of the "dark" beer. Swirl the beer and mix it until it is fully decarbonated. Then, dilute 5 mL of the beer with 5 mL of water (50%). Dilute 5 mL of the 50% beer with 5 mL water (25%). Repeat this two more times to create a 12.5% and 6.75% dilution. Then, obtain the absorbance at 430 nm of each sample, 100%, 50%, 25%, 12.5%, and 6.75%. For each value, if the absorbance at 700 nm is more than 0.039 times the value of the absorbance at 430 nm, the value should not be used. If the absorbance at 700 nm is less than 0.039 times the value of the absorbance at 430 nm, multiply the absorbance at 430 nm by 12.7. The result is the SRM value for that sample.

Make a plot of the value of SRM versus the dilution factor for the dark beer. Is there a linear relationship? If not, what is the relationship and why isn't it linear?

Then, repeat the experiment and create the plot with the "medium" colored beer. Is there a linear relationship in this case?

Finally, measure the SRM value for the two "light" colored beers.

Then, draw some conclusions about the beer analysis and what it means in terms of measurement of the color.

Molecules and Other Matters

3

3.1   The Atom

Most brewers in the pre-1900 world had never heard nor thought about the atom. But the atom was actually postulated to exist long before its discovery. Scientists in Europe developed a series of basic laws and theories about our world in the 1700 and 1800s that did include statements about the atom. For example, John Dalton in 1805 proposed what he called the atomic theory. The statements in this theory largely proved to be correct even after the atom was discovered.

The atom is the smallest unit of everything. Your desk, the pencil, and even the paper upon which you write are composed of atoms. In fact, the atom is so important to science that the periodic table (see Fig. 3.1) occupies a predominant place in nearly every science practiced on this planet. The periodic table lists each of the known elements; some are only known because they are made in the lab, and the others are naturally occurring and found in the world around us.

The periodic table is broken down into two main regions denoted by the stair-step line on the right-hand side of the table. To the left of this line are all of the elements that we call metals. Metals are shiny, malleable, ductile, and can conduct both heat and electricity. To the right of the line are all of the elements that we call non-metals. Non-metals are just the opposite of metals in that they are dull, brittle, unable to be stretched, and insulate against heat and electricity. Elements that touch the line are known as the metalloids because they have properties that are half-way between those of the metals and the non-metals.

The elements in the periodic table are arranged based upon their physical characteristics and properties, such as their reactivity with water. For example, the elements in the first group or column in the table react quickly with water to make alkaline solutions. The arrangement also accounts for the sizes of the elements; the atoms get small as you move across a row, and larger as you move down a column. Because of the way in which the atoms were placed into the periodic table, each column (or group) contains atoms that have fairly similar properties. And most of

The Periodic Table of the Elements

Z	Symbol	Name	Mass
1	H	hydrogen	1.01
2	He	helium	4.00
3	Li	lithium	6.94
4	Be	beryllium	9.01
5	B	boron	10.81
6	C	carbon	12.01
7	N	nitrogen	14.01
8	O	oxygen	16.00
9	F	fluorine	19.00
10	Ne	neon	20.18
11	Na	sodium	22.99
12	Mg	magnesium	24.31
13	Al	aluminum	26.98
14	Si	silicon	28.09
15	P	phosphorus	30.97
16	S	sulfur	32.06
17	Cl	chlorine	35.45
18	Ar	argon	39.95
19	K	potassium	39.10
20	Ca	calcium	40.08
21	Sc	scandium	44.96
22	Ti	titanium	47.88
23	V	vanadium	50.94
24	Cr	chromium	51.99
25	Mn	manganese	54.94
26	Fe	iron	55.93
27	Co	cobalt	58.93
28	Ni	nickel	58.69
29	Cu	copper	63.55
30	Zn	zinc	65.39
31	Ga	gallium	69.73
32	Ge	germanium	72.61
33	As	arsenic	74.92
34	Se	selenium	78.09
35	Br	bromine	79.90
36	Kr	krypton	84.80
37	Rb	rubidium	84.49
38	Sr	strontium	87.62
39	Y	yttrium	88.91
40	Zr	zirconium	91.22
41	Nb	niobium	92.91
42	Mo	molybdenum	95.94
43	Tc	technetium	98.91
44	Ru	ruthenium	101.07
45	Rh	rhodium	102.91
46	Pd	palladium	106.42
47	Ag	silver	107.87
48	Cd	cadmium	112.41
49	In	indium	114.82
50	Sn	tin	118.71
51	Sb	antimony	121.76
52	Te	tellurium	127.6
53	I	iodine	126.90
54	Xe	xenon	131.29
55	Cs	cesium	132.91
56	Ba	barium	137.33
57	La	lanthanum	138.91
72	Hf	hafnium	178.49
73	Ta	tantalum	180.95
74	W	tungsten	183.85
75	Re	rhenium	186.21
76	Os	osmium	190.23
77	Ir	iridium	192.22
78	Pt	platinum	195.08
79	Au	gold	196.97
80	Hg	mercury	200.59
81	Tl	thallium	204.38
82	Pb	lead	207.20
83	Bi	bismuth	208.98
84	Po	polonium	[208.98]
85	At	astatine	209.98
86	Rn	radon	222.02
87	Fr	francium	223.02
88	Ra	radium	226.06
89	Ac	actinium	[227]
104	Rf	rutherfordium	[261]
105	Db	dubnium	[262]
106	Sg	seaborgium	[266]
107	Bh	bohrium	[264]
108	Hs	hassium	[269]
109	Mt	meitnerium	[278]
110	Ds	darmstadtium	[281]
111	Rg	roentgenium	[282]
112	Cn	copernicium	[285]
113	Nh	nihonium	[286]
114	Fl	flerovium	[289]
115	Mc	moscovium	[289]
116	Lv	livermorium	[293]
117	Ts	tennessine	[294]
118	Og	oganesson	[294]

lanthanides

Z	Symbol	Name	Mass
58	Ce	cerium	140.12
59	Pr	praseodymium	140.91
60	Nd	neodymium	144.24
61	Pm	promethium	[145]
62	Sm	samarium	150.36
63	Eu	europium	151.96
64	Gd	gadolinium	157.25
65	Tb	terbium	158.93
66	Dy	dysprosium	162.50
67	Ho	holmium	164.93
68	Er	erbium	167.26
69	Tm	thulium	168.93
70	Yb	ytterbium	173.05
71	Lu	lutetium	174.97

actinides

Z	Symbol	Name	Mass
90	Th	thorium	232.04
91	Pa	protactinium	231.04
92	U	uranium	238.03
93	Np	neptunium	[237]
94	Pu	plutonium	[244]
95	Am	americium	[243]
96	Cm	curium	[247]
97	Bk	berkelium	[247]
98	Cf	californium	[251]
99	Es	einsteinium	[252]
100	Fm	fermium	[257]
101	Md	mendelevium	[258]
102	No	nobelium	[259]
103	Lr	lawrencium	[266]

Fig. 3.1 The periodic table

the groups have names that describe the key features. For example, the last group is known as the noble gases because the atoms that make up that group are fairly unreactive gases. Group 1 are the alkali metals because of their reactivity, Group 2 are the alkaline earth metals (they react with water to make alkaline solutions, too), Group 11 is known as the coinage metals (copper, silver, gold), and Group 17 are the halogens.

The atom can be thought of as a very tiny solar system. At its center where the sun would be is a collection of smaller particles known as protons and neutrons. This collection of particles is known as the nucleus of the atom. Protons and neutrons have essentially the same mass; the mass of each is approximately equal to 1 atomic mass unit (amu). Because the protons have a positive charge and the neutrons have no charge, the nucleus overall has a positive charge. Of these two particles, the proton is the most important because the number of protons dictates the specific element on the periodic table. The number of protons in an atom, in other words, determines the type of atom. That number is the whole number written in each box in the periodic table. For example, carbon has six protons and oxygen has eight.

The number of neutrons in a nucleus can vary and in most cases is not equal to the number of protons in the nucleus. When an element has more than one option, the result is an isotope. Some isotopes even have different names. For example, hydrogen can be found in three different forms in nature. One form has no neutrons in the nucleus (known as hydrogen), one has only one neutron (known as deuterium), and the other has two neutrons (known as tritium). While isotopes seem to compound our understanding of the periodic table, the good news is that even with multiple isotopes, to the first approximation, the properties of each are still the same as the other. In other words, if the nucleus has 1 proton, the atom is hydrogen regardless of the number of neutrons it possesses.

Surrounding the nucleus of the atom is a sea of particles called electrons. These very fast moving particles have a negative charge. But because electrons have a mass that is 1/1000th the size of the mass of a proton or a neutron, the overall mass of the atom is essentially equal to the mass of all of the protons and neutrons. In other words, electrons can be thought of as having almost no mass. In addition, in order to have an atom without a net charge, the number of electrons (the negatively charged particles) must equal the number of protons (the positively charged particles). Thus, the carbon atom must have six electrons circling the nucleus; there must be eight electrons around the oxygen nucleus.

> **CHECKPOINT 3.1**
> How many protons, neutrons, and electrons are in an atom of nitrogen that has a total mass of 14 amu? …in an isotope of nitrogen with a mass of 16 amu? What element would contain 11 protons, 12 neutrons, and 11 electrons?

Unfortunately, it is rare to find one of the elements of the periodic table with the correct number of electrons around the atom. This is because electrons are used to bind atoms together. When the number of electrons is different than the number of protons, the atom becomes an ion. A positively charged ion, called a cation (pronounced CAT-ION), has fewer electrons than protons. The negatively charged ion, called an anion (pronounced AN-ION), has more electrons than protons. Knowing that protons have a + 1 charge and electrons have a − 1 charge, it makes sense that a cation would have fewer electrons than protons.

What would be the charge on an ion of carbon if it had only four electrons? We could answer this question by noting that there are always six protons in a carbon atom, and with only four electrons, there would be two protons that weren't balanced by electrons. The charge would be 2+. Similarly, an oxygen ion with 10 electrons would have a 2− charge.

As it turns out, when atoms bond with other atoms they tend to gain or lose electrons based on their location in the periodic table. Metal atoms tend to lose the same number of electrons as the atoms position along the Period. This is particularly true for metals in Group 1 and Group 2. The result is a cation with a charge that is equal to the Group number, e.g., H^+, Li^+, Mg^{2+}, Ca^{2+}. Aluminum is the third element in the period, so it would have a charge of 3+; Al^{3+}. The "transition metals" in Group 3 through Group 12, however, tend to have multiple possibilities, although a 2+ charge is very common; e.g., Ti^{2+}, Ni^{2+}, Fe^{2+}, Fe^{3+}.

Nonmetals tend to form anions with the charge equal to the number of elements remaining in the Period. Nitrogen is three "boxes" away from the end of the row, so it would be N^{3-}; oxygen is two boxes from the end of the row, so it would be O^{2-}.

While this rule tends to work often, most of the nonmetals have multiple possibilities for their anion charges. The halogens almost always have a 1– charge.

CHECKPOINT 3.2
What is the charge on a sodium ion with only 10 electrons?
What is the charge on a chlorine ion with 18 electrons?

3.1.1 Compounds

Compounds are formed when two or more atoms combine. For example, water (H_2O) is a compound made from two hydrogen atoms and one oxygen atom. Salt, also known as sodium chloride (NaCl), is a compound made from one atom of sodium and one atom of chlorine. While these two compounds appear to be very similar when we write them, they are very different when we explore how the elements are combined.

The combination of a metal and a non-metal gives rise to an *ionic compound*. For example, NaCl is an ionic compound. In fact, the metal in an ionic compound (the Na in NaCl) actually exists as a cation; it has fewer electrons than protons. The non-metal in an ionic compound (the Cl in NaCl) is actually an anion containing more electrons than protons. The positive charge of the cation interacts very strongly with the negative charge of the anion and an ionic compound results. This electrostatic attraction is a very strong force and can be thought of like the attraction between the north pole and south pole of two magnets.

When the ions combine, they do so such that the total charge of the cation(s) is equal to the total charge of the anion(s). For example, in sodium chloride, the sodium ion has a 1+ charge and the chloride ion has a 1– charge. Therefore, one sodium cation must combine with one chloride anion so that the overall charge for the ionic compound is zero. In another example, when calcium cations combine with chloride anions, they do so such that one calcium cation (Ca^{2+}) pairs with two chloride anions (Cl^-) to make $CaCl_2$.

While the force of attraction between oppositely charged ions is very large, some ionic compounds can dissolve and dissociate into their separate ions in water. Sodium chloride, when added to water, dissolves into the water and becomes Na^+ and Cl^-. Calcium chloride also does this, resulting in a water solution of Ca^{2+} and Cl^- ions, though in this case there are twice as much Cl^- anions in solution as Ca^{2+} cations. But not all ionic compounds dissolve in water. As we'll see later, apatite (an ionic compound) does not dissolve in water, nor does it dissociate into its ions.

Compounds can also contain nothing but elements from the non-metal side of the periodic table. These combinations are called *molecules*. Carbon monoxide (CO) and water (H_2O) are examples of molecules. In molecules, the atoms do not exist as ions. Instead the atoms share their electrons so that the electrons encircle all of the nuclei in the molecule. The forces holding the atoms together are a little weaker than in the ionic compounds, but strong enough to keep the molecules together so that they don't dissociate into ions in water. The atoms in a molecule are said to be joined by covalent bonds.

Just like the ionic compounds, some molecules can dissolve freely in water and some can't. For our purposes, those containing nitrogen, oxygen, sulfur, or phosphorus tend to be fairly water-soluble. Those that lack these atoms, or have significantly more carbon atoms than nitrogen, oxygen, sulfur or phosphorus, tend to not dissolve very well in water. With just a couple of exceptions that we will discuss later, molecules do not dissociate into ions when added to water.

Naming Ionic Compounds Rules for constructing the names of compounds, known as chemical nomenclature, have been set by the International Union of Pure and Applied Chemistry (IUPAC). Those rules define the steps used to determine the name of almost any compound in the world. Ionic compounds are named by saying the name of the cation, and then saying the name of the anion. The ending of the anion is changed so that the entire name ends in "ide." There are modifications to this rule; oxygen as an anion is known as "oxide." Other examples include salt (NaCl), known as sodium chloride, and $BeCl_2$, known as beryllium chloride.

Note that even if the ionic compound contains multiples of either of the ions, we don't state so in the name. For example, calcium bromide has the formula $CaBr_2$. Table 3.1 lists some common ionic compounds and their names.

Polyatomic Ions Unfortunately, it is possible that one of the ions in an ionic compound is made up of a molecule (see below) that doesn't have the correct number of electrons to balance all of its protons. In these cases, a polyatomic ion is born. The polyatomic ion could be a cation, such as NH_4^+, or an anion, such as PO_4^{3-}. When encountered in an ionic compound, the polyatomic ion is not treated any differently than a cation or anion, except that it has a special name that must be used. Table 3.2 lists the names of the common polyatomic ions found in ionic compounds common in the brewery.

For example, $Ca(OH)_2$ is commonly known as slaked lime. Its chemical name is calcium hydroxide. NH_4Cl is ammonium chloride. $MgSO_4$, found in Epsom salts, is known as magnesium sulfate. And $Ca_3(PO_4)_2$ is known as calcium phosphate. Note in the formula for calcium phosphate, there are three Ca^{2+} cations and two PO_4^{3-} anions. The total positive charge is 6+ and the total negative charge is 6– in order to balance. Parentheses are used in the formula so that we know if more than one of the polyatomic ions is used in the compound. Also note that when a compound contains

Table 3.1 Common binary ionic compounds

Formula	Name	Formula	Name
NaCl	Sodium chloride	KCl	Potassium chloride
$CaCl_2$	Calcium chloride	$MgCl_2$	Magnesium chloride
CaO	Calcium oxide	MgF_2	Magnesium fluoride

Table 3.2 Common polyatomic ions

Formula	Name	Formula	Name
OH^-	Hydroxide	HCO_3^-	Bicarbonate
CO_3^{2-}	Carbonate	PO_4^{3-}	Phosphate
SO_4^{2-}	Sulfate	SO_3^{2-}	Sulfite
NO_2^-	Nitrite	NO_3^-	Nitrate
CH_3COO^-	Acetate	NH_4^+	Ammonium

these polyatomic ions, they are treated no different than if they were a single entity. This is because they are a single ion and the parentheses are there to indicate how many of these polyatomic ions are in the formula.

Multiple Charges While it seems like a further complication, chemical naming rules also provide ways to determine the name of a compound where the metal may exist as more than one cation. Remember, the transition elements likely are cations with a 2+ charge, but other charges are very possible. In those cases, the same metal may have multiple formula with the same atoms, such as in FeO and Fe_2O_3. This is very common.

To name compounds containing a transition element, the charge is placed in the name as a roman numeral in parentheses. For example, $CuCl_2$ is named copper(II) chloride and CuCl is named copper(I) chloride. Note that no space is placed between the parentheses and the name of the metal. These are actually different compounds with very different properties, and the name must contain the roman number so they can be distinguished from each other.

Sometimes we are presented with an ionic compound containing a transition metal that has a charge that we do not know. By examining the other atoms in the formula; however, it's often easy to determine the charge on the transition metal. For example, it has been shown that 2+ cations can increase the bitterness in beer by reacting with the hop acids. Would $NiCl_2$ suffice to perform this task? Since the nickel is a transition metal, we can't immediately determine the charge by considering its location on the periodic table. However, since the halogens almost always have a -1 charge, we can determine the charge on the anion Cl is 1–. With two of these in the formula, the total anionic charge would be 2–. Thus the nickel must have a charge of 2+ to balance the 2–. The name of the compound would be nickel(II) chloride, and we would predict that it should be able to increase the bitterness in beer by reacting with the hop acids.

Table 3.3 lists some common ionic compounds that require the use of the Roman numeral when writing their names. Many of these compounds are potentially found in brewery waters, are found as contaminants in ingredients, or get accidentally added to the process stream due to contact of the wort and beer with metals in the brewery. Note that all of these contain transition elements that have multiple cationic charges possible.

Table 3.3 Ionic compounds that use Roman numbers

Formula	Name	Formula	Name
$FeCl_2$	Iron(II) chloride	Fe_2O_3	Iron(III) oxide
FeCl3	Iron(III) chloride	FeO	Iron(II) oxide
AgCl	Silver(I) chloride	TiO_2	Titanium(IV) oxide
$Cu(NO_3)_2$	Copper(II) nitrate	$Mn_3(PO_4)_2$	Manganese(II) phosphate
$CuNO_3$	Copper(I) nitrate	$CuSO_4$	Copper(II) sulfate

Naming Molecules The names of ionic compounds result because of the limitations of compounds that can be made from specific cations and anions. For example, there is only one combination of calcium cations, Ca^{2+}, and chloride anions, Cl^-. The result is calcium chloride ($CaCl_2$). There are only two combinations of copper cations and chloride anions that are common (CuCl and $CuCl_2$).

This restriction doesn't occur in molecules that result from the combination of nonmetals with nonmetals. Molecules differ not only in the types of bonds that hold the atoms together, but also because the individual atoms in the molecule are not cations and anions, they can be put together in almost every possible combination. For example, the combination of carbon and hydrogen can result in molecules with formula such as CH_4, C_2H_4, C_2H_6, C_3H_8, and thousands and thousands of other formula. Thus, molecules by necessity must be named differently than ionic compounds.

For molecules that only contain two types of atoms, the process of naming is very similar to how the ionic compounds are named. These compounds, known as binary molecules, are named by adding prefixes to the names of the atoms in the formula and changing the ending of the last atom to "–ide". The prefixes we use for this are:

1.	Mono	6.	Hexa
2.	Di	7.	Hepta
3.	Tri	8.	Octa
4.	Tetra	9.	Nona
5.	Penta	10.	Deca

There are a couple of modifications we make to the name of the molecule so that it is easier to say. First, we tend to only use the first syllable of the name for the second atom and then add "–ide." Thus, oxygen as the second atom in the formula becomes "oxide," sulfur becomes "sulfide," phosphorus becomes "phosphide," etc.... Second, if the second atom is oxygen, we drop the "a" from the prefix if it has one. And third, if there is only one of the first atom in the formula, we don't add the "mono" prefix. As examples, CO becomes carbon monoxide, CS_2 becomes carbon disulfide, P_2O_5 is known as diphosphorus pentoxide, ClO_2 is known as chlorine dioxide, and N_2O_3 is named dinitrogen trioxide. Table 3.4 lists some common binary molecules and their names.

Binary molecules that contain both carbon and hydrogen are not named this way. This type of compound, because of the multiple ways that the atoms can be arranged,

Table 3.4 Common binary molecules

Formula	Name	Formula	Name
CO	Carbon monoxide	CO_2	Carbon dioxide
NO	Nitrogen monoxide	NO_2	Nitrogen dioxide
SO_2	Sulfur dioxide	N_2O_2	Dinitrogen dioxide
CS_2	Carbon disulfide	P_2S_4	Diphosphorus tetrasulfide
N_2O	Dinitrogen monoxide	N_2O_4	Dinitrogen tetroxide

requires a special set of naming rules. As we'll see later in this chapter, their names are based upon the specific arrangements of the atoms. The special rules emphasize the fact that with a larger number of atoms, many different ways to arrange the atoms become possible. Where with ionic compounds and binary molecules only one way to arrange the atoms is possible, only one name is needed to describe that arrangement. But with multiple ways to arrange the atoms, it is likely that each arrangement is a different compound with different properties. So, each arrangement needs a separate name. For example, there are two different compounds with the formula C_4H_{10} resulting from different arrangements of the atoms (butane and isobutane). The name of the molecule in these cases must be able to distinguish each of these molecules.

3.2   Laws that Govern Atoms, Molecules, and Ionic Compounds

Scientists began asking questions about the world many centuries ago. They mixed different compounds together and observed what happened. They discovered new compounds and new elements and explored their properties. In their search for explanations of the world, they discovered the underlying rules for how and why chemicals exist and react.

A Greek philosopher, Democritus (460–370 BC), came up with one of the earliest theories that helped shape the field of chemistry. Without experimentation, he reasoned that matter (the stuff around us) must be made up of small indivisible particles. For example, if you have a cup of sand and remove half of it, you would have a half a cup of sand. If you continued to divide it into halves, eventually you would get down to the point where you would have only one sand particle that couldn't be divided in half. Chemists later realized that this was correct. All matter is made up of particles (atoms), the smallest thing possible. This gave rise to our current understanding of the atom.

As scientists started experimenting with the world around them, they discovered that not all atoms were the same (in fact, we now know of 118 different elements that make up the periodic table). How those atoms combined to make different compounds became the focus of the majority of their research. The result was a series of laws that govern the modern practice of chemistry.

The first of these chemical laws makes sense in today's world, but at the time the law was being formulated, it was met with great skepticism. The Law of Constant Composition came to be understood by Joseph Proust (French chemist, 1754–1826). The essence of the law is that the formula for a compound is the same, no matter what the source of the compound. For example, water's formula is H_2O, whether it is water from a mountain spring or water obtained from a well. The formula for water is also the same whether it is obtained from natural sources or made in the laboratory. Thus "natural" vitamin C is exactly the same compound as "synthetic" vitamin C that is made in the lab. The formula for both is the same, the structures for both are the same, and the properties are the same.

About the same time that chemists understood how formula were constructed, John Dalton (English chemist, 1766–1844) proposed the Law of Multiple Proportions. The law states that atoms mix in whole number ratios as they form compounds. For the brewer, this means that a formula must contain whole numbers of atoms. For example, mixing oxygen atoms with a carbon atom allows the formation of CO or CO_2, but not $CO_{1.5}$. Again, this law seems logical to us in today's world, but at the time of its discovery this was groundbreaking information.

John Dalton expanded upon his understanding of formula and presented his atomic theory around 1806. This theory declared that (a) all matter is made up of atoms, (b) atoms are indivisible and indestructible, (c) compounds are formed by the combination of two or more atoms, and (d) chemical reactions are simply the rearrangement of atoms to form new compounds. While some slight modifications to this theory have been made since the 1800s due to our better understanding of chemistry, most of what he originally postulated in his theory is still correct today.

Let's go back to our understanding of water and look at the chemical equation below. It describes the reaction of hydrogen gas and oxygen gas to make gaseous water. Each of the compounds involved in the chemical reaction is represented by a separate formula made up of atoms. The equation also illustrates that the reaction occurs by simply rearranging atoms from the compounds of hydrogen and oxygen to make the product of the reaction, water. No new atoms are created, and none are destroyed in the reaction (there are the same number of atoms on the left side of the arrow as there are on the right side of the arrow).

$$2\ H_2(g) + O_2(g) \rightarrow H_2O(l) \qquad (3.1)$$

The equation also illustrates how we write chemical reactions. On the left side of the equation are the compounds that react together, known as the reactants. On the right side of the equation are the compounds that result from the reaction, known as the products. An arrow that can be read as "reacts to make" or "yields" separates the left and right side of the equation. The letter in parentheses after each chemical formula indicates the state of the compound. In this reaction, the hydrogen and oxygen are gases denoted with an italics "g." The water is denoted with an italics "l," indicating the water is liquid. Solid compounds are designated with an italics "s." The equation is also balanced. This means that it has the same number of each type of atom on both sides of the equation. For example, there are four hydrogen atoms on the left side (the reactants) and four hydrogen atoms on the right side (the products). The large number in front of each compound is used to help make sure the equation is balanced.

Overall, then, the equation can be read as "2 molecules of hydrogen gas and 1 molecule of oxygen gas react to make 2 molecules of liquid water." Note that the "1" is implied even though it is not specifically written in the equation.

Let's look at another example to further illustrate the use of equations. For example, when a solution of calcium bicarbonate is heated, it reacts to make solid calcium carbonate, carbon dioxide, and water (Eq. 3.2). This reaction is very important to the brewer that uses calcium-rich waters in their brewing process. It indicates that by heating the water, the amount of calcium dissolved in the water can be reduced (while also reducing some of the alkalinity of the water):

$$Ca(HCO_3)_2 (aq) \rightarrow CaCO_3 (s) + CO_2 (g) + H_2O(l) \qquad (3.2)$$

A quick look at the number of atoms on each side of the arrow tells us that the reaction is balanced without adding any numbers in front of the compounds. However, we do notice the italics "*aq*." This means that the compound is dissolved in water as an "aqueous solution." The equation can be read as "1 formula unit of aqueous calcium bicarbonate yields 1 formula unit of solid calcium carbonate, 1 molecule of gaseous carbon dioxide, and 1 molecule of liquid water." Note that because calcium bicarbonate and calcium carbonate are ionic compounds, they are referred to as formula units instead of as molecules.

Antoine Lavoisier (French chemist, 1743–1794) studied reactions like this. He noted that the weight of the compounds before the reaction began was the same as the total weight of all of the products after the reaction was complete. He noted that no weight was lost in any of the reactions that he performed. Today, we call this the Law of the Conservation of Mass. This law states that chemicals are neither created nor destroyed in a reaction; chemicals may change their identity, but the overall mass doesn't change. In the brewery, this means that if we add 10 pounds of crystal malt to the mash, we will get 10 pounds of extracted sugars and leftover grist (after drying it). While the law doesn't dictate how much of each product we will get, we know that the total mass of the products is the same as what we started with and nothing just disappears in the reaction. This law works for every chemical reaction.

In the late 1800s, scientists noted some interesting outcomes of their study of chemical formula. They realized that the formula for some chemical compounds were the same as others. For example, the formula for glucose and fructose are the same ($C_6H_{12}O_6$), even though the compounds were very different. This implied that while the formula for the two compounds was the same, the arrangements of the atoms in the molecules must be different. Glucose and fructose are examples of constitutional isomers. Their formulae are the same, but their constitution (how the atoms are arranged) is different. Thus, the arrangement of atoms in a chemical compound is very important to knowing the identity of the compound. And to carry that even further, the arrangement of atoms is important to understanding the properties of the compound.

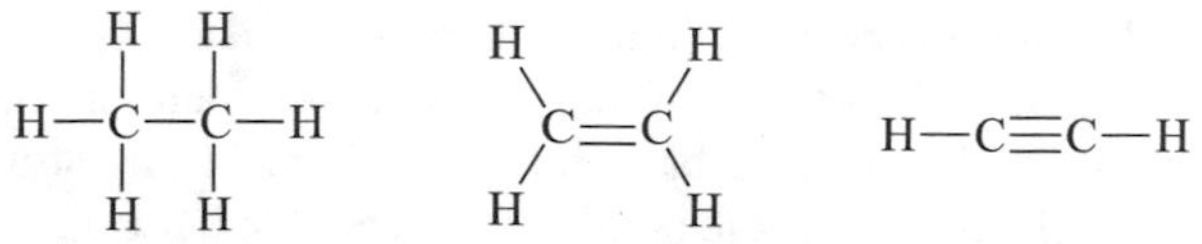

ethane   ethylene   acetylene

Fig. 3.2 Examples of molecules with multiple bonds. The number of bonds between the two carbon atoms in each structure is represented with multiple lines

Fig. 3.3 *R*-Carvone and *S*-Carvone are stereoisomers. They differ only in the three-dimensional arrangement of atoms in the structure of the molecules

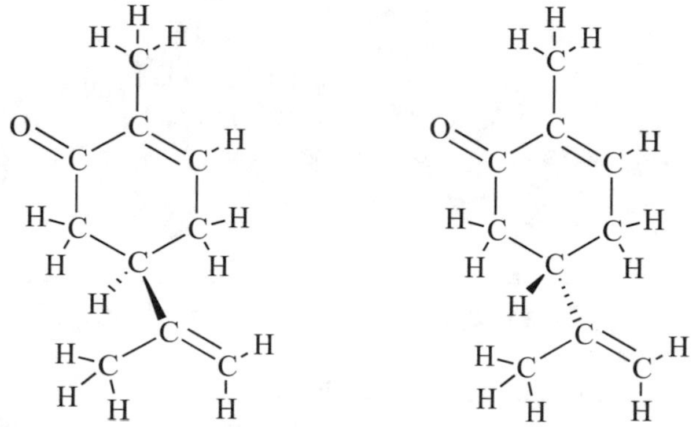

R-carvone
spearmint

S-carvone
caraway

How the atoms were arranged around each other in molecules was determined in the late 1800s. In some cases, it was determined that atoms were attached to other atoms by a single bond, such as the bonds that make up the molecule of water (H-O-H). In other cases, it was found that multiple bonds held the atoms together (Fig. 3.2). How these arrangements are constructed will be covered in the next section of this chapter.

In some cases, it was discovered that some compounds with the same formula and the same attachment of atoms still differed in the properties that they expressed. This difference was primarily found in how the compounds interacted in biological systems. For example, the active ingredient in caraway seeds and the active ingredient in spearmint are molecules with the same formula and same attachments of atoms (see Fig. 3.3). But, the taste of these two compounds is very different; one tastes like caraway and the other like spearmint. The difference in the two compounds results from the three-dimensional arrangement of the atoms. This is known as stereoisomerism. While the names of the two compounds are the same, the difference is noted by the addition of a single letter just before the name. The importance of this will be covered in greater detail in the next section.

Another example of stereoisomerism is even more striking. Glucose and galactose have the same formula and structure, but as stereoisomers, they are very different sugars. Both are sweet to the taste, but galactose is much less sweet than glucose.

Glucose is found almost everywhere (it is the sugar that makes up starch and cellulose); galactose is part of the sugar that makes up lactose found in milk. In another example, D-glucose and L-glucose have the same formula and structure, but are stereoisomers of each other. Both are similar in sweetness, but L-glucose cannot be metabolized by the body and cannot be used to make energy in the body.

CHECKPOINT 3.3

Write the formula for nitrogen disulfide, magnesium carbonate, and cobalt(II) phosphate.

Write the chemical reaction that is described by "one formula unit of solid calcium carbonate reacts to make one formula unit of solid calcium oxide and one gaseous molecule of carbon dioxide."

Based on the Law of Conservation of Mass, explain the "reaction" that occurs when noodles are cooked in boiling water.

3.3   The World of Carbon-Containing Molecules

There are millions of molecules that are known to exist in this world. And millions more are yet to be discovered. While many of these molecules do not contain the element carbon, a significant majority of them do. And many of those that do contain carbon are very important to life. In fact, an entire field of chemistry, known as organic chemistry, focuses on these molecules. Because many of the compounds found in beer contain carbon, the efforts of scientists that study organic chemistry have given the brewer, and the rest of the world, a much clearer picture of what is occurring during the brewing process.

Organic molecules containing carbon are often divided into different classes based upon the arrangement of the atoms in the compound. These classes allow the average scientist to estimate the properties of the millions of compounds without requiring personal experience with each molecule. Those arrangements that are most important to the brewer only contain the atoms carbon, hydrogen, oxygen, nitrogen, and/or sulfur. Thus, by studying those classes that contain these atoms, the brewer can get a very good sense of the reactions and properties that are possible. The arrangements of these atoms in an organic molecule are known as functional groups.

3.3.1   Basic Functional Groups in Brewing

Hydrocarbon The simplest molecules that contain carbon and hydrogen atoms are known as the hydrocarbons. These molecules contain only atoms of carbon and hydrogen, as the name implies. They are commonly found in nature as petroleum resources such as natural gas and oil, and in natural plant materials, such as waxes and oils.

The simplest of the hydrocarbons is methane (CH_4). Methane is one of the main components that make up natural gas. Methane is very useful as a fuel, and when it reacts with oxygen (combustion), the atoms rearrange to make carbon dioxide and water. The other product of this reaction is heat. It is the heat that is so important and the thing that makes methane so useful. The heat of the reaction with oxygen can be used to generate electricity, warm our houses, or heat our water. Note that based on the formula for methane, the carbon must be attached to four hydrogen atoms. This is a characteristic of carbon – it prefers to make four attachments or bonds to other atoms. This characteristic makes drawing the structure of organic molecules fairly easy. We'll cover this in more detail as we proceed through our exploration of functional groups.

Other hydrocarbons include ethane (C_2H_6), propane (C_3H_8), and butane (C_4H_{10}). Propane and butane are used a lot in the United States as fuels. Propane is the gas used in bar-b-que grills and butane is the gas used in cigarette lighters. Each of the hydrocarbons has fairly similar properties. They have fairly low boiling points and low melting points. In addition, as a class of compounds, the hydrocarbons tend not to be soluble in water in large quantities; however, very tiny amounts do dissolve. Some examples and their structures are shown in Fig. 3.4.

In some cases, the hydrocarbon is missing a pair of hydrogen atoms. This results in two options for the structure of the compound. It could be represented by a ring of carbon atoms, or it could be represented with a double bond between somewhere in the molecule. Compounds containing a ring of carbon atoms are very common in nature; the six-membered ring occurs in many compounds (see Figs. 3.3 and 3.5). If the molecule has a double bond, then it represents a new class of compound known as the alkene. Examples include ethylene and propene (Fig. 3.5). In other cases, the hydrocarbon is missing multiple pairs of hydrogen atoms. This gives rise to the alkyne class of compound. Examples include ethyne and propyne (Fig. 3.5). Both alkenes and alkynes are able to react with other chemicals to make new compounds.

Branching   While a long chain of carbon atoms can be drawn for a formula, many of the different ways to draw an organic molecule are the result of branching. This branching is the reason that a single formula results in multiple constitutional isomers. For example, the formula C_6H_{14} can give rise to a multitude of structures that represent branched compounds. Figure 3.6 lists the possible hydrocarbons with this formula. Note that each has a different name: the longest carbon chain forms the parent portion of the name. More on the naming of organic compounds is presented in Sect. 3.4.

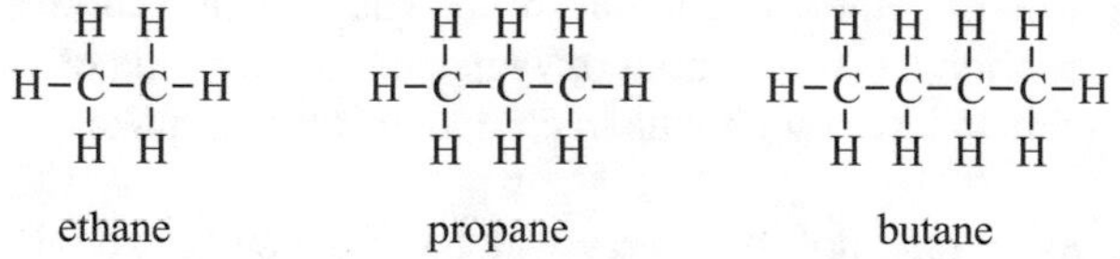

Fig. 3.4   The structure of some hydrocarbons, ethane, propane, and butane. Note that every carbon atom has 4 bonds and every hydrogen atom has 1 bond

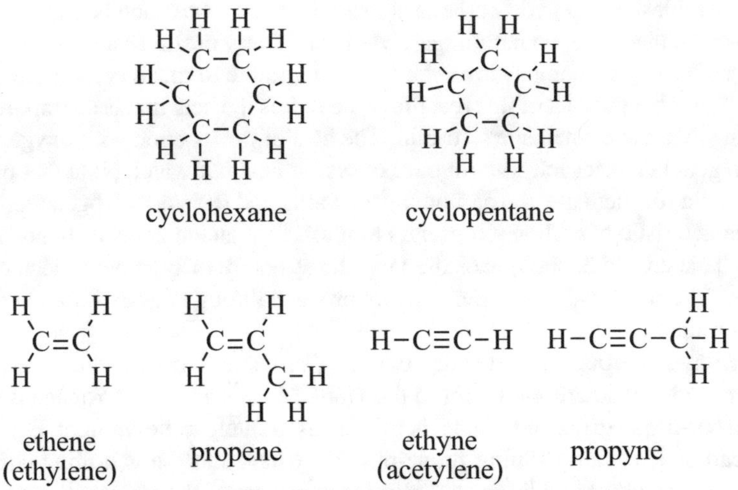

Fig. 3.5 Examples of hydrocarbons that are missing pairs of hydrogen atoms. Note that the carbon atom still has four bonds and the hydrogen atom requires one bond

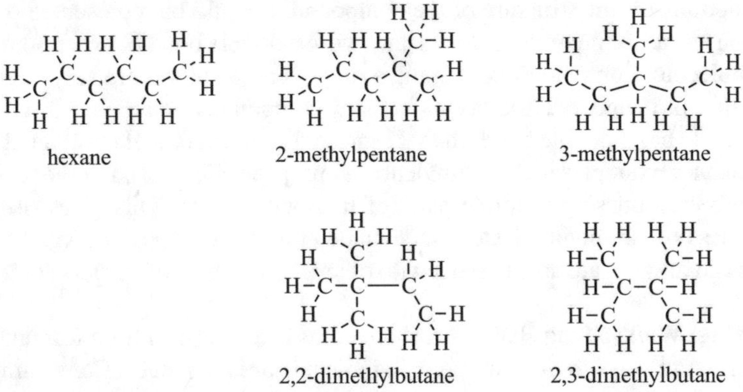

Fig. 3.6 The constitutional isomers of C6H14. Note that the longest carbon chain is the parent name (6 = hexane, 5 = pentane, 4 = butane) and that the branches are numbered along that chain

Ether If an oxygen atom is introduced into the hydrocarbon formula, there become quite a few different possible functional groups. In one arrangement, the oxygen atom is attached to two carbon atoms. This arrangement gives rise to a class of compound we know as the ether functional group. The structure of an ether, coupled with other information we know about oxygen from the formula of water, indicates that oxygen prefers to have two bonds in compounds that it makes.

As a class, ethers tend not to be very water-soluble and have fairly low boiling points (although, like hydrocarbons very small amounts of ethers can be water soluble). However, just like the hydrocarbon functional group, molecules that contain

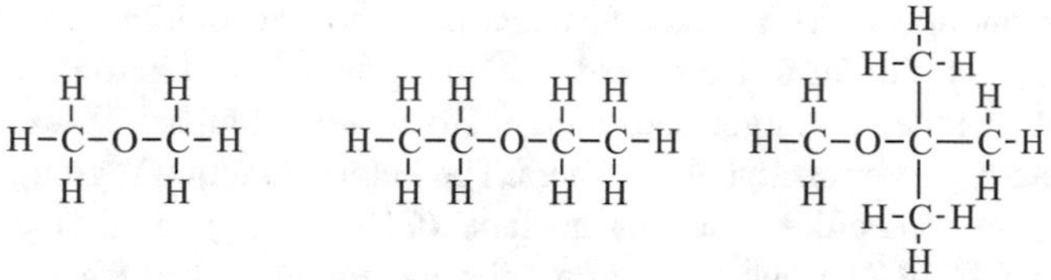

$$H-\overset{\overset{\displaystyle H}{|}}{\underset{\underset{\displaystyle H}{|}}{C}}-O-\overset{\overset{\displaystyle H}{|}}{\underset{\underset{\displaystyle H}{|}}{C}}-H$$

dimethyl ether

$$H-\overset{\overset{\displaystyle H}{|}}{\underset{\underset{\displaystyle H}{|}}{C}}-\overset{\overset{\displaystyle H}{|}}{\underset{\underset{\displaystyle H}{|}}{C}}-O-\overset{\overset{\displaystyle H}{|}}{\underset{\underset{\displaystyle H}{|}}{C}}-\overset{\overset{\displaystyle H}{|}}{\underset{\underset{\displaystyle H}{|}}{C}}-H$$

diethyl ether

methyl t-butyl ether

Fig. 3.7 The structure of some ethers. Note that the oxygen atom forms two bonds with other atoms in the structures it forms

Fig. 3.8 The structure of some common alcohols

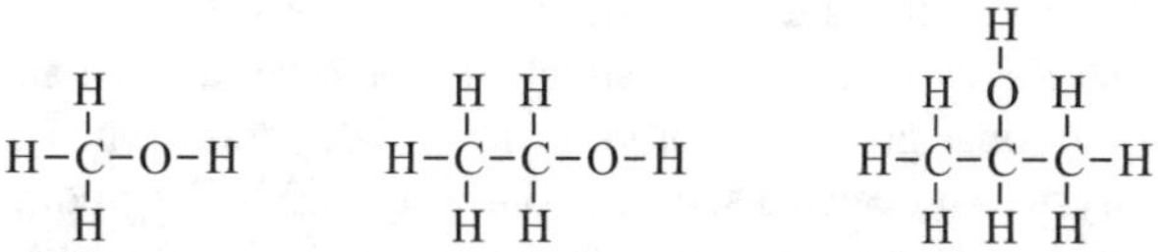

methanol   ethanol   isopropanol

the ether functional group are not very reactive. They have been used as fuel additives for automotive gasoline, and cause the octane rating to increase when added to gasoline. Examples include the oxygenated additive known as methyl-t-butyl ether (MTBE, $CH_3OC(CH_3)_3$), and the common solvent used in organic reactions, diethyl ether ($CH_3CH_2OCH_2CH_3$) as shown in Fig. 3.7.

Alcohol In another arrangement of carbon, hydrogen, and oxygen, the oxygen atom is attached to a carbon on one side and a hydrogen on the other. The result is the alcohol functional group. Yes, that's the same arrangement that we find in ethanol (CH_3CH_2OH), the alcohol in beer. Keep in mind, however, that formally, the word "alcohol" is just a word that tells us the arrangement of atoms in the molecule. While the word "alcohol" is often used as the common name for ethanol, it is a functional group name. There are many compounds that have the alcohol functional group, such as methanol (CH_3OH), isopropanol (rubbing alcohol, $(CH_3)_2CHOH$), and even glucose ($C_6H_{12}O_6$). As a class, the alcohols tend to be somewhat soluble in water and have very high boiling points. Unlike the ethers and hydrocarbons, alcohols have the ability to react with a wide variety of chemicals to make other compounds.

Ethanol is a small molecule that is infinitely soluble in water and has a boiling point of 78 °C. This doesn't seem like a very large boiling point, but comparing ethanol to propane (a molecule with about the same mass), it is much larger. Propane's boiling point is −188 °C (Fig. 3.8).

Ketone In another arrangement, the oxygen could have two attachments to the same carbon atom. This results in a double bond between the carbon and the oxygen atom in the molecule. This arrangement gives rise to a class of molecules known as a ketone (pronounced KEY-tone). Ketones have boiling points and solubility in water that is very similar to the ether functional group (small ketones tend to have some solubility in water, larger ones tend not to be very soluble). However, ketones are significantly more reactive than ethers. The ketone functional group is found in many common molecules, such as acetone (CH_3COCH_3), methyl ethyl ketone (MEK, $CH_3COCH_2CH_3$), and in more complex molecules such as fructose ($C_6H_{12}O_6$) as shown in Fig. 3.9.

Aldehyde A functional group related to the ketone is the aldehyde. This functional group requires a double attachment of an oxygen to a carbon atom, but also requires that the carbon atom be attached to another carbon and a hydrogen. The aldehyde functional group is common in carbohydrates such as glucose. Aldehydes have physical properties such as solubility and boiling point that are very similar to the ketones. However, the aldehydes tend to be fairly reactive. They are noted to undergo oxidation in air very easily. A common off-flavor in beer results from the presence of acetaldehyde, also known as ethanal (Fig. 3.10). Note the similarity of ethanol and ethanal.

Ester Another arrangement of oxygen into a carbon-containing molecule is known as the ester. This arrangement of atoms requires two oxygen atoms – one arranged in a similar fashion as an ether, and one arranged like a ketone (see Fig. 3.9). It is important that both oxygen atoms are attached to the same carbon atom; otherwise, the functional group isn't an ester. Such a molecule would contain two functional groups, the ether and the ketone.

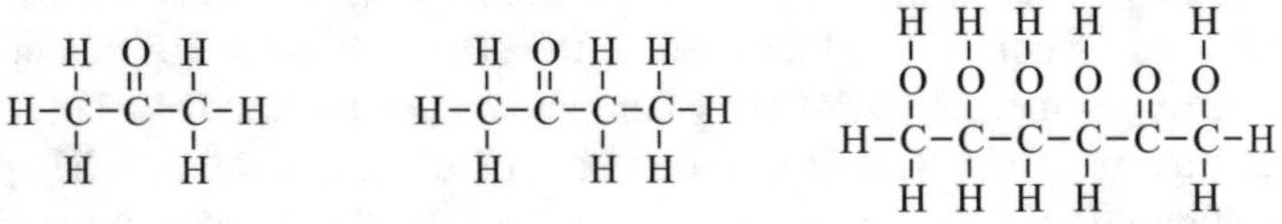

acetone   methyl ethyl ketone   fructose

Fig. 3.9 Some common ketones. Note that every carbon still has four bonds and every oxygen still has 2 bonds

Fig. 3.10 Acetaldehyde contains the aldehyde functional group

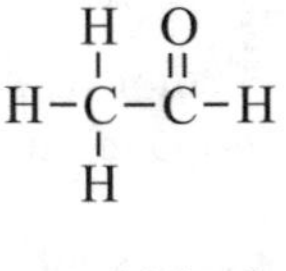

acetaldehyde

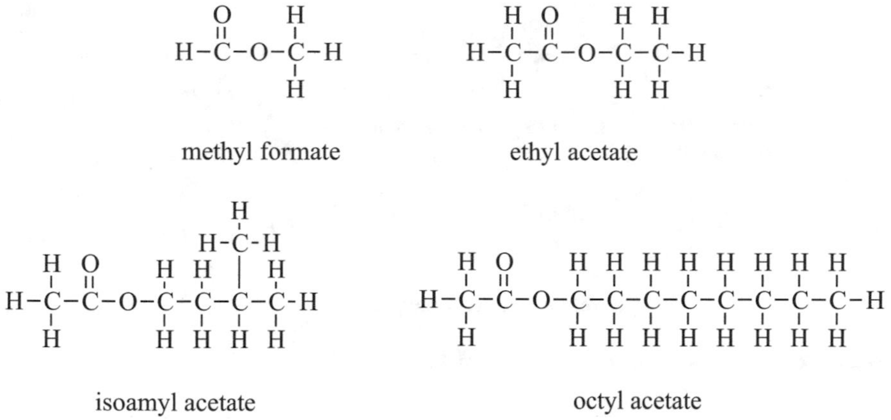

$$
\begin{array}{c}
\text{O} \quad \text{H} \\
\| \quad | \\
\text{H–C–O–C–H} \\
| \\
\text{H}
\end{array}
\qquad
\begin{array}{c}
\text{H O} \quad \text{H H} \\
| \; \| \quad | \; | \\
\text{H–C–C–O–C–C–H} \\
| \quad\quad | \; | \\
\text{H} \quad\quad \text{H H}
\end{array}
$$

methyl formate   ethyl acetate

$$
\begin{array}{c}
\text{H} \\
| \\
\text{H–C–H} \\
\text{H O} \quad \text{H H} \; | \;\; \text{H} \\
| \; \| \quad | \; | \quad | \\
\text{H–C–C–O–C–C–C–C–H} \\
| \quad\quad | \; | \; | \; | \\
\text{H} \quad\quad \text{H H H H}
\end{array}
\qquad
\begin{array}{c}
\text{H O} \quad \text{H H H H H H H H} \\
| \; \| \quad | \; | \; | \; | \; | \; | \; | \; | \\
\text{H–C–C–O–C–C–C–C–C–C–C–C–H} \\
| \quad\quad | \; | \; | \; | \; | \; | \; | \; | \\
\text{H} \quad\quad \text{H H H H H H H H}
\end{array}
$$

isoamyl acetate   octyl acetate

Fig. 3.11 Some esters

The result is a somewhat water-soluble compound that has an intermediate boiling point. In other words, the solubility in water is about half that of the alcohols and the boiling point is somewhere between the alcohols and ketones. The most interesting feature of molecules containing the ester functional group is that they tend to have a pleasant fruity odor. Examples of molecules containing this group include methyl formate, ethyl acetate, and isoamyl acetate. Methyl formate (boiling point 32 °C) has a strong plum-like odor and flavor, ethyl acetate (boiling point 77 °C) has a solventy odor and flavor, octyl acetate (boiling point 211 °C) has a distinct orange odor and flavor, and isoamyl acetate (boiling point 142 °C) has the taste and odor of banana (see Fig. 3.11).

In brewing, the esters play a very large role in the flavors of the finished product. Many are made in only very tiny quantities, but the strength of their flavor and odor are easily detected by the tongue and nose. For example, isoamyl acetate can be detected as low as 1.1 parts per million. This is equivalent to being able to taste one drop of flavor in 14 gallons of water.

Carboxylic Acid Another functional group containing two oxygen atoms and one carbon atom is closely related to the ester functional group. It differs from the ester functional group in that the combination of a ketone and an alcohol form the carboxylic acid. They tend to be fairly soluble in water and have relatively high boiling and melting points. But the most important property of the carboxylic acids is that they are acidic. When dissolved in water, the carboxylic acid functional group dissociates into an H^+ ion and the carboxylate anion. This causes the pH of the water to drop. While not as acidic as hydrochloric acid, the carboxylic acid is still an acid.

One of the important organoleptic properties of the carboxylic acids is that they have a sour taste and odor. For example, acetic acid is better known as the acid used in making vinegar (a 10% solution of acetic acid). Lactic acid and butyric acid are other examples of carboxylic acids. Figure 3.12 lists the structures of these compounds.

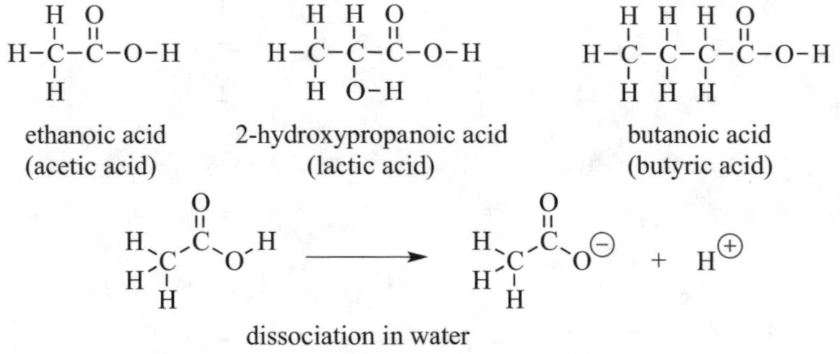

$$H-\overset{\displaystyle H}{\underset{\displaystyle H}{\overset{|}{\underset{|}{C}}}}-\overset{\displaystyle O}{\overset{||}{C}}-O-H$$

$$H-\overset{\displaystyle H}{\underset{\displaystyle H}{\overset{|}{\underset{|}{C}}}}-\overset{\displaystyle H}{\underset{\displaystyle O-H}{\overset{|}{\underset{|}{C}}}}-\overset{\displaystyle O}{\overset{||}{C}}-O-H$$

$$H-\overset{\displaystyle H}{\underset{\displaystyle H}{\overset{|}{\underset{|}{C}}}}-\overset{\displaystyle H}{\underset{\displaystyle H}{\overset{|}{\underset{|}{C}}}}-\overset{\displaystyle H}{\underset{\displaystyle H}{\overset{|}{\underset{|}{C}}}}-\overset{\displaystyle O}{\overset{||}{C}}-O-H$$

| ethanoic acid | 2-hydroxypropanoic acid | butanoic acid |
| (acetic acid) | (lactic acid) | (butyric acid) |

dissociation in water

Fig. 3.12 Some common carboxylic acids and the dissociation in water

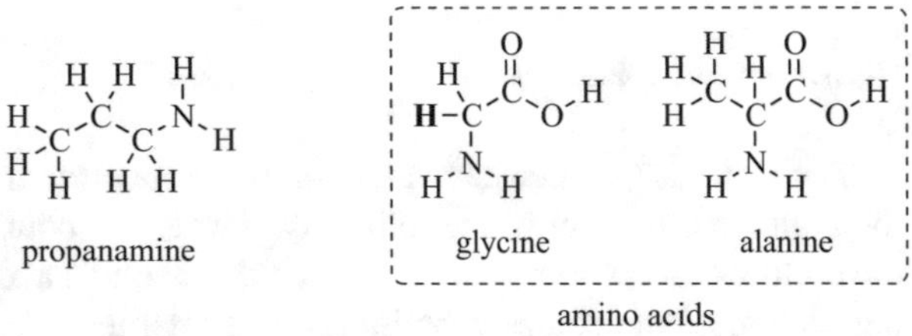

Fig. 3.13 Amine and amino acids. There are 20 amino acids used in the human body, only two are represented here. Each differs in the group that replaces the bold hydrogen in glycine

propanamine

glycine   alanine

amino acids

Amine The amino functional group is also encountered in brewing. This functional group contains a nitrogen atom attached to a carbon or hydrogen atoms. While the functional group does occur by itself in common organic molecules such as propanamine, the most common occurrence of the functional group in brewing is when it is combined in a molecule that already contains the carboxylic acid functional group. The result is the amino acid. Fig. 3.13 lists an amine and an amino acid.

> **CHECKPOINT 3.4**
> Identify the different functional groups found in carvone (Fig. 3.3).
> How many bonds do H, C, N, and O prefer in an organic molecule?

3.3.2   Amino Acid Polymers

A polymer is a molecule made up from a series of repeating smaller molecules. As we'll discover as we explore mashing, a very common naturally occurring polymer is starch. Starch, a very large molecule, is made up of glucose molecules attached together in a repeating fashion. In this example, starch is the polymer and glucose is the monomer. Cellulose is another example of a polymer made up of glucose monomers attached in a different configuration. DNA is a polymer made from nucleic acid monomers.

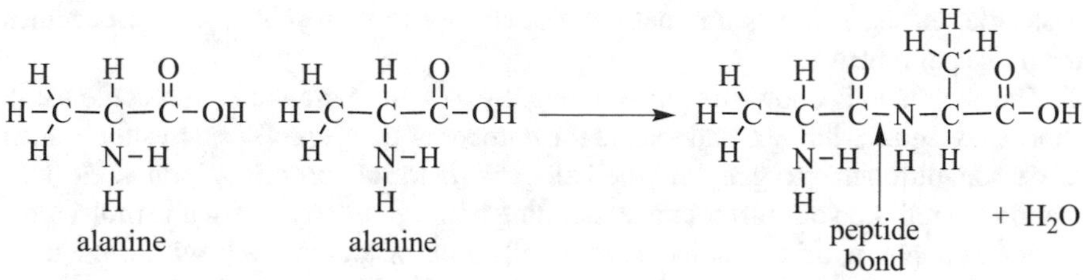

Fig. 3.14 The formation of the amide functional group in making a protein. The peptide bond in the amide functional group is noted

For the brewer, another important polymer arises from the combination of amino acids. The resulting polymer is known as a peptide if there are only a small number of amino acids in the chain. If the chain is fairly long, the amino acid polymer is known as a protein. Yes, a protein is simply a chain of amino acids that are chemically bonded together. The protein is formed in the body using enzymes and biological mechanisms, but can be simply as shown in Fig. 3.14. The product is the formation of the peptide bond and the elimination of water.

The peptide chain results from the chemical reaction of the amino end of one amino acid and the carboxylic acid end of another amino acid. The result is the formation of the amide (pronounced "AM – id" in the United States) functional group that contains the C=O bond and a nitrogen atom attached to the C=O. The new bond that is formed (the C-N bond) is known as the peptide bond. This bond is very rigid and difficult to break. For these reasons, the amino acid functional group is very stable and the proteins tend not to react with water or other compounds.

In some cases, the amino acid polymer aids reactions in the body or other biological system. The polymer in this case is known as an enzyme. On the surface, there is very little difference between a protein and an enzyme, other than the ability of an enzyme to aid the progress of other reactions. We'll uncover a lot about enzymes and the reactions that they help catalyze later in this text.

CHECKPOINT 3.5
Write the product of the reaction of glycine with another glycine.
How many different molecules (not counting the water molecule) are formed
   if a glycine reacts with an alanine?

3.3.3   Drawing Organic Molecules

Thus far we have looked at organic molecules by showing all of the atoms involved. Drawing these on paper takes a very long time, and if the molecule is sufficiently large, the clutter from showing every atom and bond makes it difficult to see the key structure or functional group in the molecule. A simpler way to draw these molecules exists, and it is the method we will use to represent organic molecules from this point forward.

The shorthand is known as line drawing. Successfully drawing a molecule using shorthand requires that we understand the number of bonds preferred by atoms such as carbon, nitrogen, oxygen, and the halogens (chlorine, bromine, iodine, etc.). In addition, it relies on our further understanding of the geometry of organic molecules.

For example, within a molecule is a collection of atoms, each with a nucleus. Those nuclei contain protons and neutrons and are positively charged. Because positive charges repel each other (just like two north pole ends of a magnet push apart), the atoms within a molecule push their nuclei to be as far apart as they can get. In other words, the atoms in a molecule do everything they can to be splayed out as far apart as possible.

The results are best represented by examples. For carbon, it is possible that it is attached to four other nuclei. In such a case, the carbon atom looks like a tetrahedron (and the angle between each adjacent nucleus is 109°). If the carbon participates in a double bond, then only three nuclei are attached. In that case, the best arrangement looks like the letter "Y" and the angles are 120°. If the carbon participates in a triple bond as it does in the alkyne functional group, then it is only attached to two other nuclei and is linear with bond angles of 180°.

Nitrogen prefers three bonds, and in the functional groups common to the brewer, it adopts a shape that looks like a three-legged stool. The bond angles in this type of arrangement are 109°. Oxygen when it has two single bonds looks like the letter "V" with bond angles of 109°.

Thus, when we draw molecules using the shorthand, we try to show the 109°, 120°, or 180° angles. As you'll see from the examples, it takes a little practice to get the drawings just right, but with that practice you'll be able to draw the molecules much faster than by showing every atom.

The rules to drawing a line drawing of a molecule are very straightforward. First, any hydrogen that is attached to a carbon atom can be omitted from the drawing. If the hydrogen is attached to an oxygen or nitrogen, it must still be represented in the drawing. Second, all carbon atoms are not drawn. Third, only the bonds connecting each atom are represented and done so to reflect the bond angles.

Let's practice drawing the line drawing for ethanol (Fig. 3.8). We start by placing our pencil on the paper (that's the first carbon), then drawing a line (that's the second carbon), nad then pointing off at about 109° we draw a line to the oxygen. Finally, we add the oxygen and the hydrogen to the drawing (see Fig. 3.15).

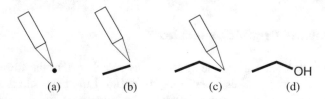

(a)   (b)   (c)   (d)

Fig. 3.15 How to draw a line drawing of ethanol (CH_3CH_2OH). We start with the pencil on the paper (**a**), draw a line to the next carbon atom (**b**), and finally to the oxygen atom (**c**). Then we add the oxygen and hydrogen atoms to the drawing (**d**)

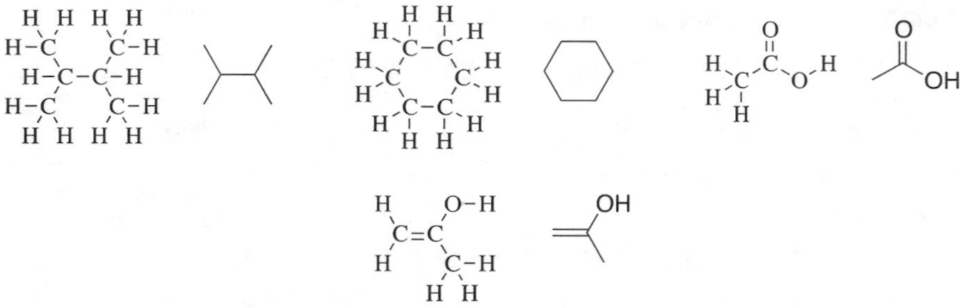

Fig. 3.16 Organic molecules and line drawings. Note the relationship between the fully expanded molecules and the simpler line drawings

Figure 3.16 shows a series of organic molecules and the line drawings that they represent. Note which atoms are omitted from the line drawings and also note the bond angles represented by the different structures. When performing the line drawings, it is very important to remember that even though some of the atoms are omitted, they are still there. This requires that we remember that carbon always makes four bonds, nitrogen makes three bonds, oxygen makes two bonds, and hydrogen and the halogens make only one bond.

CHECKPOINT 3.6

Rewrite the molecules in Fig. 3.5 using line drawings.
How many hydrogen atoms are there on each carbon in this line drawing?

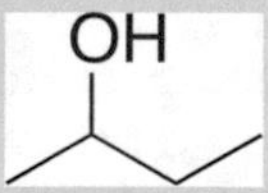

3.1.1

3.3.4 Naming Organic Molecules

While the average brewer will only be exposed to a couple hundred organic molecules (hop oils contain quite a few compounds), knowing how to systematically name the basic structures can provide a wealth of information on any molecule you might run into. Luckily, rather than being required to memorize the names of each of the million or so known organic compounds, the International Union of Pure and Applied Chemistry (IUPAC) has developed a set of rules to arrive at a molecule's name. These rules help us construct a name for a molecule if we know its structure. Unfortunately, many of the compounds in the brewery were identified and named before conventional rules were developed. We will introduce those names as well, but will focus on the use of the IUPAC nomenclature to identify these compounds.

Table 3.5 Rules for naming organic molecules

1	Identify the functional group and the ending for the molecule.			
	Alkane	ane	Alkene	ene
	Alkyne	yne	Alcohol	ol
	Ketone	one	Aldehyde	Al
	Ester	ate	Carboxylic acid	oic acid
2	Count the longest carbon chain and determine the name for that length.			
	1	meth	6	hex
	2	eth	7	hept
	3	prop	8	oct
	4	but	9	non
	5	pent	10	dec
3	Combine the functional group ending with the longest chain name. Often, two or more functional groups exist. Add both endings to the longest chain name.			
4	Number the chain to give the functional group (or branch) the smallest number. The functional group should always have the smallest number.			
5	Add the functional group to the end of the molecule name with its number.			
6	Add the branch name to the front of the molecule with its number. Branches are named using Rule #2, but end in "yl."			

Table 3.5 lists the basic steps in writing a name for a molecule. Consider we wish to provide the name of the compound shown here:

Let's use Table 3.5 to arrive at its name. First, we identify the functional group for the molecule (this compound is an alkane). Then, we count the longest chain of carbon atoms in the molecule. Using this, we write "pent" and add the ending for the functional group "ane" to arrive at the name "pentane." This method works very well for simple un-branched molecules.

For molecules that have a branch, we would then add the branch to the name. For example, consider that we have the molecule shown here:

We start by identifying the functional group (Rule #1). Here, again, this is an alkane. We count the longest carbon chain and arrive at pentane again (Rule #2, 3). However, this molecule has a branch. So, we number the carbon atoms to give the branch the lowest number. Then, we identify the group and add that, and the number, to the front of the name. The result here is 2-methylpentane.

If the functional group is not an alkane, then we must identify the carbon in the molecule that contains the functional group. Consider the molecule here as we name it:

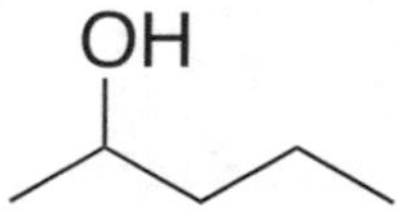

We start by recognizing this is an alcohol and an alkane. Then, we write the name for the longest carbon chain (pent) and add the functional group ending to the molecule name (pentanol). We add the number of the carbon that contains the non-alkane group. The result is pentan-2-ol. Note that we dropped the "e" from pentane when we added the "ol."

Ethers are named differently from other functional groups. The oxygen atom in these molecules is considered to be part of a branch. The ending of the branch then becomes "oxy" instead of "yl." For example, the name of these ethers follows this rule:

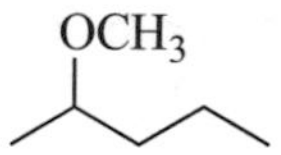

2-methoxypentane 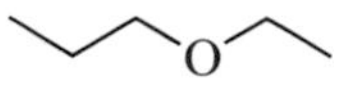   1-ethoxypropane

Esters are named very similarly to others, except that the group on the oxygen is placed in front of the name as a separate word ending in "yl." The rest of the rules apply. The names of some esters are shown here:

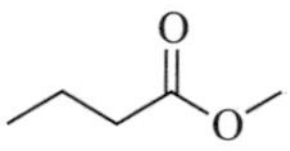

methyl butanoate 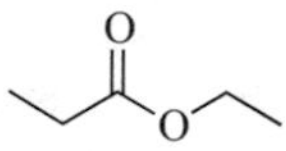   ethyl propanoate

---

CHECKPOINT 3.7
Name this molecule: $CH_3CH_2CH_2CH_2OH$

What is the name of this molecule?

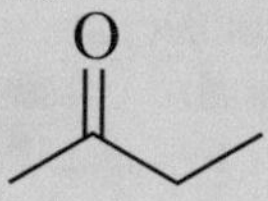

3.4   Reactions of Organic Molecules

Hundreds of different chemical reactions take place during the brewing process. The most common of these are the reaction of organic molecules with water, with oxygen, and with other molecules. These reactions result in a large number of new molecules, each of which has a significant impact on the resulting beer. In many cases, the brewer anticipates these reactions. And the flavors, colors, and properties the products lend to the finished beer are desirable. In some cases, the reactions are not desirable. Such reactions result in negative impacts on the beer.

3.4.1 Oxidation and Reduction

Oxidation and reduction are the reactions that change the number of chemical bonds to oxygen atoms within a molecule. Reduction is the process that results in a decrease in the number of bonds to oxygen; oxidation increases the number of bonds to oxygen. The brewing process is an oxidative process. In the hop plant, during the boil, the mash, and other processes, oxidation reactions predominate. This is due to the presence of oxygen. The chemical reactions during yeast metabolism are also focused on oxidation processes that convert sugars into carbon dioxide and ethanol.

Consider the following chemical reaction that occurs during the fermentation process (Fig. 3.17). If we count the number of bonds between carbon atoms and oxygen atoms in the molecule on the left (the reactant), we see six individual attachments (two single bonds and two double bonds). In the molecules on the right (the products), there are eight bonds between carbon atoms and oxygen atoms (four double bonds). This is an oxidation reaction.

3.4.2 Condensation Reactions

Condensation reactions are also common in brewing. This reaction type involves the combination of two smaller molecules to give a larger molecule and a very small stable molecule. Typically, that smaller stable molecule is water (H_2O). Condensation reactions, as a rule, result in the formation of larger molecules.

Consider the reaction of acetic acid and ethanol shown in Fig. 3.18. In this reaction, two molecules are joined together to make ethyl acetate with the elimination of a water molecule. The condensation of a carboxylic acid and an alcohol results in the formation of an ester. While esters are commonly formed during yeast metabolism, the particular condensation that occurs to form most of those esters occurs using slightly different reactant molecules.

The reverse of this reaction is also possible during the brewing process. In such cases, an ester reacts with water to form the carboxylic acid and an alcohol. Any reaction that involves water and is the reverse of the condensation reaction is known as a hydrolysis reaction. Hydrolysis (or reaction that breaks a molecule with water) reactions are quite common in the boil where hot water under acidic conditions adds to molecules and breaks them into smaller compounds. The hydrolysis reaction is one way in which larger proteins are decomposed into their individual amino acids (see Fig. 3.19).

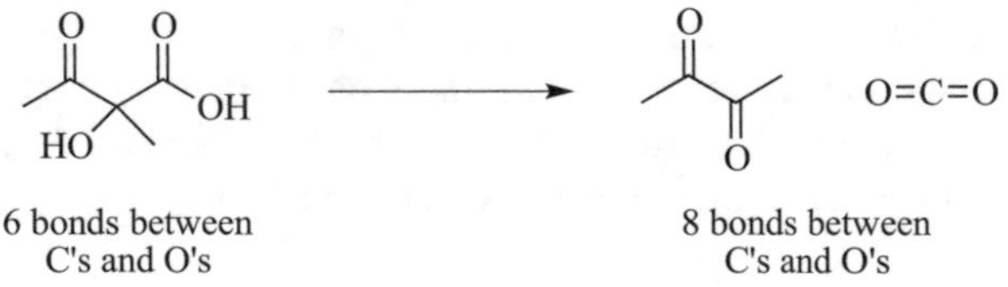

Fig. 3.17 The formation of diacetyl from alpha-acetolactate is an oxidation reaction

6 bonds between
C's and O's

8 bonds between
C's and O's

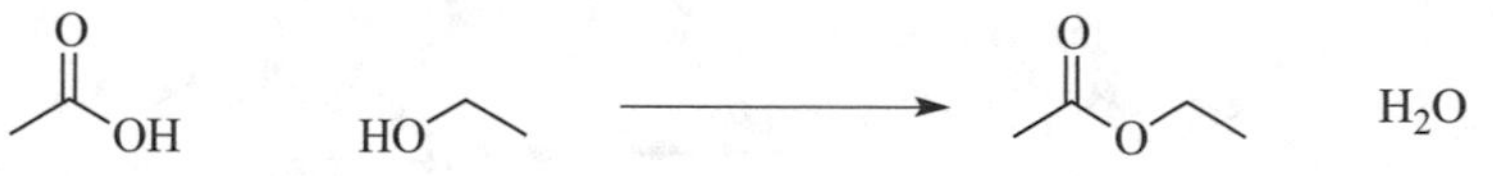

Fig. 3.18 The formation of an ester via a condensation reaction. Note the formation of water as one of the products

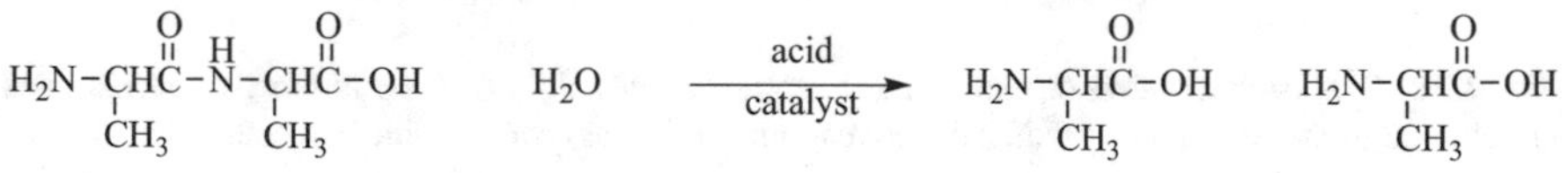

Fig. 3.19 Hydrolysis of a larger peptide or protein gives rise to the individual amino acids. Note that water is incorporated into the products

3.4.3 Isomerization Reactions

While there are not very many reactions of this type that occur in the brewing process, those that do take place are extremely significant. In the isomerization reaction, the structure of a single molecule becomes altered to form a new molecule. The new molecule ends up with properties that are different from the original molecule.

The best example of the isomerization reaction is the conversion of humulone (also known as α-acid) to isohumulone (also known as iso-α-acid) as shown in Fig. 3.20. Humulone is a molecule found in hop oil. During the boiling process, this molecule is converted via an isomerization reaction into the molecule that we associate with the bitter flavor of a hopped beer. In addition to differences in the flavor of the two molecules, there is a notable difference in the solubility of the molecules in water; isohumulone is nearly twice as soluble in water compared to humulone.

3.4.4 Radical Reactions

There are some radical reactions that take place in beer. These reactions are difficult to detect by looking at the reactants and products because the name of this type of reaction comes from how the reaction occurs. In the radical reaction, a very high-energy species, known as a radical, is formed during the reaction. This species is high energy because it contains a single unpaired electron.

In the course of the reaction, the single electron can be formed by the interaction of a molecule with sunlight. In many cases, this is the only way to tell that the reaction involves radicals. A common example of a radical reaction is the formation of the skunk-like flavor in light-struck beer, see Fig. 3.21. Note that by only considering the reactants and products, it is very difficult to determine that this is a radical reaction.

The mechanism, or process by which the reaction occurs, does show the formation of radicals. As shown in Fig. 3.22, the actual process involves the activation of a protein in beer by light. The protein then causes isohumulone to become activated.

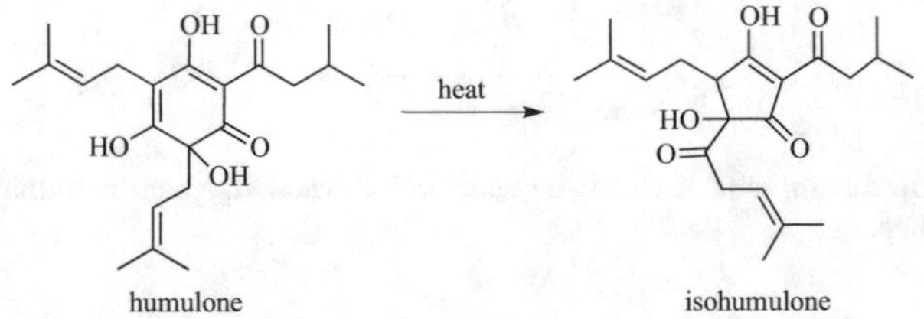

Fig. 3.20 The isomerization of humulone to isohumulone. Note that the number of atoms does not change in the reactant or product of the reaction, only the arrangement of the atoms changes

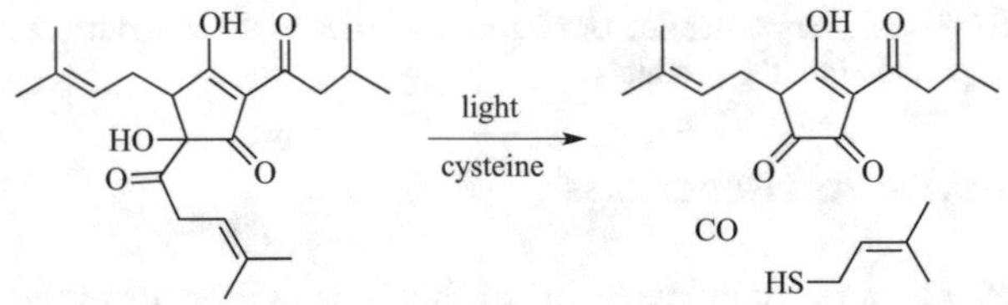

Fig. 3.21 The radical reaction of isohumulone with cysteine, an amino acid, forms the sulfur compound responsible for the skunk flavor in lightstruck beer

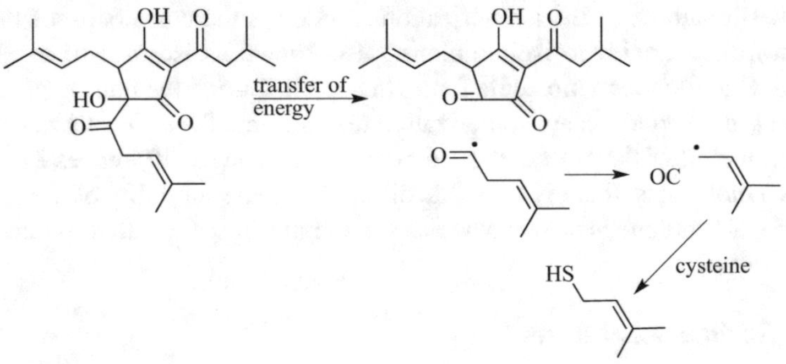

Fig. 3.22 The mechanism of the reaction of isohumulone with light. Note the formation of radicals in the formation of the sulfur-containing molecule

That molecule then cleaves into two, eliminates a molecule of carbon monoxide, and the radical reacts with a sulfur atom on cysteine to make the skunk-flavored molecule.

CHECKPOINT 3.8
Outline the key parts of the four reaction types described here.
Write the reaction that describes the hydrolysis of ethyl acetate.

3.4.5 Maillard Reactions

If we take a spoon of sugar and heat it up, we notice some very interesting changes. First, the sugar melts, then starts to brown. If we continue heating the sugar, it

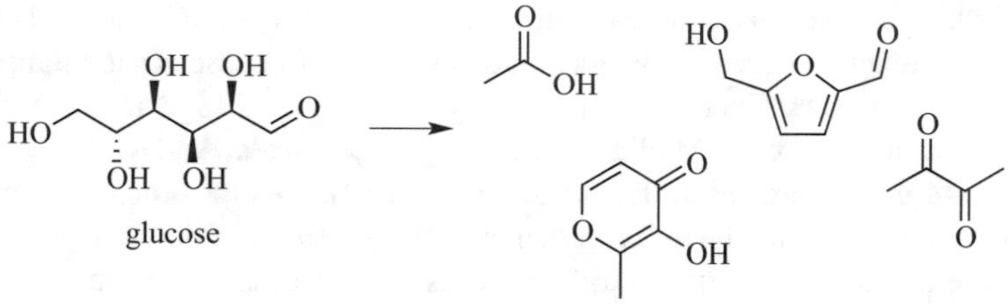

Fig. 3.23 Caramelization of sugar. This results in compounds containing oxygen, hydrogen, and carbon

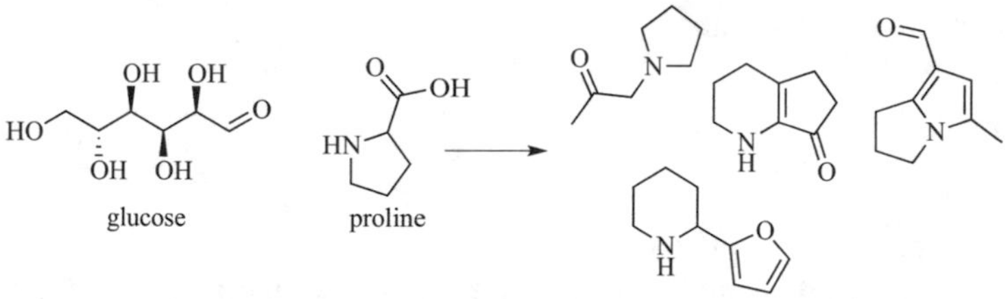

Fig. 3.24 The Maillard reaction of sugar and amino acids. Proline, an amino acid common in grain, is shown here. The products include those of the caramelization reaction (see Fig. 3.23), but also include cyclic compounds containing nitrogen

begins burning and becomes charcoal-like. These are chemical reactions that are taking place, changing the sugar molecule into new molecules. Many of these new molecules result from the loss of water, condensation reactions, isomerizations, and a host of other reactions. Some of the possible compounds formed are shown in Fig. 3.23.

This process is known as caramelization, and the resulting flavors of the heated spoon of sugar reflect that. Initially, the flavors don't change much, but once the sugars turn brown the flavors of caramel, toffee, and eventually charcoal are noticeable. These flavors are due to the larger cyclic compounds that are made in the reaction. Note that these compounds only contain carbon, hydrogen, and oxygen atoms.

Chemists studied caramelization extensively. By the 1900s, the reactions of sugar with other molecules were being explored. During his study of what happened when amino acids and sugars were heated together in the laboratory, Louis Camille Maillard (pronounced "may-yar") noted a similar caramelization. However, the result was the formation of a complex mixture of products that, while similar to caramelization reactions, contained nitrogen (see Fig. 3.24). These reactions are known as Maillard reactions. Chemists and food scientists continue to study the mechanism of the Maillard reactions to learn more about the compounds that are formed.

Maillard reactions occur when most foods are cooked because of the presence of both sugars and proteins. When we brown toast or grill a steak, evidence of the Maillard reaction appears as a brown color on the surface of the food. When malt is kilned, when wort is boiled, or when heat is added at any time during the brewing process, Maillard reactions occur.

While the mechanism of the reaction is rather complex, what is known is that the speed of the reaction between sugars and amino acids increases as the temperature of the heating process increases. If prolonged heating or high temperatures are involved, the amount of the Maillard products predominates. And with any increase in temperature, the caramel, toffee, toast, and other flavors become more and more noticeable. For example, the rich caramel and malty flavors of the bock style of beer result from the addition of heat during the mashing process (in a process known as decoction mashing).

The flavors of Maillard products are similar to those of caramelization. There is a slight difference in that the nitrogen-based compounds add an earthier, coffee-like tone to the flavors. In addition, these coffee-like flavors tend to appear earlier than in the absence of amino acids.

Chapter Summary

Section 3.1

All matter is made up of atoms.

How electrons are shared between atoms indicates the type of bond between those atoms.

Names for simple binary compounds have been developed.

Section 3.2

Laws and theories govern the rules of chemistry.

The laws and theories result in our understanding of how atoms are arranged in molecules.

Section 3.3

Molecules that contain carbon and hydrogen are classified as organic molecules. Organic molecules may contain other non-metals in addition to carbon.

The study of organic molecules is made easier by classifying them into functional groups, based on the arrangement of atoms in the molecule.

Line drawings are an easier way to represent organic molecules.

Rules for naming organic molecules exist.

Section 3.4

There are four main types of chemical reactions that occur in brewing.

Caramelization and Maillard reactions provide much of the toasty, caramel, toffee, and coffee flavors. These reactions also darken the color of the resulting beer.

Questions to Consider

1. Describe the hypothesis by Democritus in your own words.
2. According to Dalton's atomic theory, is it possible to convert lead into gold? Explain.
3. How many electrons, protons, and neutrons do F, S, Mg, B, and P have?

4. How many electrons do Na^+, N^{3-}, O^{2-}, and Cl^- have?
5. What is the likely formula for a compound made from boron and chlorine?
6. What is the likely formula for a compound made from lithium and oxygen?
7. What is the name for the compounds you described in #5 and #6?
8. What is the name for these compounds: CS_2, P_2S_3, N_2O_2, and ClO?
9. Draw all of the possible molecules with the formula C_5H_{12} and provide their names.
10. Draw all of the possible alcohols with the formula $C_4H_{10}O$ and provide the name of each.
11. Draw all of the possible compounds containing a C=O with the formula $C_5H_{10}O$ and name each one.
12. An ester has 4 carbon atoms. Draw all possible esters that satisfy this statement and name each one.
13. Show the condensation reaction that would provide each of the esters you drew in #12.
14. Based on your understanding of oxidation and reduction, arrange the oxygen containing functional groups in order of their oxidation state (most oxidized first).
15. An amide is formed between propanamine (see Fig. 3.13) and a 2 carbon carboxylic acid. Draw the condensation reaction of this reaction.
16. Explain, in your own words, the difference between a caramelization reaction and a Maillard reaction.
17. Using your own experience, explain the Law of Conservation of Mass.
18. The reaction of butane with gaseous oxygen (O2) produces carbon dioxide and water. Write this reaction out and balance it to make sure the number of atoms are the same on both sides.
19. In Fig. 3.23, one of the compounds identified as a product of the reaction is a carboxylic acid. Name it and propose a flavor for the compound.
20. Draw and name all of the ethers with the formula $C_4H_{10}O$.
21. What is the functional group in the molecule diacetyl? Propose a chemical name for this molecule.

Laboratory Exercises

Building Models in 3-D

This "experiment" is designed to familiarize you with the shapes of different molecules that you might encounter in the science of brewing. Organic molecules are not flat; they have a 3-D shape based on the number of atoms that are attached at each location in the molecule.

Equipment Needed

Organic Chemistry Model Kit – containing a minimum of 6 carbons, 12 hydrogen, 6 oxygens, and 1 nitrogen atom.

Experiment

In this experiment you will build models of each of the following molecules, draw a diagram of what you have built, and then answer the questions about each model you've built.

CH_4 – Describe the shape of this molecule in your own words.

H_2O – Does the shape of this molecule have a comparison to CH_4?

NH_3 – Compare the shape of this molecule to water and methane (CH_4).

C_2H_4 – Describe the shape of the carbon atom in this molecule.

C_2H_2 – Describe the shape of the carbon atom in this molecule. Is there a trend between CH_4, C_2H_4, and C_2H_2?

CH_3CH_2OH – Is there a relationship between the atoms in this molecule and the separate molecules of CH_4 and H_2O?

C_5H_{10} – Build a molecule that is cyclic. How many different molecules can you make?

C_5H_{12} – Build a linear version of this molecule. When stretched out as far as possible, describe the shape of the carbon chain.

$C_4H_{10}O$ – How many different molecules can you construct? Be sure to account for stereochemistry.

$C_6H_{12}O_6$ – Build a cyclic molecule containing one oxygen in the ring. This is a carbohydrate. Notice that there are many different options for arranging this molecule. Assuming that you place 5 carbons and 1 oxygen in the ring, how many different arrangements of the remaining atoms are there?

Overview of the Brewing Process

4

4.1 Overview of the Process

A good friend of mine cooks a very tasty Persian saffron rice dish. It is creamy, buttery, savory, and unlike anything this western palette has ever experienced. When I press my friend for how he makes the dish, his response is simply: *"First you plant the seeds."*

This is obviously a diversion to revealing the recipe, but there is some truth to recognizing, and revering, the beginnings of our culinary creations. The same is true for beer. We must recognize that this ancient brew begins with materials sourced from the earth. Since the beginning of the agricultural age, humans have cultivated the materials and refined the process and the art associated with the production of this beverage. Historically, the raw materials used in the production of beer depended almost exclusively on what grew in the local area. If the brewer couldn't grow a particular ingredient, buy it from a local farmer, or trade for it in the market, it wasn't included in the process for making beer. And while this statement appears to only impact grains and flavoring ingredients, it also applied to the materials used within the process to make the beer (kettles, methods to cool or heat, how the beer was stored, etc.) In fact, many of the current styles of beer can be traced back to different local materials or processes that worked satisfactorily for our ancestor-brewers.

In this section, we explore the basics of the main ingredients used in the modern production of barley-based beer. Let's start by considering the source of the sugars we'll need to make our beer.

4.1.1 Agricultural

First you plant the seeds. In most cases in the modern era, brewers did not plant, cultivate, or harvest the barley that they used. This still holds true today. But each

© The Author(s), under exclusive license to Springer Nature Switzerland AG 2021
M. Mosher, K. Trantham, *Brewing Science: A Multidisciplinary Approach*,
https://doi.org/10.1007/978-3-030-73419-0_4

brewer understood, just as we *must* understand, how barley is grown, harvested, and treated. Doing so, the brewer can recognize when the grain is suitable for brewing, and when the grain is sub-standard. After all, in order to brew beer worth consuming, one must begin with good ingredients. And it's always helpful to know where your beer ultimately comes from.

Barley (*Hordeum vulgare*) is a member of the grass family. It is likely that the barley plant was first cultivated in Mesopotamia around 10,000 BC, about the same time as wheat. Since that time, it has been cultivated, selected, and bred to produce the numerous different *cultivars* that are used currently. Each of these is grown for different purposes; barley destined for beer is often considered to be very high quality grain.

Barley is typically grown in cooler climates since it has a relatively short growing season compared to other cereal crops. In North America, it is grown in the northern continental United States and Canada where corn does not grow well. Most – about half – of the US barley produced is used as livestock feed. About one-quarter of US barley is ultimately used for brewing. Around the world, the United States doesn't even crack the top ten in barley production when compared to the production levels in Russia, Germany, and many other countries in the world.

There are three basic types of barley used when classifying the grain. Depending on how the spikelets and seeds are arranged on the central spine of the seed head, the barley is known as 2-row, 4-row, or 6-row. Two-row barley has two rows of seeds on each spike, but has one fertile floret per node. Four-row barley has four rows of seeds. Similarly, 6-row has six rows of seed on each spike, but only three fertile florets per node. Wild barley is the 2-row variety. The 4-row and 6-row varieties were historically selected after domestication and grown for different purposes. As a malted grain, 2-row barley tends to have a lower protein content. Four- and six-row barley have higher protein contents. While four-row barley is typically not used in brewing, the six-row variety has found widespread use. With modern brewing and malting processes, the slightly higher protein content is not much of an issue as we'll see in the next section (Fig. 4.1).

Once harvested, the barley is stored in silos and dried until it has less than 14% water content. After the seed reaches that level of water content, the barley can be sold. Often, however, it is stored for 1–2 months (or longer in some cases). While the reasons are not entirely understood, this storage helps reduce the dormancy of the seeds.

Prior to sale, however, the seeds are cleaned. This involves running a magnet through the seeds to pick up small pieces of metal that come from the harvesting process. The seeds also enter a cleaning drum that removes small broken pieces and stones. Finally, the seeds are sorted by size. This can be done using a screen sorter. In this machine, the seeds are poured onto a shaking screen. Seeds that are very large are shaken off of the screen and into a hopper. Seeds that are small enough to pass through the screen fall onto a second screen. If the seeds are too large to pass through the second screen, they are shaken into a second hopper. Those seeds that pass through the second screen are placed into a third hopper. In this way, the seeds are sorted into their relative sizes (large, medium, and small).

Fig. 4.1 2-row and 6-row barley. Photo by Xianmin Chang (xianmin.chang@ orkney.uhi.ac.uk)

The larger, plumper seeds are destined to be base malts for the brewer. These seeds have a large amount of starch that will yield a lot of sugar per unit weight. The medium-sized seeds are still usable by the brewer, but are converted into specialty malts instead of base malts. Unfortunately, the smallest seeds do not have enough starch to make it worthwhile for the brewer. Instead of just throwing them away, these seeds are often sold as feed for cattle.

Given that the farmers do everything to maximize their profit, production is geared toward making the plumpest seeds possible. This practice isn't always possible. Unfortunately, due to Mother Nature, the farmer sometimes struggles to correct for low rainfall amounts and temperatures that are not suitable for maximum yields. The result is that barley quality varies from year to year. The maltster and the brewer must account for this variability.

> **CHECKPOINT 4.1**
> What are the differences between 2-row and 6-row barley?
> Draw a sketch of the seed size sorter outlined in this section.

4.1.2   Malting

If we try to use freshly harvested barley to make our beer, we notice that essentially no sugar is extracted from the grain. Early civilizations probably also noted this and conducted fairly elaborate techniques (such as allowing the grain to rot, or making barley dough and bread before making beer). We now know that the required sugars are packed away in the endosperm of the seed as starch. These are energy stores

Fig. 4.2   6-row barley undergoing the sprouting process. From left to right: the seeds have been in contact with water for 0, 1, and 2 days, respectively. Rootlets begin to appear at the bottom of the barley seed

originally intended for a newly sprouting plant. To harvest that starch and convert it into sugars, the cleaned, sorted barley is sent to the maltster.

The maltster begins the process by soaking the seeds in water. This tricks the barley into thinking it is time to sprout. During this time, the "germ" part of the seed begins to grow into the acrospire, i.e., germinate (see Fig. 4.2). This process activates, and develops diastatic enzymes inside the seed. It also starts the process of unlocking starches from the endosperm. In short, the seed begins to grow by converting its stored starch into sugars. Unless halted, however, the seed will continue to grow, and consume all of the starches that are needed for brewing. Simply drying out the seed and heating it slightly halts the sprouting process. The rootlets that grew from the seed are then mechanically separated from the grain. At this point, the malted grain is ready for the brewer and is frequently called pale malt (Fig. 4.3).

Plain, pale malt accounts for up to 100% of a brewer's grain bill depending on beer style. The purpose of the pale malt in the recipe is to provide diastatic enzymes and at least some of the fermentable sugars. *Diastatic power*, a measure of the malt's ability to convert starches into sugars, is a measure of the ability of the malt to do this job. But, pale malt, by itself, doesn't provide the brewer with enough variability to make all of the styles of beer. So, to provide different flavor profiles required for those different beer styles, the brewer uses specialty malts.

Specialty malts are made in many different ways. Some are made by heating the pale malt in a kiln until it browns (Fig. 4.4). The browning or toasting of the malts occurs via the Maillard reactions we uncovered in Chap. 3. These reactions deepen and accentuate the malty-toasty flavor of the malt. As we will discover later, different degrees of heat will yield different flavor profiles for the malted grain. The only downside is that the excessive heat required to toast the grain will denature the enzymes and reduce the diastatic power of the malt.

Other specialty malts are made by soaking the pale malt in warm water. This activates the malts to begin sugar production, but since the pale malt roots and shoots have been damaged, the seed cannot grow. Instead, the sugars collect in and on the seeds. When these seeds are then dried and kilned, they result in a great source of caramel, toffee, and roasty malt flavors. The kilning process, however, also reduces the diastatic power to essentially zero. These sugary malts are known

Fig. 4.3 After 4 days in water, the rootlets are approximately the same length as the original seed. Note the fine hairs attached to the rootlets in the left panel. The acrospire is not yet visible because it grows beneath the husk of the seed

Fig. 4.4 Chocolate malt, crystal-40L, and wheat malt. Roasting the malted grain gives it deeper colors and flavors

as crystal malt (due to the presence of crystals of sugar on the surface of the seeds) or caramel malts (because of the flavors they provide.)

A brewer has a wide variety of malted and toasted grains to design different beer taste profiles. But, malting and toasting grain often requires large, dedicated areas to spread-out the grain, store the grain, and occasionally turn over the grains until the germination process is complete. Additional machines and space are needed for kilning. Brewers leave this process to specialists known as maltsters; very few breweries will take on this time, and space intensive step themselves.

4.1.3   Milling

In order to use the malt in the brewing process, the brewer has to break open the husks and expose the starch in the grain. Doing so also allows the enzymes that were formed during germination to be released once the crushed malt is added to water.

Fig. 4.5 Malted grain is gently crushed in preparation for mashing. Note the white-starchy centers of the malted grain and the largely intact husks

To achieve the crush on the malt that is desired, the brewer matches the size of the pieces of the crushed malt to the sparging process (see later in this section). In some cases, the malt is barely crushed (see Fig. 4.5). In other cases, the malt is essentially converted into flour. As a side note, the smaller the malt pieces, the faster they undergo the mashing process, but the slower they undergo sparging.

The equipment used to crush the malt is typically made of two or more large metal rollers. Malt is pulled into the gap between the rollers and crushed. The size of the gap is adjusted to provide the size of malt pieces that are desired.

4.1.4 Mashing

Once the malt is milled, it undergoes the process that differentiates this beverage from wine; mashing. Mashing can be considered an extension of malting, in that we are adding water again to the grain. This time, however, the maltster has removed the rootlets and kilned the malt. The result is that many of the enzymes needed for growth have been eliminated by the heat of the kilning process, and the lack of the root makes the seed unviable as a plant. So, the addition of water at this point doesn't result in growing a barley plant.

The malted grain is crushed to expose the white-starchy endosperm (see Fig. 4.5). The resulting *grist*, broken pieces of grain and husks, can be as fine as powder or only slightly crushed depending upon the brewer's requirements. The grist and warm water are mixed and added to the *mash tun*, giving us the *mash*. There is some acceptable variability to the water/grist ratio, but the mash typically has a consistency of fairly runny oatmeal.

In some cases, the brewer may use grains that haven't been malted and harvest the starch from them for the mash. This involves a separate step where crushed grains are added to a separate vessel known as a *cereal cooker*. Water is added to the grist and then the entire mixture is heated to the gelatinization temperature for the particular grain used (see Table 4.1). This step causes the starch in the endosperm of the grain to be released into the mixture. The entire mixture is then added to the mash tun. We'll explore this process in greater detail in a later chapter. It is important to note that the unmalted grains used in this process do not have active enzymes to convert the starches into sugars. Malt is required to provide those enzymes.

In the mashing step, enzymes developed during the germination (malting) process convert the available starches to sugars. The ultimate goal of mashing is to break down large starch molecules into sugars which can be digested by yeast, i.e., *saccharification*. The mashing process can involve several temperature rests depending on what process the brewer wishes to accomplish (create more sugars, reduce the amount of proteins, etc.). A *rest* is a period of time in which the mash is held at a specific temperature. These different rests are discussed in a later chapter.

In the modern age of brewing, nearly all malted grain available on the commercial market is fully modified – meaning that the maltster has germinated the seeds until all of the endosperm has been acted upon by enzymes. The maltster does offer some partially modified malts that require special mashing rest steps to finish the modification of the endosperm. However, for most styles of beer, the brewer acquires fully modified malt and really only needs to execute the saccharification rest. The mash temperature typically is held constant at a value between 140 °F and 155 °F where it rests for 60 min. Programmed temperature runs with multiple rests can reduce the time down to as little as 20 min, but this requires specialized equipment to accomplish. During this rest, the starches in the endosperm of the malted barley are steadily converted into fermentable and un-fermentable sugars. With appropriate starting water chemistry, a pH in the range of 5–5.5 is naturally established in the mash. This pH is essential for the saccharification enzymes to work.

There are two saccharification enzymes that remain active in the malt after kilning at the malster's factory. These enzymes become activated again during the mash. It is possible for the brewer to fine-tune the final taste profile of the beer by adjusting the temperature within the 140–155 °F range. As discussed in a later chapter on mashing, this can result in beer that is thin, alcoholic, thick, malty, or any combination of these profiles (Fig. 4.6).

At the end of the mash period, it is often desirable to halt the further conversion of long sugar chains to shorter, more fermentable chains. In this case, the brewer will execute a *mash-out* by raising the temperature of the mash just high enough to halt enzymatic activity and preserve the sugar profile. Typically, the final temperature is in the 176 °F range. This higher temperature also helps break down any proteins and glutens and helps reduce the viscosity of the mash significantly.

CHECKPOINT 4.2

What is the purpose of malting?

If an unmalted cereal was used in a recipe with the intention of converting its starches into fermentable sugars, what process would be required prior to its use?

Table 4.1 Gelatinization temperatures of commonly used unmalted grains

Grain	Temperature (°F)	Temperature (°C)
Barley	140–150	60–65
Wheat	136–147	58–64
Rye	135–158	57–70
Oats	127–138	53–59
Corn	143–165	62–74
Rice	154–172	68–78

Fig. 4.6 Crushed grain (grist) is added to warm water and allowed to sit during the Mash step. Typically the mash is held at about 155 °F for 60 min

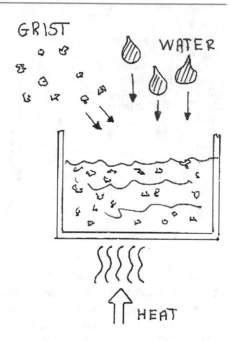

4.1.5 Lautering and Sparging

In this step, *the goal is to separate the sugar-water from the grain*. There are a variety of approaches to accomplish this task. In some instances, the entire mash (grist and sugar-water) is pumped into a different vessel known as a *lauter tun*. The water is then slowly drained from the grain while being gently rinsed with warm water (*sparging*). The lauter tun is wider and less deep than the mash tun. In addition, the lauter tun often contains a series of knives or rakes that can be rotated around the grain bed. By raising and lowering the knives, the grain bed can be gently stirred as it is being drained and rinsed. This helps ensure that all of the sugar is rinsed from the spent grains.

In some breweries, the grist has been ground into a relatively fine powder. This type of mash provides an increase in sugar production with a concomitant decrease in the time involved. One problem exists, though, in that filtering this type of mash using a lauter tun doesn't work. In this case, the mash is pumped into a *mash filter*. Inside the mash filter, bladders are used to squeeze the liquid away from the mash flour. Then, sparge water is added, and the mash flour squeezed again to remove all of the sugar water. This process, while requiring expensive equipment, is very efficient at extracting the sweet wort from the grist.

In smaller brewing operations, the mash tun and lauter tun are a combined vessel. In this case, the sparge water trickles through the deeper bed of grain and rinses it of the sugars. In both cases, the sparge water is often held at a fairly high temperature and relatively low pH. These characteristics help assure the brewer that all of the sugars are removed while reducing the possibility of extracting the tea-like

Fig. 4.7 General overview of the use of sparge water to rinse the mash. Sweet wort is separated from spent grain in the lauter/sparge step. This typically takes 60 min, or longer, to complete

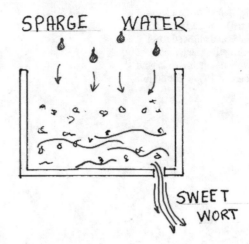

tannins from the husks. Both vessels also have a false bottom with a large number of small holes that allow the sugar-water to drain, but retain the spent grain.

The sugary, malty liquid that the brewer extracts from the mash is called *sweet wort*. It is a relatively thin liquid that contains some proteins, individual amino acids, metal cations (such as calcium, magnesium and others), and a mixture of fermentable and unfermentable sugars. If the sweet wort was just drained away from the grains, a significant amount of sugar would remain in the nooks and crannies of the grist. The problem is that the majority of the sweet wort is held in the matrix of the grist owing to a relatively large surface tension of the water, much in the same way that water adheres in a sponge (Fig. 4.7).

As we've seen, *sparging* is the process of rinsing the sweet wort away from the spent grain. Ideally, the sparge water will rinse away the concentrated sweet wort, leaving behind plain water in the spent grain. However, there are two primary issues that arise with sparging. With excessive amounts of sparge water, the final batch of sweet wort becomes much less concentrated. Unless the brewer invests the energy to concentrate the wort, the final beer will be weaker, both in taste and in final alcohol content. And, as we discovered above and will cover in more detail in a later chapter, unless the pH of the sparge water is controlled, the extraction of tannins from the *chaff* (husks of the grist) becomes a very big issue. This can result in a hazy beer with a very tea-like or astringent flavor.

For all but the mash filter process, there are variety of process methods that can used in the sparging step. Each type of sparging profile has its pros and cons and can be traced back to traditional procedures employed by different cultures. There are three basic procedures.

Continuous (or Fly) Sparge As water is drained from mash, a continuous supply of warm water is sprinkled above the grains at the same rate that sweet wort is drained. The water level is kept above the level of the mash grains. This has the potential to extract more of the sweet wort, resulting in better brew-house efficiencies over other sparging processes. The largest issue with this method is that as the sweet wort is drained from the mash, the additional sparge water causes the runoff to become less and less concentrated. This, in turn, causes the pH to rise back toward the pH of the sparge water. If the pH rises above 6, the potential exists for tannins to be extracted from the husks. The process requires an attentive brewer and must be stopped when the sugar concentration of the sweet wort drops below and/or the pH begins to rise above a predetermined point.

Cleanup of the mash tun in a continuous sparge process is somewhat involved. The excess sparge water must be drained away from the spent grains as the first step. It is possible that this water, unusable as sweet wort, could be used to mash in the next batch, assuming the brewer is interested in making another batch that day and that the pH of the water was still below 6. Then, the spent grains must be scooped out of the vessel. The water that has been absorbed by the grains makes this process very laborious. Finally, the spent grains must be removed from the brewery. Many brewers sell or donate these to area ranchers to be mixed with roughage for their animals.

No Sparge   Quite simply, the no sparge method is just that. The water from the grain bed is allowed to drain directly without addition of water. There is no risk of dilution of the sweet wort or extraction of tannins in this method. However, brewhouse efficiency is greatly reduced because a lot of usable sugars remain in the spent grains. This method is particularly useful for creating very high gravity beers, but it is fairly wasteful. Cleaning the vessel after the no sparge method is very similar to the batch sparge; however, there is no, or very little, wastewater from the process.

Batch Sparge   Similar to the no-sparge method, the batch sparge method allows the sweet wort to be drained away from the grain bed until no more liquid is removed. This batch of sweet wort is called the *first runnings* and is very rich in sugars. The first runnings, as in the no-sparge method, can be used to make a high alcohol beer. Then, the mash-lauter tun is re-filled with sparge water, mixed to allow the water to rinse all around the grains, allowed to settle, and then re-drained. The *second runnings* can be used to make a lower alcohol content beer, historically known as a small beer. The first and second runnings can be combined into one batch, or portioned out into two or more batches of different gravities. This is an improvement over no sparge approach in that efficiency improves with the second rinse. Again, care must be exercised to ensure that the pH in the second runnings has not changed to a point where tannins are extracted. In some cases, a third batch of sparge water could be obtained by adding more water, stirring the mash, allowing it to settle, and then draining off the liquid. Third runnings have very little fermentable sugars, but can still be fermented to prepare a very weakly alcoholic beer, or mixed in proportions with the other runnings to create beers of different alcohol content.

CHECKPOINT 4.3
Visit your local microbrewery. Which type of sparging system do they use?
    Describe its operation.
What is the purpose of sparging?

4.1.6 Boiling

The brewer then pumps the sweet wort into the *boil kettle*. It is then heated to boiling for a given period of time, typically 30–90 min. The exact time is determined by the outcome required by the brewer. In addition, there are several different types of boil kettles that can be used for the process, but we will cover them in much more detail in Chap. 8.

There are several reasons we boil the wort. We briefly touch on the main reasons here. Historically, the wort was boiled to kill harmful microorganisms and to allow the flavors of any other added ingredients to mix. This was important to early cultures as modern water purification procedures were nonexistent. Early beer was relatively weak, but since it had undergone this boiling step, it was safer (and healthier) to drink than the primitive water sources of the day. Interestingly, these cultures didn't understand the nature of microorganisms and didn't understand *why* boiling made the drink safer – it just did. Likely, the main reason for boiling the wort was to meld the flavors and the resulting boiled wort tasted better than the wort that was not boiled.

We now understand that boiling the wort elicits several other primary functions. It is during the wort boil that the brewer will add hops. Depending on when the hops are added to the boil, they can contribute the characteristic bitter flavor, provide the characteristic aroma, or give flavor to modern beers. Again, the brewer has many options that determine the final profile of the hops in the beer. Hops give us the crisp, citrusy flavor found in pale ales, the bitter taste of a stout, and the characteristic aroma of India Pale Ales (IPAs). Even in beers such as browns and porters where hops aren't as pronounced, the addition of hops help increase the shelf life of the final product and provide a taste balance with the alcohol in the beer. The chemistry behind the wort boil is discussed in Chap. 8.

Typically, a brewer will boil the wort with hops for 60 min in order to provide the characteristic bitter flavor we associated with modern malt-based beers. There may be other hop additions later in the boil, but at the end of the boil period no matter when the hops were added, the liquid is now called *bitter wort* (Fig. 4.8).

An important secondary effect of boiling the wort is to increase the relative sugar content as water is evaporated from the wort. This will eventually give us a higher alcohol content in the final beer, but with reduced volume. Another effect that occurs is the additional browning, the Maillard reactions. For the same reason that malts are toasted, this will give the beer a more robust, malty flavor.

At the end of the boiling time, the bitter wort is pumped out of the vessel and then back into the vessel at an angle. This creates a *whirlpool*. During the whirlpool process, which can last as long as 60 min, the proteins and other coagulated materials fall to the center of the vessel forming a cone of sludge known as *trub*. The bitter wort is then pumped away from this sludge and through a chiller. As it moves through the chiller, it is cooled to the perfect temperature to begin fermentation.

Fig. 4.8 The sweet wort is boiled with hops during the wort boil. Typically, this step takes 30–90 min

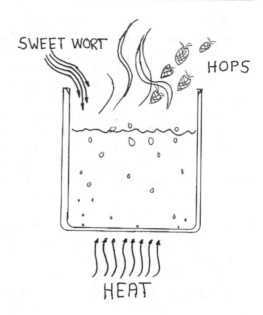

4.1.7 Fermenting

The cooled bitter wort is pumped into a fermentation vessel. As it moves into the fermenter, oxygen gas is added to the stream of fast moving liquid. The process of moving liquid around in the brewery is often called *racking*. So, using the brewer's vernacular, the bitter wort is racked to the fermenter. Then, the cool oxygenated bitter wort is inoculated with yeast, known as *pitching* yeast into the wort.

It has been said that brewers make wort, but yeast make beer. The importance of these microscopic fungi was not well understood in the early days of brewing. Even the German purity law, the *Reinheitsgebot*, did not initially recognize the importance of this ingredient in the overall process. It wasn't until the work of Louis Pasteur (1822–1895), and others, that we began to understand just how important yeast were to brewing. The biological metabolism of sugars by yeast is quite complicated and will be discussed in detail later. But we summarize the yeast's function from a macroscopic, if not empirical, point of view here. The purpose of the yeast from the perspective of the brewer is to consume the sugars in the bitter wort, and leave behind alcohol.

Again, as in every step in the process, the brewer has a wide range of strains and species of yeast to choose from. Due to subtle differences in the metabolism of each different strains of yeast, the composition of the bitter wort, and the temperatures used by the brewer, each yeast strain will yield different by-products and impart the beer with a distinct flavor. For example, the banana and clove flavors common to hefeweizen are due to minute by-products excreted from certain yeast strains during fermentation. Fermentation temperatures can also influence the final taste profile of the beer.

Fermentation can take 2–14 days for ales, and lagers may take multiple months to complete the fermentation cycle. During the process, the dormant yeast falls to the bottom of the vessel and is removed. Yeast fall out of the beer by flocculating. This process, requiring calcium ions in the beer in order to be most effective, clumps the single-celled organisms together causing them to become too heavy to stay dissolved in the beer. Different yeast strains have different propensities for doing this and are often characterized by their ability to do so. For example, highly flocculating yeasts will tend to settle and clump together very quickly after fermentation. Highly flocculating yeasts may need to be re-suspended in the beer ("roused") to complete the fermentation cycle. On the other hand, low flocculation yeast tend to remain in suspension. This is the cause of the cloudiness we see Hefeweizen and some Belgian style beers. In between, medium flocculating yeast tend to produce a cleaner beer because they stay active long enough to re-absorb diacetyl and other byproducts. They take some time, but will eventually settle to the bottom of the vessel making it easier to separate and filter from the beer.

After the fermentation cycle is complete, we have what brewers refer to as *green beer*. Green beer is drinkable, but the refined flavors haven't stabilized. So, the green beer is racked to a conditioning tank to allow that process to be accomplished. In some cases, the fermentation tanks are pressurized, allowing the carbon dioxide that is formed by the yeast to carbonate the beer. In other cases, the tanks are open to the atmosphere and the addition or adjustment of the carbon dioxide level must be accomplished during conditioning.

CHECKPOINT 4.4

Outline as many reasons as you can why the modern brewer boils the wort before fermentation.

Louis Pasteur was responsible for more than just the recognition of yeast as a requirement for brewing. What other process in food production was he responsible for (hint: the process bears his name)?

4.1.8   Maturing

At this point in the process the green beer that we have is not very tasty. The biological process of fermentation has produced a number of by-products such as acetaldehyde, diacetyl, and dimethyl sulfide. Combined, these flavors will give our beer an unpleasant "green" taste. If given sufficient time, the yeast that inhabit the beer may eventually re-absorb many of these compounds, so we allow time for the product to sit and "condition" or "mature."

During this period of rest dormant yeast, heavier and larger proteins, and other debris continue to precipitate from the beer. This helps to further clarify the beer. In the conditioning tank, the brewer also can adjust the beer. Gelatins can be added to help proteins and yeast precipitate, flavoring agents can be added to adjust the flavor, coloring agents can be added to increase the color, and reduced hop iso-acids can be added to adjust the bitterness of the beer. Each of these processes adjusts the beer so that the final product is within the brewer's specifications.

Bright beer refers to beer that has gone through this process and is ready to serve. In the case of natural conditioning, the yeast will eventually settle to the bottom of the storage vessel to make bright beer. However, even though it has been allowed to sit, bright beer will still have active yeast dissolved or in suspension (typical yeast counts of one million cells per milliliter are common).

Naturally conditioned beer requires at least a week or two for ales, depending on overall alcohol content, and up to 6 months or even more for lagers. This difference is due to the temperatures at which the yeast prefer to work. Lager yeasts grow and ferment at cooler temperatures, so the metabolism of the by-products takes longer. Because the natural process is relatively slow, the brewer often intervenes.

One way for the brewer to intervene is known as the diacetyl rest. While this is sometimes associated with the fermentation process, it is actually part of maturation. During the rest, the fermentation temperature is increased a little (an increase of 1 °C to 5 °C higher than fermentation temperature). The result speeds up the maturation process.

In Chap. 10, we explore the details of the maturation process and uncover why it is so important to the production of beer with fully developed and stabilized flavor.

4.1.9 Filtering

Once the beer has undergone the full maturation process, it can be filtered to remove any of the solids and haze that remains. The solids and haze can be made of a number of different things, but are primarily the result of excess proteins and tannins in the beer that weren't removed during the boiling or fermentation process. These compounds bind together until they become insoluble in the beer and form either haze or solids that slowly settle. Removing these from the beer improves the clarity and increases the shelf life of the beer.

But not all beer brands are completely filtered or clarified. In fact, there are many beer styles and particular brands that avoid this step in the production process entirely. For example, the Hazy IPA, also known as the East Coast IPA, is purposefully left hazy. However, in most cases, the brewer is very much interested in making sure that the beer has attained a level of *colloidal stability*. This means that the beer will not change the amount of haze or solid formation during its shelf life.

The clarification process involves a number of different options for making *bright beer*. The brewer could mechanically filter the beer to separate the yeast. There are a multitude of possibilities for separating proteins and yeast from the beer, including plate filters, leaf filters, and centrifuges. Unfortunately, mechanically removing the yeast before the yeast has had time to reabsorb byproducts could lead to beer that is permanently out of specifications. Alternatively, the brewer can add compounds and reagents to separate the solids and haze naturally by settling. In Chap. 11, we explore colloids, how they form, and how they are removed using mechanical and non-mechanical filtration systems.

4.1.10 Packaging

Packaging is the process that transfers beer from the brewery to the consumer. In some cases, this can be accomplished by placing the beer in serving tanks located at the point of sale. However, most beer is packaged in a *large pack* format (often referred to as kegs) or in a *small pack* format such as cans or bottles.

In Chap. 12, we'll re-explore how beer is carbonated and the requirements to maintain that level of carbonation as the beer moves from brewing to packaging. We'll also explore the different requirements and processes that are undertaken to package the beer in both small and large pack.

Most breweries treat their beer so that the shelf life after packaging is as long as possible. The treatment, known as pasteurization, kills any remaining yeast that are suspended in the beer. Residual yeast can continue to absorb molecules in the finished product and produce new compounds that would result in a flavor change over time. In addition, any bacteria that accidentally entered the beer are removed. Bacterial infections in a brewery are possible, and if they occur, they can impart undesirable flavors (such as sour, rotten, and bitter flavors). Removing these bacteria and yeast stabilizes the flavor and results in a product that doesn't change its flavor much, if kept cold.

In Chap. 12 we also explore pasteurization and evaluate how and why this is important to the brewer.

4.2 Cleaning and Sterilizing

Just as it is important to start with quality ingredients, it is imperative to maintain a clean brewery. This is particularly true for any equipment that is used post-wort boil, since the boil will kill any microorganisms acquired during the mashing process. The issue is that other microorganisms that are ubiquitous in nature will be just as happy consuming the fermentable sugars in the bitter wort as yeast will. The uncontrolled fermentation by foreign microorganisms will output a significant amount of undesirable off-flavors, mouthfeel, and visual appearance.

In our casual discussion of removing unwanted microorganisms, several terms tend to be used interchangeably: sterilize, disinfect, and sanitize. However, these have very well-defined, and different definitions in the brewing and food industries. To sterilize something means to completely kill all microorganisms. This generally requires excessive heat applied to the brewing equipment for an extended period. For example, a surgeon will sterilize scalpels and other operating tools in an autoclave. Unfortunately, killing all microorganisms is nearly impossible to do. In nearly every process, some microbes survive. Disinfecting means to kill all *unwanted* microorganisms which could cause spoilage. Unfortunately, this also doesn't often mean that all microorganisms are eliminated. Finally, sanitization just means we will eliminate most of the unwanted organisms.

It just isn't practical, or necessary, to completely sterilize brewery equipment of all indigenous, wild microorganisms. After all, beer has been successfully brewed long before the role of yeast or other wild microorganisms were understood. In those days, it was pure chance that a "good" strain of wild yeast would inoculate the wort and start fermenting the beer. In fact, stirring sticks from successful batches were carefully saved and reused so that a particular "house flavor" would be passed to the next brew.

Rather than sterilizing, we clean the brewery and equipment as best we can and sanitize with a chemical agent, eliminating most of the undesired microorganisms. The goal of the brewer then is to get the fermentation started with *our* selected yeast before the other microorganisms have a chance to build a significant population. Called "microbial antagonism," the yeast are able to build sufficient numbers such that it interferes with the normal growth of the other, undesirable microbes. And once the level of alcohol in the beer has reached a certain level, the alcohol itself serves as an antimicrobial agent.

Unfortunately, the dirtiest place in the brewery is the floor. The floor of the brewery is always wet, walked upon by the workers, and spilled upon with sweet wort, green beer, and other compounds used in brewing. Reduction of brewery contaminants is a result of those brewers that make sure to clean the entire space where the brewery exists. This may even include the outside walls of the brewery, and making sure that vegetation doesn't grow too close to the brewery itself.

Safety is a significant issue that all brewers must adhere to when using chemical agents to clean the brewery. For this reason, all brewers and workers in the brewery must know and understand the principles and processes for cleaning equipment. For example, some chemical agents will react with carbon dioxide. Many examples exist of cleaning agents being added to a freshly emptied fermenter and seeing the reaction of carbon dioxide and cleaning agent cause the fermenter to implode. Other examples exist of untrained workers entering a vessel to clean it using toxic chemicals, and expiring inside the confined space. Care and safety are necessary when using these agents.

For the safety of workers, most equipment is fitted with spray balls. In these cases, cleaner is cycled through the vessels using a pump that pushes the liquid through a spray ball. The result is known as *clean-in-place* (CIP). This allows the brewer to clean a vessel in the shortest amount of time possible, while reducing the exposure of workers to harmful chemicals.

CHECKPOINT 4.5

Assume the brewer wishes to NOT clean the boil kettle between batches of beer. Describe what you might expect to happen to the beers in later batches.

Outline at least three things that the brewer could adjust in the conditioning phase of the beer.

4.3   Inputs and Outputs

The most important part of brewing is making sure that the ingredients used in the process are the highest quality that they can be. In essence, a good beer can arise from good ingredients. It is very difficult to make a good beer from inferior ingredients.

4.3.1   Water

In the past, brewing ingredients and procedures were developed using local ingredients. This is still true today considering the first main ingredient; water. By percent volume of the final product, water is the most significant ingredient in beer – and, with the largest total weight, an ingredient that is difficult to ship from place to place. So even today, brewers source their main ingredient locally. Since clean water by itself has very little taste, it is easy to overlook the contribution of water as an ingredient. But just as one must "plant the seeds" to start with quality grain, one must also consider the importance of water. We will look more closely at water and its makeup in Sect. 5.4.

Early brewing culture drew water from the ground (well water) or from the surface (lakes, streams, springs, rivers, etc.). As it happens, ground water contains a significant amount of dissolved salts and minerals from the surrounding bedrock – all of this depending on the geology of the area. It is quite fortuitous that many of the naturally dissolved minerals contributing to the hardness of ground water are essential for brewing.

Owing to the different geology of different areas, brewing evolved differently depending on the water chemistry in the area. For example, regions with higher levels of bicarbonate, such as London or Dublin, are known for their darker beers. High levels of bicarbonates tend to make the pH of the water too high, effecting both mash efficiency and yeast metabolism. Adding highly roasted grain – darker roasted malt – tends to lower the pH back to the optimum acidic values. So, by accident or by trial and error, the brewers with this type of water who brewed darker beers such as porters and stouts noticed that the beer tasted very good. This style of beer is *appropriate* for the local water chemistry. On the other hand, the water from wells near Burton-on-Trent is high in gypsum (calcium sulfate). And coincidentally, this type of water is very efficient at enhancing hop aromas and flavors. Thus, the early brewers in Burton-on-Trent developed a very characteristic Pale Ale.

Early brewers brewed beers appropriate for their local water chemistry. Today we have the option to start with purified water, obtained by either distillation or reverse osmosis (RO), and adding measured amounts of minerals and salts appropriate for the beer to be brewed. In the brewery, this means that city-provided water is often treated prior to its use. Or, a brewer can start with water that is available, either from a well or from a municipality, and make small corrections. However, an accurate accounting of the starting water chemistry is required. One step is clear, though. City water supplies that are treated with chlorine, fluoride, or other compounds must be treated at the brewery to remove these agents. In addition to their ability to alter

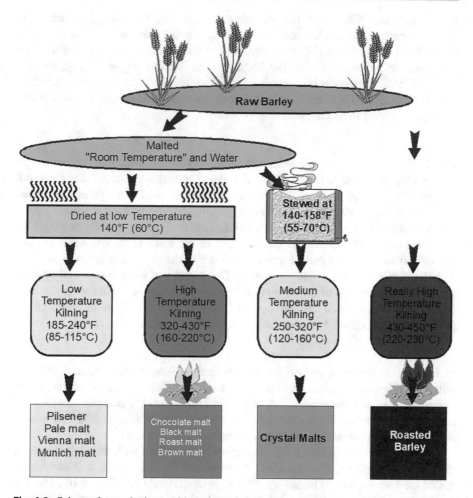

Fig. 4.9 Schema for producing a wide variety of malts from a single type of grain

the flavor of the finished product, the action of some of the disinfecting agents during the brewing process can result in the formation of potential cancer-causing agents. No brewer wants to make beer with that on the label.

4.3.2 Grains and Malts

The maltster provides the brewer with a large selection of malts to use in designing a beer. The variety of flavors available to the brewer is primarily due to the process used to dry and roast the malts, and to some extent the species (2-row or 6-row) and cultivar of grain used. We'll consider the basic process differences and the type of malt each yields in this section. In Fig. 4.9, the overall processes and the products that are produced by each are outlined.

Roasted Barley   The first distinction comes by examining Fig. 4.9. Roasted barley is simply barley that was toasted, almost burnt, without executing the malting process. The kilning at very high temperatures not only denatures the saccharification enzymes, but browns and burns the starches originally in the grain. Roasted barley is not malt; this ingredient is used sparingly for adding a dry roasted, distinct coffee-like flavor. It also adds a significant amount of color to the final beer. Roasted barley is frequently used in Stouts and some Porters.

Pale Malt   On the other hand, pale malt has undergone the basic malting process by germinating the grain in the presence of water. The malting process starts the process of saccharification, as well as the development of needed enzymes. The growing process is halted by drying the malt. Typically, the malt is dried at around 100 °F (40 °C) with very high levels of ventilation. This drying step rapidly removes the water but does little to damage the enzymes in the malt. Once the water content reaches about 10%, the malted grain is cured by raising the temperature to about 175 °F (80 °C) with continued ventilation. The drying process is monitored by measuring the temperature of the air pushed into the grain (*air on*) and comparing it to the temperature of the air that passes out of the grain (*air off*). Once the temperature of the air on and air off becomes the same, the malt is considered cured. Typically, this leaves about 4% moisture in the malt.

Different cultivars, slightly different temperatures, and the amount of ventilation will yield different malts such as Pilsner malt, Vienna malt, etc. The main point here is that the temperature used to dry the grain is low enough that the scarification enzymes are not denatured. The grain is said to have sufficient diastatic power needed to convert starches into sugars during the mash process.

A grain bill must contain a significant amount of grain with diastatic power to execute the starch-sugar conversion. A grain used for this purpose is frequently called the *base grain* and makes up 70%–95% of the grain bill (Fig. 4.10). Munich and Vienna Malts are initially dried at slightly higher temperatures with much lower amounts of ventilation. Then, they are cured at slightly higher temperatures than pale malt. This process helps protect most of the enzymes, but the higher temperatures increase the number of Maillard reactions. They have a darker, richer flavor and are an ideal base grain for Bock or Oktoberfest styles. Since they are dried at a higher temperature, they do not have as much diastatic power as pale malt, but enough that they can still be used as the base grain.

Dark malts, such as brown malt and chocolate malt, have been kilned at higher temperatures. Again, a variety of temperatures are responsible for the biggest differences in these types of malts. The freshly dried pale malt is first cured at the typical pale malt temperature, then kilned typically in a drum roaster at higher temperatures. The higher temperatures increase the rate of Malliard reactions inside the malt and begin to change the flavor of the grain.

Fig. 4.10 2-row pale malt

Malliard reactions, as we discovered in Sect. 3.3, are non-enzymatic browning reactions initiated by the condensation of amino acids and sugars in the grain. The products of the reaction are increased amounts of nitrogen and oxygen heterocycles and the formation of melanoidins. The process gives very distinct enhanced flavors, and a significant increase in the intensity of brown colors.

While the elevated temperatures enhance the malty, biscuit-like flavors, this also denatures the enzymes and makes them unable to perform the reactions needed in the mashing process. Thus these malts have very little diastatic power. This is primarily why these malts are not used as a base grain. Furthermore, a beer made exclusively with brown or chocolate malt would have a very strong, and unbalanced taste profile (Fig. 4.11).

Crystal malts initially skip the drying step in malting. As soon as the germination of the barley is complete, the wet grain is placed in a rotating drum and heated to about 150 °F (65 °C) for 30 min. During this initial heating, additional water is sprayed onto the malt. This is the normal temperature for conducting a mash, so it's like conducting a miniature mash right inside the barley seed. The result is that the enzymes in the grain become activated and convert the starches of the malt into sugars. Once the process has completed, the temperature is increased to 300 °F (150 °C) and the malt is ventilated for an hour or more. This higher temperature caramelizes some of the sugars and increases the flavors of toffee, caramel, and brown sugar. This also increases the Maillard reactions that further enhance the flavor and color of the malt.

The amount of time that the malt is left at 300 °F (150 °C) determines the level of browning that occurs. In this way, the maltster can prepare lightly colored crystal malts (known as Crystal 10 L) or even very dark crystal malts (such as Crystal 90 L) (Fig. 4.12).

Fig. 4.11 Chocolate malt. The name of the malt arises from the color of the malt, not from the flavor, although the use of this malt does increase the chocolaty flavors in a brew

Fig. 4.12 Crystal 40 L malt. Note the exposed center of the malt has been caramelized. Compare this color to the white-starchy interior of pale malt (Figs. 4.5 and 4.10)

Crystal malts cannot be used as base malts. Even though the enzymes were initially active in the malt to convert the starches in the malt, the kilning process at 300 °F (150 °C) sufficiently denatures the enzymes such that none remain. These malts have no diastatic power. However, because the starches in these grains have been converted into fermentable sugar, crystal malts do not need to be mashed. The brewer often does add them to the mash, but this is often just done to help increase the permeability of the filter bed when the mash is sparged. It is entirely possible that the brewer could simply steep the crystal malts in warm water to extract the sugars and add that directly to the sweet wort.

CHECKPOINT 4.6
Is it possible for the temperature of the air off to be lower than the temperature
   of the air on?
Crystal malts do not need to be mashed. Do they need to be ground into grist
   in order to be used? If not mashed, how are they used in the brewery?

4.3.3 Hops

Early beers did not use hops. These historical recipes, however, often used additives
to spice up the brew so that it wasn't so one-dimensional in flavor. Many different
plants, herbs, fruits, vegetables, and spices found their way into the recipe. The list
is quite extensive, but includes ingredients such as heather flowers, bog myrtle,
dandelion, and spruce needles. As we noted in Chap. 1, the use of gruit was common
in Europe. And while hops were used in beers as early as the 800s, it wasn't until the
thirteenth century that hops saw consistent use as a flavoring agent. Yet, in Britain,
hops were condemned as a "wicked and pernicious weed" and ale, a drink made
exclusively from water, malt, and yeast, was the preferred beverage. It wasn't until
the mid-fifteenth century that beer was accepted as a drink in England; and until the
mid-eighteenth century, the words ale and beer meant entirely different drinks. The
prevalence of the use of hops in beer is tied to both the pleasant flavor it produces
and to the preservative power of the hops (beer would last longer).

The hop plant (*Humulus lupulus*) is a climbing, herbaceous perennial, in the family Cannabaceae. Although it looks like a vine, it is actually a *bine*. The difference
is in how the stem of the plant attaches itself to structures as it grows. A bine grows
helically around its support and has downward pointing bristles to help it climb,
while a vine uses tendrils and suckers to attach to the support as it climbs.
Incidentally, hops are very closely related to cannabis (marijuana) in the relatively
small Cannabaceae family. The hop plant is also *dioecious* meaning that it has distinct male and female plants. The flower that the female plant produces is the most
desirable for brewing. But once that flower has been pollinated by the male plant, it
begins to produce seeds. The production of seeds greatly reduces the utility of the
flowers for use in brewing, so, the hop farmer works to make sure that only the
female plant is grown in the field. While seeds do allow the plant to propagate, so
too does the growth of the underground rhizome.

The rhizome of the hop plant is essentially a stem that grows beneath the surface
of the soil. From the rhizome are perennial roots that can survive throughout the
year, and in the spring, buds on the rhizome send annual shoots above the soil.
Cuttings from the rhizome allow a genetically identical hop plant to be produced
without the use of pollination. Each year, the annual shoots rapidly grow from the
rhizome during the spring and once they reach a given length, convert their growth
spurt into the production of strobiles, also called cones.

The female hop cone contains lupulin glands, which produce the active bittering
agents (Fig. 4.13). Hop resins, from the lupulin glands, contain Alpha acids and beta
acids. There a large variety of chemical compounds found in these resins and their

Fig. 4.13 Hop cone cross section

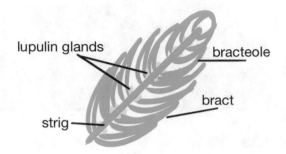

lupulin glands

bracteole

bract

strig

effect on the final taste of the beer depends on a) the relative proportion of compounds and b) the length of time these compounds are allowed to boil in the wort.

When hops are added to the boil, the components of the hop oils such as myrcene, humulene, and caryophyllene begin to effuse into the wort boil immediately. However, these lighter compounds will be quickly boiled away if allowed to boil for a significant amount of time. It is for this reason that if the brewer wishes to retain those compounds in the finished beer, hops are added at the last 5–10 min of the boil. Hops intended for this step are usually called the "aroma" hops. Different types of hops are specifically cultivated to have higher levels of the lighter compounds – and thus intended as an aroma hop.

Other compounds found in the hop oils also dissolve in the wort when the hops are added. Many of these compounds are less volatile, and are not carried away so quickly. Hops added at this stage, about 20 min before the end of the boil, are called "flavor" hops. Flavors and aromas from these two steps include descriptors such as "citrus," "piney," "grapefruit," "floral," and sometimes "earthy" depending on the type of hop used.

Hops added near the beginning of the boil do not contribute much in the way of flavor and aroma since the flavoring and aroma compounds that are volatile will have evaporated away during the course of the boil. Hops added at very early, "bittering hops," are intended to add the bitter flavor to the beer. As we uncover in Chap. 8, the alpha acids (humulones) isomerize in the hot wort. A hop with a high alpha-acid content is intended as a bittering hop.

A hop intended for bittering would not necessarily be suitable for flavor or aroma – or vice versa. For example, Cascade hops are commonly used in pale ales and IPAs because they have a larger proportion of the lighter compounds, such as myrcene and humulene, which gives these beers a characteristic "citrus" or "piney" flavor. But Cascade hops have a relatively low alpha acid content, compared to a bittering hop such as Chinnook that has up to three times more alpha acids. Chinook hops might be a good choice for bittering.

CHECKPOINT 4.7

Is it possible to use a "bittering" hop for aroma only in a beer?

Use the internet to research at least two varieties of hops that are grown in New Zealand. Is each of these a bittering, aroma, or flavor hop?

4.3.4 Yeast

Yeast is a single-celled eukaryotic organism, and is technically a fungus. A eukaryote cell contains a nucleus and other organelles enclosed with a membrane. Most yeasts reproduce asexually by mitosis, but the species of yeast most common to brewing reproduce by an asymmetric division process called budding. We'll explore this in much greater detail in Chap. 9.

Brewing yeasts can be loosely divided into "top-fermenting" and "bottom-fermenting." They are classified into these categories based on where they are typically active. The most common strain of top-fermenting yeast is *Saccharomyces cerevisiae*, and typically most active at a temperature between 60°–70 °F. This yeast is commonly used for ales and is colloquially known as "ale yeast." On the other hand, *Saccharomyces pastorianus*, is a bottom-fermenting yeast. This yeast, "lager yeast", *can* ferment at higher temperatures but, unlike *S. cerevisiae*, it remains active at lower temperatures, as low as 40 °F. Further, *S. pastorianus* is also able to hydrolyse melibiose into more fermentable monosaccharides – something that *S. cerevisiae* cannot do. This leads to more consumption of the available sugars for a dryer, crisper flavor. Lager yeast can be used at higher temperatures, giving the characteristic flavor of "steam beer," or California common beer.

Yeast have a significant impact on the final flavor of the beer. Their selection and use is another variable over which the brewer has control. Yeast, via its metabolism of sugars, will also produce small amounts of byproducts. Some of these compounds, such as acetaldehyde, diacetyl, or dimethyl sulfide, are not typically desirable in large quantities. Individually, these compounds give the beer a green apple, buttery, or cooked corn taste respectively. These undesirable compounds will be eventually re-absorbed by the yeast, which is why the natural conditioning step takes so long – and is so important.

On the other hand, some by-products are desirable in certain styles of beers. These compounds can impart a clove, banana, or fruit flavor such as one might encounter in a Hefeweizen. Ale yeast, because of the higher temperatures, synthesize a host of flavor compounds, such as esters, during fermentation. Esters are the compounds that give fruits their characteristic flavor. In a Hefeweizen, the banana-like flavor comes from an ester called isoamyl acetate. The yeast also produces other esters such as ethyl acetate, which tastes like a flowery solvent. So how do we control which esters end up in the beer, and in what quantities?

We'll discover more about the production of esters and other byproducts in Chap. 9, but as an overview, yeast do produce a variety of alcohols including ethanol. These fusel alcohols can be converted into esters through a process called esterification. The type of ester that is produced depends on the alcohol that starts in this process. So, if we start with ethanol, we might end up with ethyl acetate after the esterification. Or, if we start with isoamyl alcohol, we would expect to end up with the banana-like isoamyl acetate. To control the production of these esters, the brewer may select a specific strain of yeast known for its particular ester production. Temperature also has a large impact on the production of esters. Generally speaking, increased temperature will increase the amount of esters produced.

4.3.5   Finished Product

Once the beer has been conditioned, it is typically ready for packaging or serving. The flavors of the beer have been set and the action of the yeast has reduced itself to almost nothing. The brewer has had time to adjust the color, the flavor, and the clarity of the product in the conditioning tank. As we'll see in Chaps. 10 and 11, clarifying agents may have been added, the level of carbonation has been adjusted, and the beer is verified as being within specifications for what the brewer had intended. In some cases, this has taken months to achieve from the malted barley. In other cases, the process has only taken a week or two.

Whatever the case, the brewer now decides the way that the beer will be consumed. Will the customer buy it in a bottle, a can, or in a keg? Or, will the customer consume the beer at the brewery's taproom? These decisions determine the last few steps in the overall process.

Once to the consumer, the beer undergoes the true test. It is the consumer that determines if the brewery makes good beer. As we'll see in Chap. 13, the measure of quality isn't determined exclusively by the brewer or a tasting panel. It does not matter if the beer is "within spec" or "true to style." What matters most is that the customer enjoys the product and wants to consume it. And for that reason, it is important that the beer fits with what the customer believes is what a beer should look, taste, smell, and, yes, even feel like. If the beer strays too far from the public's perception of quality, sales will decline or stagnate. And the brewery will suffer.

The perception of a quality beer starts with how it is served. Whether it be from a bottle, a can, or a tap, the beer must "look" like a beer. In the United States, this means that a beer must be carbonated appropriately, have a thick head that lingers as long as possible, and be relatively clear. Obvious exceptions to this exist, and those exceptions depend upon the particular style.

In the taproom, bar, pub, or tavern, the way that the beer is handled is important. The glass should be tipped as the beer is first poured into it. And once the volume of beer has reached an appropriate level, the glass is held upright while the remaining beer is poured right into the center. This method of pouring allows the bulk of the beer to be added to the glass without a great loss of carbonation. Then, for the last few ounces, the beer is agitated with the pour into the center. This causes the head to rise and form on the top of the beer.

One of the more important indicators of good beer handling is how the beer is served. Servers should carefully present the glass to the customer so that the head still remains on the beer. Did the bartender spill it all over the glass and stick the receipt to the outside of the glass? Or did the bartender carefully pour the perfect pint with as much care as a newborn infant? Did the beer get served to you right after pouring, or did the server go on vacation right after you placed your order.

The perfect pour should come to you while it still has the majority of the foam still remaining. It should be served in a glass that is appropriate for the style, and dry to the touch. And the beer should still be at the preferred serving temperature (see Table 4.2).

Table 4.2 Suggested beer serving temperatures

Style	Temperature	Style	Temperature
American Light Lagers	32–40 °F (0–5 °C)	Pale Lagers Pilsners	38–45 °F (3–7 °C)
Belgian Ales Abbey Ales	40–45 °F (4–7 °C)	Wheat-based beers Lambics	40–50 °F (4–10 °C)
American Pale Ales IPAs	45–50 °C (7–10 °C)	Stouts Porters	45–55 °F (7–13 °C)
Strong Lagers Cask and Real Ales	50–55 °F (10–13 °C)	Barleywine Wee Heavy	50–55 °F (10–13 °C)

Data compiled from Randy Mosher, Tasting Beer: An Insider's Guide to the World's Greatest Drink, Storey Publishing: 2009

If these steps are correctly accomplished, the customer has the best chance to enjoy the beer to its fullest potential. And if the brewer has done the task to provide the best product possible, the customer will repeat their business. That alone is a sure sign that everything was well done and done well.

Chapter Summary

Section 4.1

Brewer's barley is available in 2-row and 4-row varieties, and in a wide variety of cultivars.

Barley is sorted by size; the plumper seeds are sold to the maltster for making malt. The seeds are then stored to reduce their dormancy and enable sprouting.

Malt is made by germinating the seeds until all of the starch has begun to be utilized by the seed for growth.

The malt is then dried and kilned to prepare a variety of different malt products ranging from pale malt to black malt.

The brewer mashes the malt using warm water until the starch is converted into fermentable sugars.

The sweet wort is boiled. This allows the brewer to add flavoring agents and hops while at the same time sterilizing the wort.

The hopped wort is cooled, oxygenated, and fermented using yeast.

The green beer undergoes maturation to adjust the flavor, carbonation, and colloidal stability.

Bright beer results from the use of mechanical filtration or settling using the addition of compounds.

The finished beer is then packaged in large pack or small pack.

Section 4.2

Brewing beer requires constant cleaning and sanitizing.

Safety and care must be taken into account when working with any cleaning or sanitizing agent.

Section 4.3

The four main ingredients used in beer production are water, malt, hops, and yeast. While the overall process for making beer is relatively uniform, the options when using each of these ingredients gives rise to an almost infinite number of flavor options.

The end result of the process depends entirely upon consumer approval.

Questions to Consider

1. For the brewer, what is the most important part of the hop plant?
2. What is the difference between a vine and a bine?
3. Distinguish between Bittering, Flavor, and Aroma hops with regards to a.) Time of addition to the wort, b). effect on bitterness, and c.) effect on the finished beer.
4. What is diastatic power?
5. What is meant by the term "base" grain?
6. Outline the steps involved in brewing beer from malt.
7. Why is a plumper barley seed considered better for use as a base grain?
8. Is barley the only grain that can be malted? Explain your answer.
9. Why do some breweries utilize a cereal cooker?
10. Although beer is made with just four ingredients, the brewer/malster has significant control over the final taste of the beer. List as many things the brewer can change *in the process* to produce the beer which will influence the final flavor.
11. Why would it be important to make sure that the brewery floor is free of bacteria if the production of beer is inside sealed vessels?
12. Explain how crystal malt is prepared from barley?
13. What purpose would the addition of crystal malt have in beer manufacture? Why is this malt used instead of just adding sugar?
14. Why would a brewer use chocolate malt?
15. List some of the compounds that yeast can produce.
16. Use the Internet to determine the appropriate glass and temperature for serving a Belgian Ale.
17. What is meant by the term sparge?
18. If a brewer batch sparged, how many different "strengths" of beer could be made?
19. Why would a brewer prefer to use rain water over ground water?
20. What is meant by these terms: lauter, sparge, mash, condition.
21. What is the difference between green beer and bright beer?
22. Why are hop yards (hop fields) made up of only female hop plants?
23. How are hops propagated?
24. What is meant by the statement: "the customer is always right"?
25. A brewer wishes to use lager yeast and ferment hopped wort at 75 °C. Based on the information in this chapter, what would you expect to be the outcome of this process?

26. If two brewers utilize the same exact recipe, will they make beer that has the same flavor profile? Explain why or why not.
27. Why is it not possible to make beer from unmalted barley? What would a brewer need to do to use unmalted barley as one of the ingredients in the production of beer?

Laboratory Exercises

Sketch the Overview

This laboratory assignment requires the students to view videos from the Internet and then create a flowchart outlining the overall brewing process. The assignment can be made as a homework assignment, or can be a class assignment that requires students to discuss their outline and compare it to the rest of the class.

Sources for the video are easily found as brewery tours of local breweries. These tours are often uploaded to the Internet and can be found in YouTube, VIMEO, Google Play, and other sites. The students should watch one or two video tours of breweries that illustrate each of the processes found in the brewing of beer. Then, the students, working in small groups, should produce a sketch of the brewing process using the correct terminology. If each group produces a sketch of the process using a different video tour, comparison of the processes used in regional and local breweries can be made.

Research on Barley

This laboratory assignment requires the students to evaluate the commercially available varieties of barley to determine the key characteristics of their use in malting and brewing. The assignment can be given to groups of students for presentation to the entire class, or multiple barley varieties can be assigned to each student to complete compare-contrast reports.

Students are given a list of barley varieties to examine. The students then research the varieties on the Internet and in the library to determine key features of the varieties they've been assigned. Specific information that works well in this exercise includes a.) parents of the specific cultivar, b.) location where the cultivar was first grown, c.) percent of protein, starch, etc., d.) typical growing period, e.) information about performance in malting.

Malting and Water

5

5.1 Biology of Barley

Barley is likely one of the oldest plants to be domesticated by human civilization. Evidence of its domestication has been found in Mesopotamia that dates from 8500 BC. Archaeologists have uncovered an ancient village near the Sea of Galilee in northeastern Israel. Because it appears that the village burned down and was then covered by rising waters from the Sea of Galilee. Silt and clay then covered the entire site and preserved evidence of over 100 different types of seeds.

It is likely that this plant was domesticated much earlier than this. The wild barley plant, in fact, is native to the area and ranges from the Nile River into Tibet. Domestication of the wild barley (*Hordeum spontaneum*) gave rise to the barley plant that we know today.

The modern barley plant is a member of the family of grasses. The grass grows vertically and at the top of the stem rests a spike along which the barley seeds grow. When the plant matures, the spikes separate into *spikelets* containing the seeds. Two main ways in which the seeds grow along the spikelets give rise to two very important species of barley, two-row barley (*Hordeum distichon*) and six-row barley (*Hordeum vulgare*). In both varieties, the seeds grow opposite each other in long straight rows. A layer of hair-like structures known as *awns* protects the seeds.

The brewer regards the two species differently. Two-row barley seeds (also known as *corns*) tend to be larger and, because of that, they contain slightly less protein per corn than the six-row barley. However, the crop yield of the six-row species tends to be a little higher than two-row barley. In the United States, the higher protein level was desired in the production of beers with high levels of un-malted adjuncts. Thus, the six-row barley was, and continues to be, the main barley grown in the United States. Two-row barley, however, produces more starch per corn and tends to be favored as the primary source of starch in the craft beer industry.

5.1.1   The Barley Corn

The barley seed (see Fig. 5.1) consists of three main structures: the *embryo*, the *endosperm*, and the *pericarp-testa*. The *proximal* end of the barley corn contains the embryo. This structure contains the *acrospire* and *coleoptile* that develop into the above ground growth of the plant. In addition, pointing in the opposite direction are the *rootlet* and *coleorhiza* that become the roots of the plant. The embryo occupies less than a third of the total volume of the corn.

Separating the embryo from the endosperm is the *scutellum*, cells that serve to absorb nutrients from the endosperm as the seed develops and to produce enzymes that can be used by the growing seed.

At the *distal* end of the barley corn is the endosperm. This collection of cells contains the starch needed by the growing seed and by the brewer. Each cell in the endosperm is made up of walls that contain a significant amount of β-D-glucan (see Chap. 2). Inside those cells, surrounded by a protective layer of proteins are the granules of starch. Those proteins include the large hordein (~35%) and glutelin (~30%) storage proteins that are used in the production of amino acids during growth. Also included are the albumin (~5%) and globulin (~30%) proteins that are the source of the enzymes needed by the seed during growth.

Finally, the *pericarp* – a semipermeable membrane that surrounds the corn – and *testa* – a thinner membrane that contains much of the compounds that result in the haze found in beer – are fused together to form the outer protective layers. Just outside of these layers is the *husk*. The husk is a collection of dead cells that contain a significant amount of silica. It is for this reason that the husk is both abrasive and hard. Along the dorsal side of the corn, the husk is known as the *palea*. The husk along the ventral side of the corn is known as the *lemma*.

Just inside the pericarp-testa lies the *aleurone* layer of cells as shown in Fig. 5.2. When the aleurone layer hydrates, it produces enzymes that help the seed utilize the endosperm's starches as food during growth. At proximal end of the corn, the husk

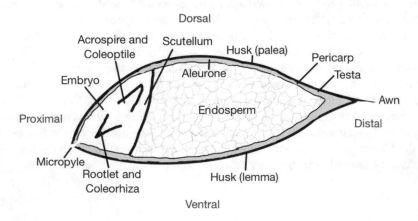

Fig. 5.1 The barley seed

Fig. 5.2 A cutaway of the layers that make up the barley walls

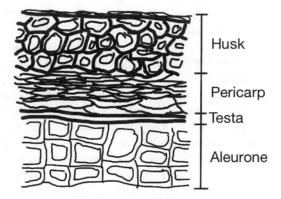

has a *micropylar* region that is permeable to water. Because the aleurone layer is a little more permeable than the starchy endosperm, water that has entered the seed can travel along the outside edges of the seed.

CHECKPOINT 5.1
Describe the differences between two-row and six-row barley.
What is the difference between the coleoptile and the acrospire?

5.1.2 Barley and the Farmer

Barley has been grown nearly everywhere; however, it prefers cooler drier climates. As a relatively drought-resistant crop, it can be planted either as a winter crop (planted in the Fall) or as a spring crop (planted in early Spring). The winter crop is harvested in July; the spring variety is harvested in August. Based on where the farmer lives, it may be possible to plant two crops in 1 year.

The world's largest producer of barley is Russia, followed by Germany, France, and Canada. Over 155 million metric tons of barley was produced worldwide in 2008. After a dip in production in 2010, the level of production has risen back to its current level of about 145 million metric tons. The United States is a relatively minor producer of barley. In 2015, the United States produced only 4.7 million metric tons of barley. This level of production is only 67% of what was produced in the United States in 2000. The downward trend in barley production is due to some increasing prices for corn. In the current economy in the United States, as corn prices fall, so do the prices for barley.

Approximately 75% of the barley used in beer production is grown in only five states in the United States; Idaho, Minnesota, Montana, North Dakota, and Washington. The remainder of the barley grown in the United States is distributed across a large number of states, primarily Utah, Colorado, South Dakota, Wisconsin, Wyoming, Oregon, and California. The influx of craft maltsters and microbreweries across many other states has driven some of the production of this valuable crop.

The farmer typically harvests when the moisture level in the seed is below 18%. The barley is usually ready for harvest when it has less than 25–30% moisture, but if it were threshed with this level of moisture, it would severely damage the seeds. So, the barley is either dried in the field or in the dryer attached to the silo until it is 12% moisture.

Drying the barley after harvest occurs by either a batch dryer or a continuous dryer. One such batch dryer is known as the tower dryer. In this device a large quantity of wet barley is conveyed onto a screen and then warm air applied. The air temperature is controlled to be about 40–50 °C (105–122 °F) as it passes through the grain. It is very important to monitor the temperature of the "air on" – the air that is passed into the grain – and the "air off" – the air coming out of the grain. This is done to make sure that the temperature of the grain itself does not rise above the 35–40 °C (95–105 °F) range. If it were allowed to get that hot, it could severely damage the barley corns. While the temperature of the air is much warmer than this, the evaporation of water from the corns helps to keep them cool. After a given amount of time on the screen, the barley is conveyed down to the next screen below and the process repeated. The series of screens are arranged in a tower pattern so that the air on at the bottom of the tower passes through multiple screens before exiting the tower. This type of dryer is relatively expensive, but efficient in drying the grain without damaging it.

An alternative to simply dropping the grain through the stream of warm air is the continuous dryer. Many different designs exist, but are essentially a long vertical tube where the grain is fed into the top and air blows horizontally across the grain. The dry grain is removed from the bottom of the tube. Improvements to this device have been made. One such improvement is a slanted conveyor (see Fig. 5.3). Grain is added to the conveyor and dried by blowing warm air through the conveyor. The air on is kept in the 55–60 °C (130–140 °F) range and as the grain passes through the device, the air off tends to stay around 25 °C (77 °F). The grain is cycled through the drier until it reaches the 12% moisture level and can be stored, although one pass is likely all that is needed.

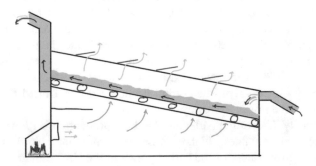

Fig. 5.3 Continuous barley dryer and air monitoring. The orange arrows indicate the flow of hot air (air on from the firebox, air off out the top of the device). The black arrows indicate the flow of the barley into the dryer and along the conveyor to the auger at the end of the device

In fact, barley must be stored before use. This allows the barley to "break dormancy." Barley seeds, just like most seeds that are harvested, must be dried to about 12% moisture and stored at around 35–40 °C (95–105 °F) to help break the dormancy before they can germinate and grow. If the freshly harvested barley seeds were planted without this resting stage, they would not grow. But after a few days at elevated temperature (depending upon the variety of barley), the seeds do sprout and grow when planted. And if not used immediately, they can be stored cool until they are needed.

There are many different cultivars of the barley plant that are used for making beer. And, there are a large number that are specifically used for feed. The American Malting Barley Association is one such source for a list of the different cultivars. This organization makes recommendations to the farmers as to which cultivar of barley is likely to better grow and sell well in the coming year. They also track which cultivars were actually used and how many acres are planted in each.

5.1.3 Barley Diseases and Pests

Most farmers also have to worry about diseases and pests in their barley fields. In addition to reducing the yield of barley per acre, diseases and pests can cause damage to the product they make that can significantly reduce the price they get. The farmer knows these pitfalls very well and spends time watching for them. After all, good beer can be made from good ingredients. It is hard to make good beer from inferior ingredients.

The pests that inhabit barley in the United States are very damaging to the crops. One of the more common of these pests are aphids. There are a large number of species of these small insects, each causing its own specific damage to the barley plant. In addition to eating portions of the plant, aphids tend to cause the spread of sooty mold that can further injure the plants.

In addition to aphids, armyworms and wireworms can cause significant damage to the barley in the field. Wireworms tend to eat the young plants as they first grow. Armyworms feed on the leaves of the plant stunting its growth. The damage is not only noticeable, but treatable with pesticides and other pest management practices.

Fungi can also cause damage to the crops and to the finished beer made from infected barley. One of the most notorious fungal infections is caused by the genus *Fusarium*. These fungi can result in "common root rot" that stunts plant growth or Fusarium Head Blight (FHB) that makes the seeds of the barley plant look like they are dried out. The fungus responsible for FHB, *Fusarium graminearum*, also produces a mycotoxin in addition to causing damage to the development of the barley plant. *Fusarium graminearum* releases deoxynivalenol (DON; also known by its common name: vomitoxin). This toxin is well regulated in the food supply with maximum levels set at 1 ppm. However, many farmers, maltsters, and brewers set even lower levels of this toxin.

Bread and beer made from grains containing DON cause nausea and in some cases vomiting, hence the common name for this compound. In addition, when barley infected with this compound is used to make beer, the mycotoxin forms crystals at the bottom of the bottles after bottling. If the level of DON is great enough, the crystals can get large enough to cause issues. When those bottles are opened, a significant amount of gushing occurs, due to the sharp crystals of DON that provide nucleation sites for the formation of bubbles.

CHECKPOINT 5.2
Given that barley has a density of 609 kg/m^3, determine the pressure exerted on the bottom of a barley storage vessel that is 10 m tall and full of grain. What would be the difference in the mass of 100 kg of barley that has 12% moisture, versus the same number of barley corns that have 18% moisture?

5.1.4 Sorting and Grading

The brewer only wants the plumpest, most uniform, barley corns that they can obtain. And the farmers do their best to grow those. The maltster is the intermediary in this process. They are the ones that purchase the grain from the farmer, convert it into malt, and then sell it to the brewer. But they only buy those barley corns that are appropriate for the brewer, and they only pay the premium price when the barley corns are perfect and plump and free of debris.

So, the barley from the field is graded and sorted before being sold. The barley is first passed along a screen containing air jets and magnets. The air jets whisk away any light debris that might be present from the barley harvest. The magnets pick up any metal pieces. Metal shavings from the farm equipment, combines, augers, and other equipment can end up in the barley harvest. The magnet is needed to remove these shards. The screens on which the grain is moved allow miniature particles to fall through. These are typically small stones and other items that should be removed.

The barley then moves into a cleaning drum (Fig. 5.4). The drum contains a number of miniature scoops oriented around the inside. The drum is slowly turned and any small barley corns, broken corns, or other foreign material are picked up by

Fig. 5.4 Cleaning drum. The drum rotates and carries dockage upwards. It drops in the center chute that conveys it out of the barley

the miniature scoops. This material, known as *dockage*, is damaging to the malting process. As the drum rotates, the broken material moves along the perimeter of the barrel until it reaches the top. It then falls from the scoop and lands in a tube that runs out of the barrel. In this way the barley is cleaned so that only corns remain.

In the final step, the barley can be graded by further sorting. This step is very important to ensure that the barley used in the brewing process (or at least at the malting stage) is of the same size. Barley corns of the same size are needed because they will germinate at the same rate. If the batch of barley is not graded, the final malt that the brewer receives will be of varying stages of modification (see Sect. 5.2), which will result in an uneven mash, and more importantly, an unpredictable result of mashing.

Grading takes place when a sample of the barley is placed in a hopper and then dropped onto a set of vibrating screens (Fig. 5.5). The top screen holes are 2.78 mm (0.109 in, 7/64") wide, the second screen has holes that are 2.38 mm (0.09375 in, 6/64") wide, and the third screen is 1.98 mm (0.078 in, 5/64"). Kernels that stay behind on the first and second screens are considered plump. Barley corns that pass through the second screen but stay on top of the third screen are considered "thins." The kernels that fall through the third screen are considered "thru" or dockage.

Malting barley in the United States is graded based on the standards set by the USDA (United States Department of Agriculture). There are four grades available for malting barley based on the type of barley (six-row versus two-row) and on the amount of plump, thins, and thru. Table 5.1 lists the maximum amount of thins that can exist in any of the grades.

The percentage of plump kernels in any grade can be requested in an analysis of the grain sample. The higher percentage of plump kernels is highly desired in a quality malt sample, irrespective of the USDA Grade. From Table 5.1, the maximum percent thins indicates that the best Grades (1 and 2) require more than 90% plump kernels. Samples that don't conform to these standards are typically graded in the "barley" category and are used as food or feed.

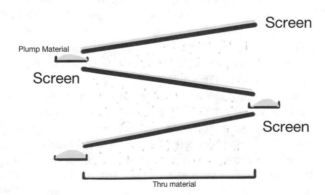

Fig. 5.5 Grading screens. The vibrating screens are held at an angle

Table 5.1   Maximum percent of thins allowed in US graded malting barley

USDA grade	Type	Maximum thins (%)	Minimum unbroken kernels (%)
1	Six-row	7.0	97
2	Six-row	10.0	94
3	Six-row	15.0	90
4	Six-row	15.0	87
1	Two-row	5.0	98
2	Two-row	7.0	98
3	Two-row	10.0	96
4	Two-row	10.0	93

5.2   Malting Barley

Once the barley arrives at the maltster, the workers use the natural machinery of the barley itself to convert it into malt. There are three main processes involved in malting the barley. These are steeping, germination, and kilning. We'll explore each of these processes from both the aspects of what goes on inside the barley corn and what equipment accompanies each of these processes.

5.2.1   Germination of Barley

In the malthouse, the incoming barley is steeped in water. It is very important that the water be high quality because it will be absorbed by the barley and carried forward into the finished beer much later in the process. When barley is soaked in water, some of the water enters the seed through the micropylar region. The water then hydrates the embryo and the aleurone. The influx of water causes the activation of the biological machinery inside. The husk also begins to absorb water, passing it into the pericarp and testa by capillary action (Fig. 5.6).

The water moves slowly through the barley seed from the proximal to the distal end and much faster along the dorsal edge. The vascular structure of the seed along the dorsal side means that water moves by capillary action faster. The seed becomes hydrated from the embryo through the scutellum and into the aleurone layer. Then the endosperm becomes hydrated from the outer edges toward the center. The movement of water is fairly slow through the endosperm, such that the entire seed is only fully hydrated when it is soaked in water.

The goal for the maltster is to make sure the barley increases to about 45% moisture. Too little moisture and it will not germinate fully. Too much moisture and it will rush too quickly through germination. Neither of these two options is useful for the maltster. Typically, it takes between 30 and 50 h to reach this level of moisture.

Once hydration begins, the embryo and scutellum begin to produce *gibberellic acid*. Gibberellic acid is a plant hormone, and just like human hormones, it triggers other actions inside the seed. Specifically inside the barley seed, it diffuses into the aleurone layer and signals the production of enzymes needed by the seed to grow (Fig. 5.7). Because of the way the seed hydrates, the gibberellic acid moves to the

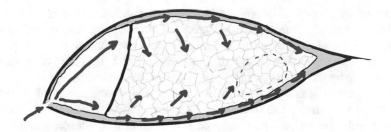

Fig. 5.6 Water movement as a barley seed hydrates. The dotted blue circle is the last place to become hydrated

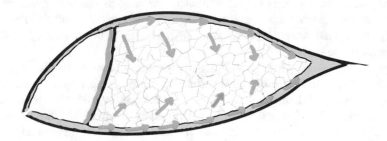

Fig. 5.7 Activation of enzymes by gibberellic acid. Gibberellic acid (represented by the orange color) is made in the scutellum and passed into the aleurone layer. From there it migrates through the aleurone activating enzymes. Those enzymes then diffuse into the endosperm

dorsal side of the seed and then moves to the distal end. Enzymes are produced wherever it goes, and the pattern of their production means that the seed "grows" in an asymmetric pattern. The last portion of the seed to be affected by the enzymes is located near the distal end on the ventral side (it is the same location as the last place that gets hydrated).

The enzymes that are activated include those that are already produced by the seed and are awaiting hydration. These are known as zymogens and only require water to become active. Gibberellic acid also signals the biological machinery in the aleurone layer to make enzymes from scratch as well. The enzymes generated by the growing barley seed include:

- α-amylase – converts starches into useable sugars
- β-amylase – converts starches into useable sugars
- limit dextrinase – breaks starches into smaller pieces
- glucanase – breaks down the cell walls in the endosperm
- pentosanase – breaks down the cell walls in the endosperm
- protease – breaks down the proteins that make up the cell walls and surround the starch granules in the endosperm
- phytase – releases phosphate from phytin and lowers the pH

Each of these enzymes is covered in greater detail in the chapter on mashing. However, as we can tell from the listed actions that each performs, the main goal of the production and release of these enzymes is the conversion of the starches in the endosperm into sugars and the proteins into amino acids. This soup of nutrients is then taken up by the scutellum and given to the growing embryo.

The degree to which the endosperm is *modified* by the action of water and enzymes is very important. If only a portion of the endosperm is modified, the resulting malt is considered *partially modified*. If all of the endosperm is modified, the malt is considered *fully modified*. Fully modified malt is squishy and spongy to the touch – a good indicator that the endosperm's cell walls have been broken down. Partially modified malt contains a hard nib inside the malt that isn't broken down by the enzymes. This becomes very important in mashing as the fully modified malt can be easily converted into fermentable sugars. The partially modified malt must undergo additional processing during the mash in order to be completely converted into fermentable sugars.

More importantly, partially modified endosperms have a significantly larger amount of β-glucan remaining. This carbohydrate that makes up a portion of the endosperm cell walls causes some problems for the brewer unless it is taken care of during malting. Of primary concern is that it makes the grain bed during mashing much less permeable, a problem that can give rise to a stuck mash. In addition, the large amount of β-glucan in the mash can translate into haze in the finished beer.

The proteins that make up the cell walls and surround the starch granules are broken down into amino acids by proteases. This contributes the amount of free amino nitrogen (FAN) available in the grain. FAN is useful for the fermentation process because the yeast need these small amino acids and amines as nutrients. Some proteins are only released and not completely degraded into amino acids. Four different protein classes have been identified: albumins, globulins, glutelins, and hordeins. The albumins and globulins are smaller proteins that tend to be more water soluble than the glutelins and hordeins.

High levels of protease activity initially seem like they would be a good thing to have happen because higher levels of FAN would be beneficial to the yeast later in the process. However, this is not the case. If the level of FAN is too great, the amount of Maillard Reactions (the heat driven reactions of amino acids and sugars) increases. This can result in significantly more browning and caramelization than desired. In addition, if too many of the proteins are broken down, the result is a loss of the head on the beer. The combination of proteins, tannins, and other compounds are needed to support a full head of foam.

If the level of protease action is too low, many of the proteins won't break down. This results in problems with haze and foam later in the process. It also results in a lower than usual conversion of the starches into fermentable sugars during the mash. High protein levels can result in a more viscous wort during mashing, increasing the risk of a potentially stuck mash. Finally, the foam generated during the initial stages of the boil can be quite significant meaning additional care must be taken prior to the hot break. In short, it is very important to have adequate protein degradation.

The end result of all of these steps is the growth of the barley seed. Initially, the maltster notes the emergence of the *chit* from the bottom of the seed. This will eventually develop into the roots as the seed continues to grow. The maltster also notes the growth of the acrospire that eventually will become the shoot for the plant. The acrospire grows from the embryo under the dorsal husk of the seed toward the distal end. Basically, it isn't seen during the initial stages of growth – but if a seed is pulled apart, the acrospire can be readily observed.

The maltster stops the growth of the seed as soon as the acrospire is approximately the same length as the entire seed. By this time, the roots are about two-times the length of the seed. This amount of growth is required for the endosperm to become fully modified. If stopped too early, the malt would be partially modified. If the maltster doesn't arrest the growth of the seed when the acrospire is as long as the seed, the level of starch in the endosperm would continue to drop as the seed grows and the utility of the seed to be used in making beer would be reduced.

To stop the growth, the maltster applies heat and dries out the seeds slowly until they are almost free of moisture. The dehydrated seed cannot continue growth and the enzymes are unable to continue to act on the starch and cell walls. The enzymes are, for the most part, still active. They just lack the water to allow them to do their work. At this point, the malt can be kilned to add color or flavor to the malt itself.

5.2.2   Equipment Used in Malting

In the previous section, we uncovered what happens inside the barley seed as it undergoes the malting process. The maltster uses some specialized equipment to ensure that the process is efficient, rapid, and uniform across an entire batch of barley. Special attention is placed on making the process uniform. Deviations from uniformity result in partially, fully, and over modified malt. Such malts would be difficult to handle in the brewhouse and would lower the mash efficiency for the brewer.

The first step in the process after cleaning, sorting, and grading the barley is to steep it in a vessel. This can simply involve soaking the barley in a vat for a period of time. To ensure that the barley doesn't "drown," oxygen bubbles through the mixture. The action of the gas in the vat helps to stir the mixture. Unfortunately, this leaves areas in the vat that have significantly more oxygen exposure than other areas. The stirring action of bubbling gas tends not to mix very efficiently either.

So, the maltster has developed a more efficient steeping tank. This tank looks very similar to the cylindroconical vessels (CCVs) that are used as fermentation tanks. However, the interior of these vessels belies their difference. Figure 5.8 illustrates a drawing of a cutaway view of the modern steep tank. The sloped bottom and internal baffles force the slurry of water and barley to mix thoroughly. This provides adequate water and oxygen uniformly to the barley seeds.

The steeping of the barley is adjusted by the maltster to allow even hydration. Often this includes periods of soaking in water, periods of bubbling oxygen through the water, and periods of resting (where the water is drained). It is very important

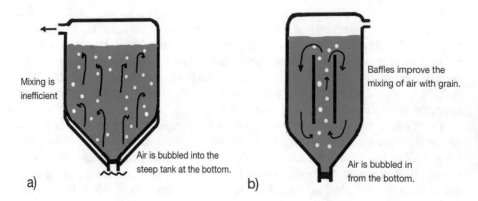

Fig. 5.8   Steep tank designs. (**a**) Traditional design with poor uniformity. (**b**) Modern design with good mixing and uniformity

that the soaking cycles include cycles of resting. Resting allows the removal of CO_2 and waste products from the barley as it grows. Resting also is important because it allows the barley to be washed of foreign substances and bacteria.

The barley seeds generate heat as they absorb water and swell in size. Initially, the maltster starts the steep by adding water that is between 15 and 20 °C (59–68 °F). The heat generated by the germination process can raise the temperature of the water as high as 25 °C (77 °F) before it is drained and refilled. This might not seem like a significant change that we'd worry about, but it is. As the temperature increases, the speed of the germination increases and is less able to be controlled. Remember that the maltster is working hard to accurately control the entire process in order to make sure that every barley seed in the batch has the same modification. Higher temperatures make this harder to accomplish. More importantly, if left unchecked, the temperature could get high enough from the mass of barley to damage it during the germination step.

Once the water content in the barley seeds rises to about 45%, the barley is moved to a vessel and allowed to sprout. There are many different types of germination vessels. Traditionally, the hydrated barley was spread out onto the floor of a building. This process, known as floor-malting, is still practiced. Some brewers and maltsters believe it provides a very uniform modification and a hand-crafted taste to the malt. After spreading it out into a layer on the floor, the barley is turned over multiple times until the germination is complete. Turning can be accomplished by hand (where the maltster uses a hoe or fork to turn the barley) or by machine. Typically the entire germination process takes between 3 and 5 days.

Turning the grain as it germinates has the same effect as cooling the grain. Reduction of the heat is just as important during germination as it is during steeping. If left unchecked, the hot grain supports the growth of bacteria. In addition, as the grain gets hotter, it begins to become uneven in its rate of germination. Warmer seeds begin to *bolt* – germinate very quickly. And the entire germination process becomes uneven.

Fig. 5.9 The Galland drum. Air enters through the tubes along the perimeter and exits through the central tube

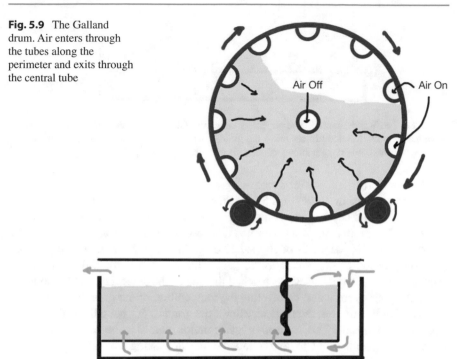

Fig. 5.10 The Saladin box. Air, represented with the blue arrows, can be recycled and the humidity and temperature adjusted. The screws move along the length of the grain bed turning it over gently

Other alternatives to floor-malting include drum-based malting systems. In the 1880s, Frenchman Nicholas Galland developed a drum to germinate barley (Fig. 5.9). Its use in malting barley was an immediate success. The Galland Drum is essentially a large drum with perforated tubes that run down the inside of the drum's perimeter. The axis of the drum is also a perforated pipe. Air is blown into the tubes along the perimeter and exits through the center pipe. Periodically, the drum rotates to turn the grain.

Charles Saladin, a French contemporary of Galland, invented a germinating vessel that bears his name. First introduced in the 1890s, the Saladin Box (see Fig. 5.10) is a rectangular cement box with a false perforated bottom for air flow. Modern versions of the Saladin Box are round instead of rectangular. The addition of a rotating arm to the top of the round box allows the grain to be evenly added to the germinator. Steeped grain is placed into the box to a depth of 0.9–1.2 m (3–4 feet). Along one end of the box is the air intake. The intake also includes humidifiers to keep the air moist and prevent the growing seeds from drying out. Air is passed into the intake and up through the false bottom. It exits the top of the box. In some systems the entire box can be covered so that the exiting air can be recycled. At the top of the box is an arm containing large screws that rotate and move along the length of the box. The screws gently turn the growing grain.

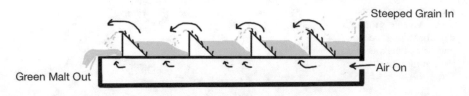

Fig. 5.11 The Wanderhaufen. The air enters under the false bottom and blows up through the germinating grain. Slanted turners move the grain along the device. Unlike the Saladin box and other germinators, this system allows continuous germinating of the grain

The Wanderhaufen was a germinating system invented by the founder of the Carlsberg brewery in 1878 (Fig. 5.11). Steeped grain is placed at one end of the device at a shorter depth than in the Saladin box and fully modified malt exits at the other end. The grain rests on a perforated false bottom and in a manner similar to the Saladin box, air is pushed through the bed of grain. Slanted turners move the grain from one portion of the device to the next. This movement turns the barley as it grows. By the time the barley has reached the end of the device, it is fully modified.

As the grain germinates, controlling the temperature of the grain becomes more difficult. For this reason, the maltster allows the grain to dry out a little in a process known as *withering*. The control of germination can also be accomplished by increasing the amount of CO_2 that is added to the air passing through the grain. This slows the respiration of the grain and reduces its growth rate. A slower growth rate allows the maltster to more carefully gauge when it is finished.

Once the grain has reached the modification level desired by the maltster, it is known as *green malt* and is pumped into the *kiln*. This sounds like a very hot oven, but just the opposite is true. The kiln is instead a process by which warm air is passed through the malt until the malt's moisture content is reduced to about 4%. The green malt is pumped into a vessel with a perforated false bottom. Air is blown under the false bottom and up through the malt bed.

The effects of kilning reduce the moisture in the malt and inhibit the further growth of the barley seed. Control of the removal of moisture is very important, just as is every other step in the malting process. That control comes about in the kiln and how it operates. If the malt is dried and kept cool, most of the enzymes that are in the seed will remain "alive." Just as in the dry barley seed though, they won't have any water present, and so will not be active. The kilning process involves the following steps:

1. Free drying
2. Forced drying
3. Curing

In the *free drying* step, warm air (50–60 °C, 120–140 °F) passes through the grain with a high flow rate. The moisture in the seeds slowly evaporates and is removed from the malt. Because evaporating water requires energy, the malt is cooled as the moisture content drops. This initial stage takes about 12 h and results in the reduction of the moisture content of the malt to about 25%.

The maltster pays attention to the humidity of the air off. It is very desirable for the air off to have 90–95% humidity. However, if the humidity reaches 100%, the moisture in the air might condense on the kiln itself. This causes a severe problem with the malt. Since the air cannot support additional water (i.e., humidity) the air on slowly cooks the malt. This increases the color and flavor of the malt (see Sect. 5.3), but also destroys the enzymes inside. This detrimental effect is known as *stewing*.

Once the free drying is complete, the malt enters the *forced drying* stage. This stage occurs when the air off humidity levels start to decline and the temperature of the air off rapidly begins to rise. This point is known as *breakthrough*. The decline in humidity indicates that the moisture in the outer part of the malt has evaporated. What remains is the water deep inside the malt. Getting this out requires higher temperatures (about 70 °C, 158 °F). The process doesn't take long, but eventually the moisture in the malt has dropped to about 10%. Because the amount of moisture in the malt is significantly reduced, there is little cooling effect during this stage. The result is that the temperature of the air off begins to rise.

Once the air off and air on get close to each other, the malt enters the *curing* stage. In this stage, the temperature of the air on is increased to about 85 °C (185 °F) for lager malts and to about 100 °C (212 °F) for ale malts. In this stage, the moisture content is reduced to about 4%. The curing stage only lasts about 2–3 h. Figure 5.12 illustrates the temperatures of the air on, air off, and moisture level of the malt during kilning.

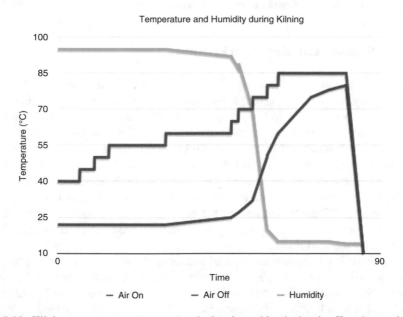

Fig. 5.12 Kilning temperatures. The red line is the air on, blue is the air off, and green is the moisture content of the malt itself

As soon as the malt is cured, the temperature of the air on is dropped quickly to cool the malt. At the end of the entire process, the malt is stored cool to allow the remaining moisture in the seed to redistribute evenly across the seed. Storage times of up to a month may be required for certain malt cultivars.

Kilning actually requires a significant amount of energy, as the air on must be constantly heated. Traditionally, the energy has been provided through the burning of wood, coal, or fossil fuels (such as natural gas). The heat from the burning material was passed directly through the false bottom of the kiln. Modern kilns heat water that passes through a radiator. The air on is fed through the radiator to be warmed and then through the malt. This reduces the flavor impact of burning wood, coal, or fossil fuels on the malt.

Given the cost of construction, many malting plants (aka *maltings*) are designed in a tower format. Barley is conveyed to the top of a multi-story building. There it enters the steeping process. Upon completion of steeping, the grain is conveyed to the next floor down and enters the germination process. It may be moved additional floors and go through additional germination steps. Finally, it enters the kiln in the bottom floors. In this way, the amount of land required to operate a maltings is reduced. The entire facility is known as a *tower maltings*.

Once kilned and dried, the malt may be further kilned at elevated temperatures to provide malt with more color and flavor. Simply kilning at an elevated temperature gives that result. In other cases, the malt is quickly re-steeped and then placed back in the kiln. The air on is re-humidified to 95–100% and the temperature increased to 66 °C (150 °F). This causes the malt to begin to mash (see Chap. 6) and the sugars begin to form. After an hour, the humidity is reduced and the malt kilned until dried back to about 4%. This results in crystal or caramel malt.

5.2.3   Problems Arising from Malting

Malting barley requires the use of hot air. The traditional method of applying heat directly to the air on through the use of burners causes the production of NO_x. NO_x is a general formula for a series of compounds that include NO and NO_2. These compounds are gases that when exposed to the germinating barley can react with amines (such as those found in amino acids and proteins) to form nitrosamines. Of particularly notorious reputation is N,N-dimethylnitrosamine (NDMA), a carcinogen that is highly regulated. For example, the World Health Organization limits NDMA in drinking water to no more than 0.1 parts per billion. The use of indirect heating of the air on, use of the lowest temperatures possible to accomplish the malt drying, and increased air on flow rates have significantly reduced this issue.

In germinating malt, the amino acid methionine undergoes a reaction to form S-methylmethionine (SMM). When the malt is kilned, the heat causes the decomposition of SMM into dimethylsulfide (DMS). The warm air on drives off most of the DMS, but a small amount becomes oxidized into dimethylsulfoxide (DMSO).

DMSO is less volatile and can remain in relatively large amounts in the finished malt. Increased heating reduces both DMS and DMSO during the kilning process, but this comes at the expense of darker malt with increased Maillard reactions.

DMS imparts a creamed corn or canned corn flavor to the malt. Unless this is removed or reduced significantly, this flavor will be found in the finished beer. While this flavor may be desired in some beer styles such as the Pilsner style, it is not desirable in many. DMSO has a garlic-like flavor that is not desirable in most beer styles.

CHECKPOINT 5.3

Smoked malts are usually kilned over a wood fire. In addition to the flavor of the smoke, what would you predict would also exist in the malt?

Describe the differences between fully modified and partially modified malt.

5.3   Maillard Reactions

When heat is applied to the combination of sugars and amino acids, the result is their combination and reaction to form flavor and color compounds. The reaction, described in Chap. 2, is fairly complex, but the amounts of flavor and color are directly related to the time and temperature of the system.

The first step in the reaction sequence involves the condensation of an amino acid and a reducing sugar as illustrated in Fig. 5.13. The reducing sugars that exist in malt are the same as what exist in wort and include glucose, maltose, maltotriose, etc. In malt, these compounds are not as prevalent as those in wort, but exist in

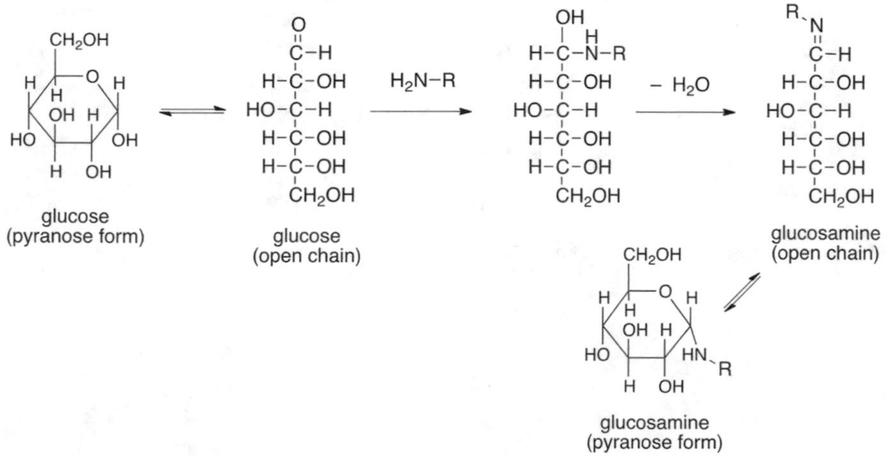

Fig. 5.13 The first step in the Maillard reaction. The amine group (H2N-) is part of a larger molecule, such as an amino acid, protein, or other compound. The "R" stands for the rest of the molecule to which it is attached

quantities sufficient enough to result in the browning and flavor that we attribute to kilned malts. The amino acids that exist are the result of the action of the proteases on the starchy endosperm.

While Fig. 5.13 illustrates the reaction with glucose, the action of maltose with amino acids produces a similar compound. This first reaction is catalyzed by heat. The initial addition of the amine functional group of an amino acid or protein to the carbohydrate is relatively unstable and eliminates water to give rise to glucosamine (or maltosamine, etc.). Glucosamine and maltosamine can then re-cyclize into the pyranose form that is relatively stable.

The second step of the reaction is illustrated in Fig. 5.14. This step, named after the Italian chemist Mario Amadori, is known as the Amadori rearrangement. This reaction is proposed to be acid and heat catalyzed. It involves reopening of the ring and then rearrangement of the double bond to the more stable carbonyl (C=O). This compound, an aminoketose, can also re-cyclize into the pyranose form.

It is at this stage that the Maillard reaction can go one of three ways. In the first pathway (Fig. 5.15), the aminoketose can dehydrate and lose a single water molecule. That results in dicarbonyl compounds, many of which require a rearrangement to more stable compounds. The products of this first pathway are very similar to the products of caramelization (the reaction of carbohydrates with heat). Words used to describe the flavors of these compounds are caramel, toffee, and sugary.

The second pathway (Fig. 5.16) involves the extensive dehydration of the amino-ketose. The product of the initial dehydration then undergoes the Strecker degradation (named after German chemist Adolf Strecker who discovered it in the 1860s). The product of the Strecker degradation is an aldehyde (known as the Strecker aldehyde) and an aminoketone. These products alone have nutty, buttery, or butterscotch

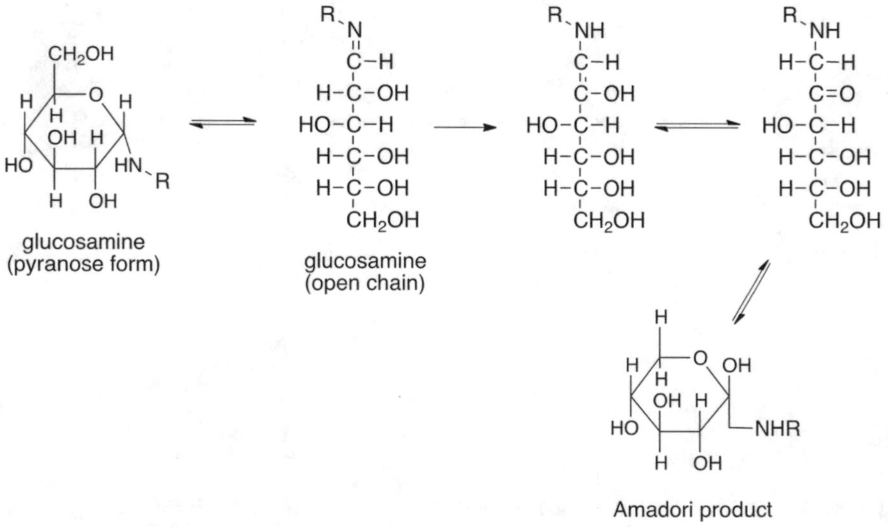

Fig. 5.14 The second step in the Maillard reaction – the Amadori rearrangement

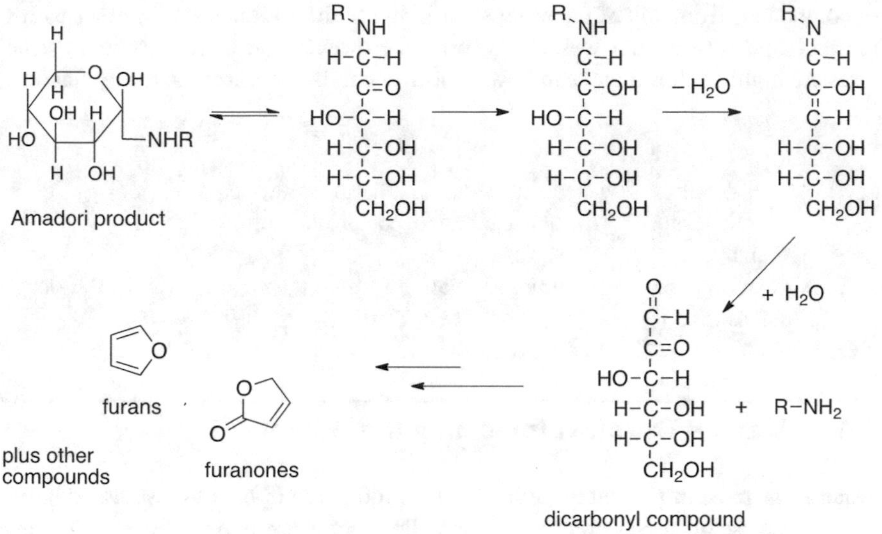

Fig. 5.15 Dehydration gives rise to dicarbonyls that can make compounds with caramel flavors

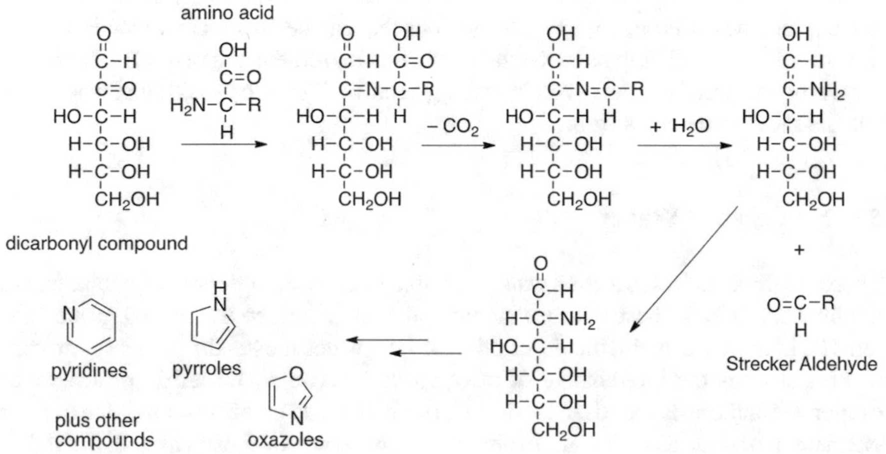

Fig. 5.16 The Strecker degradation and related products. The amino-carbonyl product of the Strecker degradation can make compounds with burnt, astringent, toasty, and nutty flavors

flavors and can contribute to the aroma of the malt. Further reaction of these compounds gives rise to cyclic compounds such as maltol and isomaltol. In fact, kilning of Munich malt imparts a malty flavor to the finished beer. That flavor is the result of relatively large amounts of maltol and isomaltol.

The third pathway gives rise to the melanoidins. Extensive heating or high temperatures form these compounds. Melanoidins are very complex and arise from multiple dehydrations and then combinations of other amino acids, carbohydrates, or compounds from any of the other steps in the Maillard Reaction. In other words, these compounds are not well characterized. However, the large polymeric structures are highly colored and can have an astringent, bitter, burnt, or roasty flavor.

CHECKPOINT 5.4

Given the structure of glucosamine that results from the reaction of an amino acid with glucose, draw the structure of the product of the same amino acid with maltose.

Can malt or other compounds containing sugars and amino acids under Maillard reactions occur at room temperature?

5.4   Water – The Most Important Ingredient

Water is the most important ingredient in the production of beer. As we noted in this chapter, the use of steep water requires that the water is purified or treated. The same is true for the next steps in the brewing process. The properties of mash water, also known as hot liquor, can significantly impact the quality of the finished beer.

In this section, we will uncover this ingredient, discover where it comes from, and explore the different compounds and ions that can be present in water. Knowing this information will help us understand how this ingredient can result in changes to the flavor and processes involved in making beer. Let's start by looking at where this valuable resource comes from.

5.4.1   Types of Water

There are three main sources of water available to the brewer depending upon where the brewery exists. These sources include rainwater, surface water, and groundwater. The hydrologic cycle, the description of how water moves through our environment, describes the interchange of these sources of water. Initially, surface water evaporates and forms clouds. Clouds are primarily made up of water but due to their exposure to the gases of the atmosphere can have some of these gases dissolved in the droplets of water. The gases found in clouds include carbon dioxide, methane, sulfur dioxide (from volcanoes and the burning of fossil fuels), and NO_x (from forest fires and the burning of fossil fuels). Clouds that form over oceans and other salty waters can also contain very small amounts of ions such as sodium, chloride, potassium, etc.

Clouds can traverse many miles from where they are formed before precipitating as rain or snow. And since the rain comes directly from the cloud, the rain and snow contain the same dissolved gases and ions as were found in the clouds. Due to the compounds dissolved in the rainwater, the pH tends to be around 5.5. In the not too distant past, when the regulations on industrial emissions were much more lax than they currently are, the pH of rain was routinely in the 2.0–4.0 range. In fact, the lowest recorded pH of rain occurred in West Virginia in 1978. That rain had a pH just under 2.0. As a point of reference, stomach acid has a pH of 2.0.

Once on the ground, the rain and snow melt into streams, rivers, and eventually end up in lakes and oceans. Water on the surface of the earth comes into contact with plants, soil, rocks, and pollutants from human and other sources. Prolonged exposure to rocks and soils is required for any significant quantities of ions to be present. However, the acidity of the rain can greatly reduce the time required for the water to "pick up" dissolved ions such as calcium. Plants and animals can also greatly impact what is dissolved in the water. Tannins from decaying leaves, bacteria, algae, and other compounds from dead animals can be part of the stream of water. Coupled with the interaction of living creatures with the surface water, this organic material significantly changes the perceived and actual quality of the water.

There are two "spurs" on the hydrologic cycle. One results from the location where snow collects after it has fallen to the ground. Snow can fall in locations that rarely melt, such as at the poles or in glaciers. Ice and snow eventually do melt and are returned to the surface waters of the hydrologic cycle. The other spur results from the permeation of water into the ground. This water can flow just like streams and rivers and can re-enter the surface waters via springs. Often it takes years for the water to return to the surface.

Only one important sink occurs in the cycle. Storage of water outside of the hydrologic cycle can occur when the water permeates down into the soil into underground lakes that do not return to the surface at some other location. It can take hundreds of years for the water to enter these aquifers, providing significant time and pressure to dissolve ions from the surrounding rocks. Because the water must filter down through the soil and rocks, it is often less contaminated by the organic material found in surface waters.

5.4.1.1 Aquifers

There are more than 64 principle aquifers in the United States alone, according to the United States Geological Service. Particulars on each of these aquifers can be found by visiting their website (water.usgs.gov); however, there are five basic types of rock that line these aquifers. Knowing these basic types will give us a good background on what we can expect from our water:

- Sand and gravel
- Sandstone
- Sandstone and carbonate
- Carbonate
- Igneous and metamorphic

Sand and gravel deposits that line aquifers are permeable, and recharge fairly quickly. Prior to the withdrawal of water via wells, most of the water in these aquifers was able to flow into adjacent aquifers or other groundwater sources (such as oceans or rivers). Because the water can flow readily within, into, and out of these aquifers, the water tends to have slightly higher quantities of organic material.

Runoff from agriculture and industry can pollute the water fairly easily. In addition, the water within the aquifer can be relatively high in ions that characterize the location of the aquifer. For example, central California is home to a large sand and gravel aquifer. Water from this aquifer can contain relatively high levels of ions such as iron, sodium, boron, arsenic, and chloride. Many of these ions are the result of intrusion of seawater into the aquifer.

Sandstone lined aquifers have fairly small pores and fractures within the rock. As such, the beds are permeable, but flow of water is restricted to mostly local areas unless the fracturing is fairly extensive. In places where the fracturing is extensive, contamination of the water from agricultural or industrial sources is possible. The pores within the sandstone do a good job of filtering organic materials, but the fractures do not. The intimate contact of the water with the rocks means that the water tends to have fairly large concentrations of calcium, magnesium, and bicarbonate. And, because of the presence of fractures, the ground water can have up to intermediate levels of organic solids and other materials dissolved within it. Western Colorado contains an example of the sandstone aquifers. This particular aquifer has extensive fracturing as indicated by the intermediate level of organic material and high to very high levels of *hardness* (calcium and magnesium).

Carbonate aquifers often have large caves, pipes, and other openings within the rock in which the water rests. Flow of the water within the aquifer is relatively restricted, except in those areas where the caves and openings resulting from dissolving of the rocks are rather extensive. Contamination of the aquifer in these areas can be an issue because of the flow of the water. The amount of organic material in the aquifers is variable, also due to the presence of the openings along the rocks within the system. Calcium, magnesium, and bicarbonate tend to predominate the ions that are found in the water making the water hard to very hard. While other ions tend not to be an issue, sulfate concentrations can be higher in those regions where deposits of lead and zinc exist. For example, southern Missouri is home to a carbonate aquifer and, as a visit to the area will confirm, there are a significant number of caves and fresh-water springs in the area. The springs arise from the fact that the aquifer is very close to the surface of the ground in some areas. This allows the aquifer to discharge directly to the surface, become replenished easily with rainfall, and, unfortunately, be easily contaminated with agricultural and industrial runoff.

Sandstone – Carbonate aquifers contain a mixture of sandstone and carbonate rocks in which the aquifer lies. While pore size is fairly small, the movement of groundwater in these aquifers can be quite large. In fact, the most productive wells in the United States are located in the city of San Antonio, Texas, located on the Trinity-Edwards aquifer (an example of sandstone-carbonate aquifers). These wells can produce more than 16,000 gallons of water per minute. Organic solids tend to be low to intermediate in concentration. But, just like the individual sandstone and carbonate aquifers, the concentration of calcium, magnesium, bicarbonate, and carbonate ions tends to be fairly large. The concentrations tend to be high enough that the water is considered very hard and relatively basic (pH > 7) due to the carbonate and bicarbonate concentrations. Many of these aquifers tend to be located fairly close to the surface of the soil, so they are easily replenished from surface waters and contamination can exist.

Igneous rock aquifers are the least permeable of the aquifer types. These rock systems result from depositing molten rock onto the ground and then over time becoming buried in the ground. The result is a dense rock formation that is often crystalline or fused in nature. Water permeability, then, is limited to fractures within the rocks. Often, multiple layers of igneous rocks occur together, with fractures between the layers. Water moves through fissures in the layers and between the layers to fill the aquifer. The result is limited filtering of the water and few ions from the dissolution of the surrounding rocks.

5.4.1.2 Brewery Water

For the brewer, there are three main places to get water. The first involves collecting it as rain. This is actually illegal in some locations in the United States so it's important for the brewer to understand the local laws about water. Because the compounds that dissolve in rain are variable based on the makeup of the atmosphere at the time it rains, a problem exists in the use of this water. To accurately understand what is being used, the brewer must analyze each and every sample before its use in brewing. More importantly, unless the brewery is located in a rainforest, the quantity of rainwater is likely not sufficient to serve as the sole source of water for the brewery. Even pilot batches of 1 to 2 bbl would require significant quantities of water that could not be routinely supported solely by rainfall.

The second place the brewer can obtain water is from a well. Wells can be dug deep enough to reach into an aquifer, or may be shallow so that they depend solely on groundwater below the water table. While the aquifer option is likely the best, it may not be possible to dig that deep. So, some of the wells in use only dip into the ground water supply under the water table. Water quality of these shallow wells is better than the surface waters surrounding, but in some cases not much more.

For the deep wells that take water from the aquifer, the ions and species in the water are a result of the type of rock associated with the aquifer. For example, a brewery located in the southeastern Kansas region would be using water from carbonate rocks. The water would be very hard, with high levels of calcium and magnesium. The pH of the water would likely be at or above 7.0 due to the high levels of carbonate ions. The brewer would want to periodically test the water for contamination from agricultural sources and for organic materials as these could periodically become large enough to damage the flavor of the finished beer. While unusual, the brewer may also want to check for the presence of bacteria or other microorganisms indicating further contamination.

The third, and most common, place to obtain water for brewing comes from a municipal supply. In fact, most brewers won't have an option and be forced to use water supplied from their town. Across the United States, cities and towns get their water from surface and ground water supplies as needed. Some towns pull water from nearby lakes, some from rivers and streams, some dig wells, and others use reverse osmosis to grab water from oceans. In fact, larger towns and cities may obtain water from multiple sources. It all depends on where the closest water supply exists.

Thus, the water delivered to a brewery is highly dependent upon the location of the brewery. In addition, if the city uses multiple sources for their water, the source may be different from season to season. Of greater impact, however, is that the water obtained from municipalities is often treated to ensure that no harmful pathogens or ions exist. To ensure a safe water supply, most municipalities add chemicals.

Water purification often begins by adding compounds such as aluminum sulfate or iron(III) chloride. These compounds react with water to make aluminum hydroxide or iron(III) hydroxide. The hydroxide salts coagulate and entrap organic solids as they precipitate from the water. The sulfate and chloride ions that remain in the solution raise the overall concentration of these ions a little, but the removal of the suspended matter is more beneficial.

The purification plant then adds a disinfectant to the water to reduce or eliminate the presence of pathogenic organisms. This can be done by either adding chlorine gas (Cl_2) or sodium hypochlorite (NaOCl). In water, both are sources of the hypochlorite ion (OCl^-). These species are very powerful oxidizers and destroy the cell walls of the microorganisms. One problem exists with these disinfectants, though. They do persist in the water supply and can be found in measureable quantities in the brewery water. These oxidants can also cause damage to the malt if the water is used without removing them.

The other issue is the presence of byproducts that result from the disinfection process. These oxidizers can react with organic material in the water and form trihalomethanes and haloacetic acids. These compounds are potentially carcinogenic and chronic exposure to them causes other health issues. Their presence in the water is highly regulated by state and federal authorities. For the brewer, however, these compounds impart a taste to the finished beer that is perceived as an off flavor. So, at the very least, they must be removed to protect the flavor of the beer. Their removal is accomplished by passing the water through a charcoal filter. The charcoal absorbs most organic compounds and the disinfection byproducts. It should be noted, however, that after passing the water through the charcoal filter, it has lost all of its disinfectant (i.e., bacteria can again grow in the water if there are any sources of contamination.)

CHECKPOINT 5.5
Use the information in this chapter to draw a diagram illustrating the hydrologic cycle.
What is the likely formula of the trihalomethane that results from the addition of chlorine gas (Cl_2) to water contaminated with organic matter?

5.4.2   What's in the Water?

Water's chemical formula is H_2O. As we discovered in Chap. 2, water is a polar molecule that can dissolve other polar substances. Gases, such as CO_2 and O_2, tend not to be polar and, as such, tend not to dissolve in water. Many of the larger organic molecules tend to be only sparingly soluble in water as well. On the other hand, ionic compounds have variable solubility. Some are quite soluble and others are very insoluble.

5.4.2.1 Cations in Water

The typical ions that are found in drinking water include calcium (Ca^{2+}) and magnesium (Mg^{2+}). Both come from the dissolution of rocks that are permeable by water. Calcium and magnesium tend to occur together, with calcium as the major ion in water. Other ions are also possible.

Calcium and Magnesium   Calcium and magnesium contribute to the hardness of the water sample. Two forms of hardness exist. *Temporary hardness* is the result of calcium and magnesium in the presence of carbonate or bicarbonate. If the water has temporary hardness, boiling it for a few minutes will reduce the level of dissolved calcium in the water as shown in Fig. 5.17.

   Water with *permanent hardness* is the result of the presence of anions of sulfate, nitrate, chloride, and others. These anions do not form solids with calcium and magnesium when heated and instead remain in solution. This is somewhat desirable as both calcium and magnesium are both necessary at different stages of the brewing process. For example, flocculation of yeast during the cold crash stage of fermentation is sluggish when the calcium level is low in the water.

   At nearly every concentration level, calcium tends to be beneficial. Too high of a level, however, can cause the formation of beer stone (calcium oxalate) on the vessels and kegs in the process. Magnesium acts very similarly to calcium but at levels greater than about 15 ppm can cause some issues. Above this concentration, a bitter taste becomes evident. In addition, digestive issues such as a laxative effect can be noted in those that consume this concentration.

Iron   Iron ions can be found in some water supplies. And sometimes these occur from poor plumbing systems (either in the municipality or within the brewery). If iron is found in the water, it can exist as one of two forms (Fe^{2+} and Fe^{3+}). The ferrous ion (Fe^{2+}) is typical in iron-containing waters that have not been aerated. Water that has been aerated typically contains the ferric ion (Fe^{3+}). When the levels of iron are high in water, rust stains appear on fixtures, the water takes on an orangish or rust-colored hue, and the flavor can be very metallic. But even low concentrations of iron above 0.5 ppm are harmful in the brewery. It is toxic to yeast at this level and causes any of the tannins in the beer to oxidize faster, imparting poor flavors to the resulting product.

$$Ca(HCO_3)_2 \, (aq) \rightleftharpoons CaCO_3 \, (s) \; + \; H_2O \, (l) \; + \; CO_2 \, (g)$$

Fig. 5.17   Boiling water with temporary hardness reduces the calcium and magnesium content

Copper   Copper ions can enter the water supply when they leach into ground water. Copper can also enter the water if the pH of the water supply is acidic and copper pipes are used. At levels greater than 10 ppm, copper can be toxic to yeast. In addition, it can speed the oxidation of tannins and cause permanent haze. Humans, on the other hand, can react to levels as low as 1.3 ppm. This can result in gastrointestinal distress or, in some cases, kidney or liver damage.

Sodium   This ion enters the water stream naturally from surface and ground waters. It also gets into brewery water if the water used is conditioned using an ion-exchange conditioner. While moderate and low levels have a minor impact (high sodium levels can affect yeast growth), they do affect the flavor of the finished beer. At levels above 150 ppm, the beer will taste salty. At levels less than this, the sodium imparts a perceived sweetness to the beer.

Potassium   This ion also arises in brewery water from natural sources. It has effects that are similar to sodium. High levels of this ion in beer can cause digestive problems. Very high levels can affect cardiovascular function in humans.

Other Metal Cations   Many other metal cations may find their way into the water that the brewer uses. For example, if lead pipes are part of the supply of water to the brewery, it may be possible to have some lead dissolved in the water. Most of these cations are not desirable at anything more than trace levels, where they can be useful for yeast health. Beyond those levels, they can be toxic to the yeast.

5.4.2.2 Anions in Water

The typical anions found in brewery water are carbonates and bicarbonates. These arise from natural sources and have effects on the brewing process, the health of the yeast, and on the flavor of the finished beer.

Bicarbonate and Carbonate   As we will discover throughout this text, these ions are very important to the brewing process. They arise naturally in the water from its exposure to air (see Fig. 5.18). The result of dissolving carbon dioxide in water is

$$H_2O + CO_2 \rightleftharpoons H_2CO_3$$

$$H_2CO_3 \rightleftharpoons HCO_3^- + H^+$$

$$HCO_3^- \rightleftharpoons CO_3^{2-} + H^+$$

Fig. 5.18   The formation of bicarbonate and carbonate from dissolution of CO_2. The second reaction predominates in water. The third reaction occurs only in very alkaline (pH > 10) water

$$CaCO_3 \rightleftharpoons CO_3^{2-} + Ca^{2+}$$

$$CO_3^{2-} + H_2O \rightleftharpoons HCO_3^- + OH^-$$

Fig. 5.19 The formation of bicarbonate resulting from the dissolution of carbonate containing rocks in water. Note that the reaction also generates hydroxide ions

the formation of carbonic acid. Carbonic acid then decomposes into the bicarbonate anion and a proton. The proton lowers the pH of the solution as we'll see in the next section. The bicarbonate anion can further decompose into the carbonate anion and release another proton (and further lower the pH); however, this last reaction only occurs when the pH of the solution is already rather high.

Bicarbonate and carbonate anions can also be added to water through its contact with carbonate containing rocks, such as in carbonate aquifers. Figure 5.19 illustrates the reactions of the carbonate anion in water. The reaction of carbonate anions with water gives rise to the bicarbonate anion and an anion of hydroxide. The hydroxide results in the increase in the pH to more alkaline values.

Water that has been exposed to both carbonate containing rocks and air (containing carbon dioxide) has both carbonate and bicarbonate anions in it. In addition, these waters tend to have pH values that are greater than rainwater, but less than well water from a carbonate aquifer.

Chloride Chloride in the brewery water occurs naturally. At levels up to about 350 ppm it can impart a beneficial effect on the fullness of the flavor of beer. At levels above 500 ppm, it can interfere with the flocculation of yeast.

Fluoride Many municipalities in the United States add fluoride to the drinking water and some natural deposits can increase the level of fluoride in ground waters. The addition of fluoride has been very useful in the prevention of dental caries (cavities) in both adults and children. The U.S. Environmental Protection Agency (U.S. EPA) maximum contaminant level for fluoride is 4.0 ppm. Consumption of water containing fluoride above this level can cause pain and tenderness in bones.

Nitrate and Nitrite These ions enter water naturally from deposits, but also can indicate contaminated water. In oxygenated water, the level of nitrite is usually quite low as it becomes oxidized to nitrate. The U.S. EPA maximum contaminant level for nitrate is 10 ppm and for nitrite, it is 1 ppm. Water containing more than these levels causes blue-baby syndrome in infants. Nitrite is toxic to yeast and both nitrite and nitrate can form carcinogenic compounds during the process to make beer. Whenever possible, water containing nitrates and/or nitrites should be avoided.

Sulfate Sulfate in brewery water can be a very useful anion. At low levels it can be beneficial in creating a drier flavor. It can also help enhance the bitter hop flavor. At levels above about 250 ppm, it can begin to impart a slightly salty flavor. When those levels get above 400 ppm, the sulfate can cause gastrointestinal distress. The flavors of the traditional Pale Ales made in Burton-on-Trent, England, are considered well enhanced by the high levels of sulfate (>600 ppm) in the well water used. Note that this level is far above the concentration that might cause stomach issues. In the production of beer, the greatest impact is that it can be converted to SO_2 and H_2S by yeast or other microbes. These compounds can add poor off-flavors; H_2S smells like rotten eggs.

5.4.2.3 Reactions in Water

There are many reactions that take place in water. And since beer is mostly water, it is useful to understand how these reactions take place. In Chap. 3, we explored many reactions that occur in beer. While we focused on the reactions that occur with organic molecules, such as the condensation of an alcohol with a carboxylic acid to make an ester, there are many other reactions that occur in water.

When a reaction occurs in water, we say that the reaction is an *aqueous* reaction. This doesn't mean that water is one of the reactants in the reaction, although it can be. Instead, this means that the water is involved in dissolving the components of the reaction.

Water is a polar molecule that has a region of partial positive charge around the hydrogen atoms and a region of partial negative charge around the oxygen atom. If a cation is present, the water can orient itself so that the oxygen atom interacts with the cation. Similarly, water interacts with anions by orienting its hydrogen atoms toward the anion. The result, an intermolecular force of attraction due to proximity of opposite charges, is more stable than the individual anions and cations alone. In other words, water can stabilize cations and anions. It is this stabilization that makes the anions and cations soluble in water.

If the interaction of water and the anion or cation is not strong enough, the anion and cation associate with each other to make an inorganic compound. The inorganic compound without the water interaction tends to attract more anions and cations to make a larger association of ions. The result becomes so large that it cannot be supported by the water and a solid forms. When the solid is large enough, it becomes visible to the naked eye (and forms a haze in the beer). And if the solid grows even larger, it drops out of solution and forms a sediment at the bottom. This is known as *precipitation*.

Precipitation reactions are common in beer. For example, a precipitation reaction of importance to the brewer is the reaction of calcium ions with oxalate anions. Calcium, one of the ions common in water, is necessary for many enzymatic and metabolic processes during the mash and fermentation. Oxalate, present in very small quantities in malt and hops, has been suggested to also form during yeast metabolism. Typically, the quantities are fairly small (~ 10 ppm), but calcium oxalate is so insoluble in water that it forms a precipitate. In a large vessel, it results in a large amount of solid. The overall reaction is shown in Fig. 5.20.

$$Ca^{2+} (aq) \quad + \quad C_2O_4^{2-} (aq) \quad \rightleftharpoons \quad CaC_2O_4 (s)$$

Fig. 5.20 The reaction of calcium cations and oxalate anions forms calcium oxalate, a precipitate

$$K = \frac{[products]}{[reactants]}$$

$$K = \frac{[CaC_2O_4]}{[Ca^{2+}] \, [C_2O_4^{2-}]} = \frac{1}{[Ca^{2+}] \, [C_2O_4^{2-}]}$$

Fig. 5.21 The equilibrium constant and an example of the constant for the calcium oxalate reaction. The concentration of each component in the reaction (denoted with square brackets) is included in the equation. Based on the equation outlined in Fig. 5.20, the equilibrium constant for calcium oxalate is written as shown

Note that the calcium cation and oxalate anion are written with an (aq). This means that they are soluble (they are in an aqueous solution). The product is insoluble in water and is written with an (s) to indicate a solid. This solid, calcium oxalate, is also known as beer stone. It is a brownish solid that forms a sediment at the bottom of a vessel and can stick to the sides of the vessel where it can be difficult to remove.

Most reactions are written with two arrows pointing in different directions. This emphasizes that the reaction can proceed in both the forward direction and the reverse direction. The result is known as an *equilibrium*. In an equilibrium, the reactants are constantly forming products and the products of the reaction are constantly reforming the reactants. The rate at which the reaction moves forward to make products and the rate of the backward reaction to make reactants are not the same initially. But at some point, the rates become the same and the reaction reaches a steady state. At that steady state, the concentration of the reactants and products is constant (although the reaction is still running forward and backward at the same time.)

Some reactions reach a steady state with higher concentrations of the product than the reactant; some equilibria favor more of the reactant than the product; and others lie more toward a fairly uniform distribution of reactants and products. This distribution is referenced for each reaction as a constant known as the equilibrium constant (K). The value of K can be calculated using the equation shown in Fig. 5.21. The only compounds that are listed in the equilibrium constant equation are those that are aqueous solutions or gases. If a compound is indicated as a solid or a liquid in the equation, it is not included in the equation. This is because the concentration, a measure of the amount of a substance per unit volume, doesn't change for a solid or a liquid.

Fig. 5.22 The acid
equilibrium constants for
the equilibria from
Fig. 5.18. Neither reaction
goes to completion; instead
the reactions make only a
small amount of product

$$H_2CO_3 \rightleftharpoons HCO_3^- + H^+$$

$$Ka = 4.5 \times 10^{-7}$$
$$pKa = 6.35$$

$$HCO_3^- \rightleftharpoons CO_3^{2-} + H^+$$

$$Ka = 4.7 \times 10^{-11}$$
$$pKa = 10.33$$

If the value of K is large, this indicates that the reaction favors making products. If the value of K is small, the opposite is true and the reaction doesn't really make much products at all. When K is equal to 1, the reaction essentially goes half-way and results in lots of starting materials and products. For the calcium oxalate reaction, the value of K is 3.94×10^8. This large value of K means that almost all of the calcium and oxalate present in solution will react to make the solid beerstone.

The use of the equilibrium constant can be helpful in determining whether a reaction makes products or not. However, the constant is often a fairly large or small number that is often written in scientific notation. To make things a little easier to write and relate to other brewers, the value can be written as the negative log of the equilibrium constant. This is known as the pK. Note that mathematically a large K would result in a small pK, and vice versa.

$$pK = -\log K \tag{5.1}$$

When the reaction in question relates the acidity of a compound, the equilibrium constant becomes the acid equilibrium constant, Ka. Similarly, the negative log of the constant would be the pKa. For example, Fig. 5.22 illustrates two acids and their equilibrium constants.

5.4.3 pH

As we saw in our discussion of equilibria, pKa values express the degree at which compounds that are acids make products. The product mixture of an acid reacting with water includes a proton (H^+). This is the species that makes a solution an acid. And measuring the acidity of a solution requires one of two methods. The first option is to know the concentration of the acid and its pKa value. Then, a calculation can be performed to determine the acidity of the solution. The second method is more widely used and is very familiar, but it requires that the solution be in water. The acidity can be directly measured by determining the concentration of the H^+ in solution.

As with the values for the equilibrium constant, the concentration of H^+ is reported as the negative log of the concentration, the pH. In an aqueous solution, the

$$H_2O \ (l) \ \rightleftharpoons \ OH^- \ (aq) \ + \ H^+ \ (aq)$$

$$Ka \ = \ [OH^-] \ [H^+]$$

$$1 \times 10^{-14} = (1 \times 10^{-7})\ (1 \times 10^{-7})$$

$$pH = -\log (1 \times 10^{-7}) = 7$$

Fig. 5.23 The dissociation of water into H+ and OH–

measure of the concentration of H^+ results in a measure of the acidity of a solution. Figure 5.23 illustrates that water can ionize by itself to make H^+ and OH^-. The equilibrium constant is very small, indicating that water doesn't make very much of the H^+ or OH^- ions. In a solution that is neither acidic nor *alkaline* (aka *basic*), the concentration of H^+ is 1.0×10^{-7} M. Thus, the pH is 7.0. When the concentration of H^+ is greater than this, the concentration of OH^- is proportionately less, and the solution is acidic. This means that the pH would be less than 7. Conversely, if the concentration of H^+ is less than 1.0×10^{-7} M, the solution becomes alkaline or basic.

5.4.3.1 Residual Alkalinity

Residual alkalinity is a measure of the concentration of alkaline ions that will impact the pH of water after all of the carbonate and bicarbonate have complexed with the available calcium and magnesium. For the brewer, residual alkalinity is a bad thing. Any of these anions that remain will increase the pH of the water. And when the brewer uses such water to mash, the pH of the system may cause the extraction of tannins from the grain.

Luckily, the type of malt that is used in the production of beer can absorb some of the residual alkalinity. It turns out that a darker colored malt has more absorbing power than a lighter malt. So, the brewer can "get away" with a residual alkalinity that is a little high if they're planning on using dark malts. Further adjustment to the residual alkalinity would definitely be needed if the brewer isn't using dark malts.

Kolbach, in 1953, recognized that calcium and magnesium react with the alkalinity in water. On a per mass basis, 1.4 equivalents of calcium can react with 1.0 equivalent of alkalinity. In addition, 1.7 equivalents of magnesium can neutralize 1.0 equivalent of alkalinity. Thus, the residual alkalinity of a sample of water is:

$$\text{Residual Alkalinity}\,(\text{ppm}) = \text{Total Alkalinity}\,(\text{ppm}) - \left(\frac{\text{ppm Ca}}{1.4} + \frac{\text{ppm Mg}}{1.7} \right) \quad (5.2)$$

The residual alkalinity can be adjusted by the addition of acid (to neutralize some of the total alkalinity). If such a method is used, the acid chosen should be neutral to the brewing process. Thus, the use of hydrochloric acid (HCl) is not a good choice. The use of phosphoric acid (H_3PO_4) is a more common choice. The phosphate ion is useful in the brewing process and is even produced naturally by the mashing of malt. Other ways to adjust the residual alkalinity include the addition of calcium and magnesium to the water. This would increase the concentration of ions that can react with the total alkalinity.

Managing the residual alkalinity is very important in the brewing process. In addition to making sure that enough calcium and magnesium are present in the water to take care of all of the alkalinity, it is vital that additional calcium and magnesium remain. Those ions cause the isomerization of hop acids, aid in the flocculation of yeast, help form foams for the head, and the list goes on. If the water used isn't treated or adjusted to reduce the residual alkalinity, the finished beer (and the overall brewing process) can suffer.

CHECKPOINT 5.6

Where does nitrate in ground water come from?

If a water sample has 200 ppm chloride, how many ppm sodium would it have? (assume that the only cation in the water is sodium).

What is the pH of a solution that contains 4.8×10^{-5} M H⁺ ions? Is this solution acidic or alkaline?

How would adding more calcium reduce the residual alkalinity of a water sample?

Chapter Summary

Section 5.1
    The barley plant is a member of the grass family.
    Barley is dried to about 12% moisture before being sent to the malthouse.
    Sorting and grading provide information about the quality of a barley harvest.
Section 5.2
    The barley seed germinates when water rehydrates the corn.
    Gibberellic acid triggers the production of enzymes that convert the endosperm
        into usable sugars.
    The process of malting includes steeping, germinating, and kilning.
Section 5.3
    Maillard reactions are complex but involve three main steps.
    The products of the reactions increase the color of the malt and impart caramel,
        toasty, malty, or burned flavors to the malt.
Section 5.4
    Water is the most important ingredient in the brewhouse.
    The makeup of cations and anions in the water is a result of the source of the
        water and any added contamination.
    pH is the measure of acidity of a solution.
    Residual alkalinity must be managed to avoid extraction of tannins and to ensure
        the adequate concentration of calcium and magnesium during the brewing
        processes.

Questions to Consider

1. Why is water considered the most important ingredient?
2. Use the Internet to look up reasons why some believe that barley was likely one of the first grains grown.
3. If a barley corn's husk is damaged, will this change how the seed hydrates?
4. To take question #3 further, assume that the damage to the barley corn is a hole through the husk, pericarp, and testa. Will this change how the seed hydrates?
5. Use the Internet and identify countries in Europe that produce barley.
6. Use the Internet to visit the American Malting Barley Association website. Which cultivars are recommended for next years' crop?
7. What is the minimum mass of a bushel of Grade 1 2-row malting barley?
8. Why is the air on during kilning slowly ramped up, rather than being set at the initially warm temperature?
9. One of the compounds produced during the initial stages of the Maillard Reaction is diacetyl. Why does the flavor of this compound decline as the malt is further kilned?
10. A water sample is reported to have a total alkalinity of 100 ppm as $CaCO_3$. If the sample contains 50 ppm Ca and 12 ppm Mg, what is the residual alkalinity of the sample?
11. What is the benefit to the malt in a floor maltings? What are the disadvantages for this method of preparing malt?
12. Use the Internet to look up how your city or town gets its water. How is it treated prior to being delivered to your home?
13. What is the likely fate of the roots that grow on the barley seed as it is modified?
14. Would sulfuric acid be a good choice to decrease the alkalinity of a water sample that was to be used in the brewhouse? Why or why not?
15. Can wheat, another member of the grass family, be malted? Why or why not?
16. What would a brewer need to do to use rainwater as the water in the brewhouse?
17. Why is rainwater typically acidic, but well water from a carbonate aquifer typically alkaline?
18. Rank the pH values of water from a carbonate-sandstone aquifer, surface water, and rainwater.
19. What would you predict to be the effect of adding gibberellic acid to a sample of barley that is about to become malted?
20. Estimate the total time to convert a bag of barley that has been freshly harvested into a bag of malt that is ready to be used by the brewer.
21. How many milligrams of calcium (as calcium carbonate) are there in 1.0 bbl of water with 50 ppm calcium (as calcium carbonate)?
22. If a 1.0 bbl water sample has 80 ppm total alkalinity, 50 ppm calcium, and 10 ppm magnesium, how many milligrams of calcium (as calcium carbonate) must be added to give a residual alkalinity of 0.0 ppm.

Laboratory Exercises

Germination of Barley

This experiment is designed to allow you to see the changes that barley seeds undergo during the germination process. The entire experiment takes 3–7 days to complete, but analysis of the seeds along the way is very helpful in learning the key names of the parts of the seed as well as observing the growth of the seeds.

Equipment Needed

Barley – seeds that are ready for planting
Magnifying glass
Paper towels
100 mL beaker
Water

Experiment

Each student group should obtain 20–30 barley seeds for this experiment. The seeds are placed into the beaker and then room temperature water is added. The seeds are gently stirred and left to sit in the water for an hour. The water is then removed by decanting it from the seeds. Then, slightly warm water is added to cover the seeds. The seeds are again stirred and left to sit for another hour. Then the water is decanted and the seeds placed onto a paper towel that is wet with water.

The paper towel is folded over and water is added to wet the towel completely. The paper towels are placed in a cool dark place (such as a drawer), and left until the next day. Periodically over the next 7 days, the paper towel is observed. It should remain damp during the entire period.

Observations should be recorded and, using a magnifying glass, drawings of the seeds should be made. At each observation, one of the seeds should be pulled apart (if possible) and the interior of the seed drawn.

Make a table containing the following headers:

- Time since first steeped
- Total Number of seeds
- Number of seeds that have chitted
- Number of seeds with roots that are 10 mm long
- Number of seeds with acrospire that is visible

Plots of the data in the table can be used (time versus percent that…) to graphically observe the same data.

Milling and Mashing

6

6.1 Milling

Milling is the first of the steps that the brewer takes when a batch of beer is being made by the brewer. The purpose and the results of the milling process can't be underestimated. The quality of the milling process sets the stage for the entire process that follows. If that quality is lacking, the resulting beer that comes out of the final step in the brewery will also be lacking.

Milling involves crushing the malt so that the starchy endosperm is exposed to the environment. In some cases, as we'll discover, a slight crush that just breaks each malt kernel is the best option. In others, the best option is to crush the malt into a fine powder.

6.1.1 Purpose of Milling

As we noted, the purpose of milling is to expose the starch in the endosperm to the environment. This allows the enzymes within the mash to reach all areas of the endosperm and convert the entire endosperm into sugars. If the malt kernels are only slightly broken, the action of the enzymes may take longer as the pieces of the endosperm will be large. If the process requires that the malt is ground into flour, the small particle size undergoes the mashing reactions much faster.

While making the endosperm available for the mashing enzymes is the primary goal, the brewer has a secondary goal for the mill. That goal is to make sure that the malt husks are as intact as possible. Larger pieces of malt husk, which are essentially untouched in the mashing process, can help form a filter bed in the mash or lauter tun. These husks help to separate the particles of endosperm so that they don't form a cement-like mixture. The result: the wort can slowly drain from the spent grains/husks at the end of the mashing process.

M. Mosher, K. Trantham, *Brewing Science: A Multidisciplinary Approach*,
https://doi.org/10.1007/978-3-030-73419-0_6

If the process requires that the malt is ground to a flour, the husk material usually doesn't survive intact. Then, when the flour is added to the mash mixer, the resulting slurry of water and malt must be constantly stirred so that it doesn't set up. This ensures that all of the endosperm is hydrated with water and exposed to the enzymes. If the stirring stops or slows, pockets of the mixture settle out and form a cement or dough-like ball. The exchange of water and enzymes into and out of those regions is very slow and the mash suffers because of it.

A problem with grinding the malt into flour for the mashing process is that the husk material has also been ground up. Thus, all of the husk is exposed to the mash water. And, if the pH, ions in the water, or heat isn't just right, the mash water can extract the tannins (polyphenols) from the husks. This can result in the addition of an astringent finish to the beer. The brewer is well aware of this and takes extra cautions to ensure that the mashing and sparging water chemistry is as perfect as possible.

When the malt is milled exactly to the specifications of the mashing process, the enzymes rapidly and efficiently convert the starch to sugars.

6.1.2   Equipment Used in Milling

There are a wide variety of different mills that can be used in the brewery. In general, they fall into one of three categories.

Roller Mills Just as the name suggests, these mills are made up of pairs of cylindrical rollers that are similar in shape to the rolling pin used to flatten out dough for biscuits. The rollers have a rough serrated surface that grabs onto the malt kernels and forces them into the gap between the rollers. The width of the gap determines how much "crush" is imparted to the malt kernel.

The roller mills come in multiple configurations. Typically, there are either 2-roll, 4-roll, or 6-roll versions that contain either 1 pair, 2 pairs, or 3 pairs of rollers. The version with the most control over the resulting size of the crushed malt (also known as grist) is the 6-roll version. It is also the most expensive. The first pair of rollers is set to a fairly wide gap to gently crush the malt into relatively uniform, but too large, size. The second pair of rollers has a slightly smaller gap that makes everything a little smaller. And the third pair of rollers fine-tunes the size of the pieces. The benefit to having multiple rollers is that any dust or flour that is formed during the process can be directed to a separate bin so that it doesn't impact the mashing process.

Wet Roller Mills This type of mill is essentially identical to the first category, except the malt is first mixed with water before it enters the mill. This hydrates the husk slightly so that it doesn't shatter as it is crushed. More importantly, no dust is formed during the milling process. This increases the safety of the milling operation by reducing the amount of dust in the air.

As a side note, dust is a significant hazard in manufacturing. Finely divided particles that are dispersed in air can be explosion hazards. And any spark could set them off. Brewers are well aware of this issue and take precautions to make sure that dust is eliminated or greatly reduced. Unfortunately, that's not the only hazard associated with dust. The other hazard is to the workers that are within the area. Breathing dust can cause significant health problems due to inhalation of the dust. So, if an area of the brewery might generate dust, the workers are often found wearing breathing masks to limit their exposure.

Hammer Mills   The noisiest of the mills is the hammer mill. This device is essentially a large horizontal drum containing a set of movable hammers. Malt is added to the drum and as it turns, the hammers crush the grain. Control of the size of the kernels that result isn't as easy as it is with the roller mill and the wet roller mill. In fact, the only option that produces good results is to convert the malt into flour.

Just like the dry roller mills, hammer mills generate dust. And since their goal is to make flour, they make a lot of dust. So the brewer has to have dust controls in place for the hammer mill. Workers have to wear hearing protection in addition to breathing protection when working near a hammer mill.

6.2   Purpose of Mashing

Mashing is necessary as a step in the brewing process. Essentially, it is a continuation of the germination process begun in the malt house. However, the difference between germination and mashing is that the barley seed does not continue its growth after being kilned. In mashing, the enzymes inside the barley seed become reactivated. Once hydrated and at the correct temperature, the enzymes go to work and convert the starch into fermentable and non-fermentable sugars. They also decompose the proteins and other biological structures holding the starch in the endosperm and release it into the water. They convert the proteins into smaller pieces and individual amino acids, and also help lower the pH of the overall system.

The fermentable sugars are very necessary in the next step of the brewing process. These sugars include maltose, glucose, maltotriose, and a host of others that can be consumed by yeast during fermentation. While fermentable sugars are the primary product of the mashing process, mashing also creates non-fermentable sugars. These sugars are not consumed by yeast during fermentation and remain essentially unchanged by the end of the entire process. In other words, they remain in the beer after fermentation to lend a sweet taste.

Mashing also reduces the size of some of the proteins that are extracted from the malt. In some cases, the individual amino acids are cleaved from the proteins. Many of these are essential as nutrients for the yeast during fermentation. The amount of protein degradation during mashing has a direct impact on the mouthfeel and the properties of the head of the final beer. As we'll uncover later, other enzymatic activity can even result in a lowering of the pH of the wort to the perfect level for other enzymatic activity.

6.3   Equipment Used in Mashing

There are many different types of mashing vessels available on the market. In the
microbrewery, cost and efficiency often result in the use of a mash tun that can
double as a lauter tun (see the next chapter for more information on this step in the
brewing process). In the large-scale brewery, when performing mashes on large
batches, or when using malt that has been milled using a hammer mill, the use of a
separate vessel for separating the spent grains from the wort is necessary.

Let's take a look at the process in a stepwise fashion. The first process in the
mashing process is to combine the grist with hot liquor. First, we add *foundation
liquor* to the mash mixer or mash tun. This hot liquor serves multiple purposes and
is a vital step in the process. It pre-heats the mash vessel and provides a buffer for
the grist that will fall into the vessel. Without the water at the bottom of the vessel,
the grist could fall with such force that it would be driven into the holes of a false
bottom or even further break the grist or the husks into smaller pieces. Finally, the
foundation liquor helps reduce the amount of air that is mixed into the mash and
results in a reduction of oxidation during mashing.

Once the foundation liquor is in the vessel, we can add the *grist* through the top
of the vessel either directly into the vessel or after a pre-mash mixing with additional hot liquor. The Steel's Masher, a common pre-mash mixer, was developed in
England in 1853. It hasn't changed much since then because it mixes the grist and
hot liquor. This device, see Fig. 6.1, admits dry grist through a grist case and into a
horizontal arm. Hot liquor is injected into the grist at this point. Then, an auger
moves the grist/water mixture and mixes it before directing it into the mash vessel.
The use of a Steel's Masher gives the brewer control of the temperature of the initial

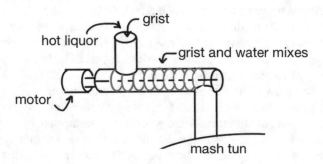

Fig. 6.1 Steel's Masher

mash-in and mixes the mash sufficiently such that it doesn't have to be mixed much, if at all, during an infusion mash.

Another common pre-mash mixer that we find in the brewery is the vortex masher. This device can be used with dry and wet milled grist. The grist is added to a funnel at the top of the mash vessel while the hot liquor is also added. Some designs allow the hot liquor to be sprayed into the grist to aid in the reduction of dust. Other designs simply add the grist to a stream of hot liquor. The vortex masher adds the two streams together in such a way that they swirl around each other as they enter the mash vessel. This mixing constitutes the start of mashing, known as the mash-in. The design of the masher is such that the grist is fully hydrated by the time it enters the mash tun.

Alternative pre-mash mixers have also been developed. In some cases, these involve adding hot liquor to the grist and using a positive displacement pump to transfer the wet slurry into the mash tun via the bottom of the vessel. And, as we'd expect, there are some breweries that simply add the grist directly to the mash tun containing the appropriate amount of hot liquor.

In the homebrewing world, this is the process that takes place. The mash tun (for the homebrewer, this is often a cooler) is filled with hot liquor whose temperature is just a little warmer than the desired mash temperature. Then, the grist is added and the mixture stirred with a long paddle or spoon. Stirring is continued until all of the grist has been wetted (the dough balls are broken up). In fact, the authors have even seen some homebrew setups where the grist is placed in the cooler first, and then the hot liquor is added with a LOT of stirring.

While the processes at the homebrewery level are not ideal, they work. When translated to the larger scales of microbrewing, they can still work – but cost a lot in terms of time and elbow grease. For the small, startup microbrewery, the addition of dry grist to a mash tun full of hot liquor can be a way to complete the mash in. It isn't efficient, but it works and is a little less expensive than the use of a pre-mash mixer. The most effective part of the simpler process is that a brewer can easily get the correct ratio of water and grist during the mash. Compared to the use of a vortex mixer, the rate of addition of water versus the rate of addition of grist must be closely monitored to arrive at the correct mass of both in the final mash.

> **CHECKPOINT 6.2**
> What is the purpose of the Steel's Masher? What benefit would it or another pre-mash mixer have on the outcome of the mashing step in brewing?

6.3.1   Cereal Cookers

A cereal cooker is a separate mash vessel needed when an unmodified adjunct is added to the mash. Unmodified cereals that are commonly used in brewing include corn and rice. Unmalted barley, oats, wheat, and rye also find their way into the brewing recipe. Initially, the use of unmodified grains was common as a way to save expense in the manufacture of beer. Today, the cost is a small part, the main reason for the use of unmalted cereals is that they are an integral part of the recipe. In other words, the use of adjuncts may be desired for the final flavor or the beer.

If these adjuncts are used, we have to "modify" them so that their starch is available for fermentation. That modification is essentially the same thing that happens when barley is malted, with one large exception. None of the enzymes will be produced or survive the process. To do this, the brewer mills the cereal to break open the grains and then adds them to the cereal cooker. The addition occurs in the same ways that malt is added to the mash vessel. The slurry of cereal grist and hot liquor is stirred while the temperature of the vessel increases. The final temperature of the mixture is based upon the type of cereal used (see Table 6.1). This temperature allows the starch in the cereal to become available for enzymatic action during the mash. Stirring aids in this process and also keeps the mixture from scorching on the side of the heated vessel. It is important to note that while the process makes starch available for enzyme action, no enzymes are activated in the cereal cooker.

Cereal cookers are required for unmodified adjuncts because those grains have not been malted. For example, if we add a handful of milled barley to hot water and waited a few minutes, we would easily see that this is true. The "barley tea" we just made would not be sweet. The lack of sweetness in the water indicates that none of the starch has been converted into sugars. Because the cereals have not been malted, none of the enzymes needed to convert the starch into fermentable sugars are present. In other words, cereal added to hot liquor will not mash. However, if the starch is made available by breaking down the cell walls and protein coats, the resulting slurry can be added to an existing mash. That slurry would be rich in starch and available to be mashed. The enzymes in the existing mash made with fully or partially modified grains can convert all of the starch from the cereals into fermentable (and non-fermentable) sugars.

Once the cereal has maintained its gelatinization temperature for a specified amount of time (usually 20 min or so), it is pumped while still hot into the mash mixer with the malted grist. The malted grains and the cereal grains are mixed to make sure that the enzymes have access to all of the starch in the slurry. This mixing results in raising the temperature of the mash overall because the slurry from the cereal mixture has a higher temperature. A typical temperature profile for the result

Table 6.1 Gelatinization temperatures of unmodified cereals

Cereal	Gelatinization temperature
Barley	52–59 °C (126–138 °F)
Corn	62–72 °C (144–162 °F)
Oats	53–59 °C (127–138 °F)
Potato	56–71 °C (133–160 °F)
Rice	68–77 °C (154–171 °F)
Rye	57–70 °C (135–158 °F)
Sorghum	68–75 °C (154–167 °F)
Wheat	58–63 °C (136–147 °F)

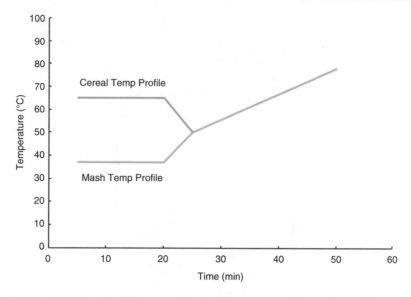

Fig. 6.2 Gelatinized cereal addition to mash raises temperature of the mash. Note that this mash profile holds at 37 °C for 20 min, then the temperature is raised to 50 °C by the addition of the gelatinized cereal, then ramped up to 78 °C over the next ~25 min. This ramp allows the brewer to hit each of the mash rests and provides the flavor profile that the brewer is looking for in this case

is shown in Fig. 6.2. Note that the result of adding both the gelatinized grits and the malted mash results in the final saccharification temperature where the starches are converted to sugars. While the heated cereals could be added with the initial mash-in and then treated just as the mash is treated, the starch in the cereals would not be available for the enzymes in the mash. And if that was done, we would only have sugars that result from the malted grains, plus a bunch of flavor from a hot extraction of unmalted grains. In addition, adding them later in the mash steps saves energy associated with warming the overall mash.

6.3.2   Mash Mixer and Mash Kettles

The typical mash mixer, fitted with the Steel's Masher or vortex mixer, contains rotating paddles at the bottom to stir the mash, see Fig. 6.3. The slurry of grist and water is stirred during the entire mashing process to ensure an even temperature, an even distribution of all of the enzymes, and provides some protection from scorching the mash if the vessel is heated. Stirring must be carefully monitored during the mashing process, because the speed of stirring is directly related to the amount of shear stress transferred to the mash. A low shear rate, and a slower stirring speed, improves the quality of the finished beer. If the mash is stirred too violently, the high shear stress on the mash can result in lower enzymatic activity, smaller protein fragments, and fracturing of the grist that decreases the particle size (which may result in a stuck mash during the next step in the process.)

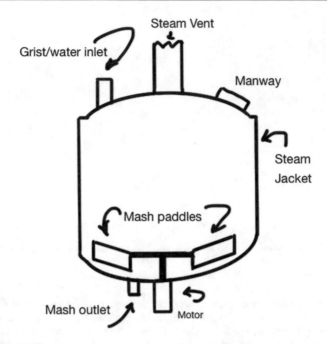

Fig. 6.3 Mash mixer. Note the asymmetric bottom on the mixer that improves mixing without increasing shear forces. The vessel is also steam jacketed to allow temperatures to be maintained or adjusted to match the desired temperature. Not shown is the CIP system that is almost always included in the mixer

Modern mash mixers are not symmetrical in their design. The offset stir paddles increase the mixing of the mash without increasing the shear stresses. The mixing blades are also positioned so that a space exists between the blade and the sides and bottom of the vessel. This space is necessary so that the grist isn't further ground by the blades.

Steam jackets surround the sides of the modern mash mixer. Typically, the jackets provide different heating zones inside the mash vessel. In this way the mash can be slowly or differentially heated. Alternatively, additional hot liquor or a gelatinized cereal can be added to adjust the temperature.

Mash kettles can be used in the mashing process, and are required if the brewer's recipe requires a *decoction mash*. The mash kettle itself looks nearly identical to the cereal cooker. In a decoction mash, a portion of the mash (both liquid and grist) is pumped from the mash mixer into the mash kettle. Then, the temperature of the mash kettle is increased, typically to boiling. This allows Maillard reactions (the reaction of amino acids and sugars) to take place and increase the browning and caramel-like flavors. The heated and stirred mash is then transferred back into the mash mixer – the addition of which raises the temperature of the mash.

An example profile for a double-decoction mash using a mash kettle is shown in Fig. 6.4. In this particular mash, the temperature of the mash-in results in a

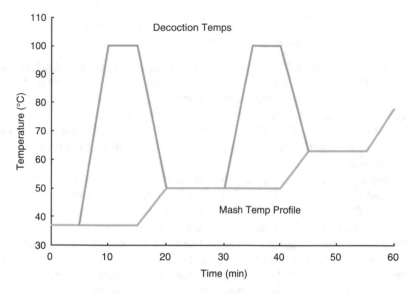

Fig. 6.4 One possible double decoction temperature profile. Note that when some of the mash is withdrawn, boiled, and then added back into the mash, the temperature is raised to the next rest temperature. This example shows the decoction pulled from the dough-in temp, returned to make the protein rest at 50 °C, pulled and returned to give the saccharification rest. The entire mash is then raised in temperature to the mash out temp

temperature of 37 °C (the dough-in rest, see Sect. 6.4). The temperature is held constant for a while and then approximately 1/3 of the entire mash is removed to the mash kettle. The removed mash is then heated to boiling and kept there for about 5 min. When the boiled mash is returned to the mash mixer, it raises the temperature of the entire mixture to 50 °C (the protein rest, see Sect. 6.4). The temperature of the mixture is maintained for about 10 min and then another 1/3 of the entire mash is removed and heated to boiling in the mask kettle. Again, after returning this to the mash kettle, the temperature of the entire mixture is raised to 63 °C. This is the saccharification rest where the starches are converted into fermentable and unfermentable sugars. After 10 min, the entire mash temperature is raised to the mash-out temperature of 78 °C.

A more typical decoction is to withdraw 1/3 of the mash after the protein rest starts (50 °C) to raise the temperature of the entire mash to 63 °C. After a 20–30 min rest, the second decoction is used to bring the entire mash to the mash-out temperature.

Can we calculate the new temperature at each step when we do a decoction mash? In short, yes. Using the standard principles of heat transfer, the process is relatively straightforward. Because we know the temperatures that you want and the temperatures that you have, the calculation is most easily done by determining the amount of the mash to withdraw for the decoction. For example, assume that we start with a dough-in at 37 °C and want to raise the temperature of the mash to the protein-rest stage at 50 °C. We use the formula:

$$\frac{\text{Temp change of mash}}{\text{Temp change of decoction}} = \frac{50-37}{100-37} = 0.20$$

So, we would remove 20% of the mash, decoct it, and then return it to the original mash. The temperature of the mash would then raise from 37 to 50 °C. The formula is very useful and works for any decoction (as long as the specific heats of the mash and the decoction do not change during the process).

Each time a portion of the mash is removed and heated to boiling, the enzymes become denatured at the elevated temperature. In fact, many of the proteins in the boiling decoction are degraded. This results in thinning of the mash (making it easier to transfer) and destroys any further enzyme activity. However, since the boiled mash is returned to the unboiled portion, it mixes again with more enzymes. So, the overall end result is that the mash gets thinner, some of the proteins become degraded, and additional caramelization is added to the wort.

Because the mash is being transferred from one vessel to another, whether the mash kettle is used or not, a fairly thin mash is needed. Typically, the ratio of water to grist is 3–5 L/kg (~0.35–0.60 gal/lb). This means that temperature and pH control are very important to the brewer. The enzymes in this thin of a mash can easily be destroyed by small variations in the temperature or pH that deviate from the ideal. Thick mashes tend to buffer the enzymes from small changes in the temperature or pH because of the relatively large amount of hydrated grist to water ratio. As the amount of grist drops, its ability to buffer the changes in temperature and pH also drops, resulting in a loss of enzymatic activity in the mash.

CHECKPOINT 6.3

Figure 6.4 indicates one possible double decoction profile. Draw the profile that would exist from the more "typical" double decoction outlined in this section.

6.3.3 Mash Tun

In some operations, the mash tun is the preferred, and only, vessel for performing the mash operation. The mash tun is an insulated or jacketed vessel that is operated similarly to the mash mixer. First, hot liquor is added to the vessel to pre-warm it and provide protection to the grist as it is added. Addition of the grist can then occur via a pre-mash mixer. Alternatively, in the most basic of mash tuns, the dry grist is added directly to a tun filled with hot liquor. In these cases, the hard work of the brewer to mix the grist and the hot liquor determines the quality of the mash. The end result in both the basic tun and that fitted with the latest gadgets is to end the mash-in with a temperature that starts the enzymatic processes.

Variability exists in mash tun design. As noted above, the simplest design is just an insulated vessel with a metal cover (that may or may not be used). More complex designs may include stirring paddles, mash rakes, or heating jackets; the simple vessel contains none of these, requiring the brewer to stir the mash with a handheld mash paddle to ensure even distribution of hot liquor and grist. Some designs even include a special rotating paddle that aids the removal of the spent grain from the vessel after mashing has finished.

The key in the use of the mash tun is that the vessel must be able to maintain the desired temperature. An even and steady temperature during the mash rest ensures the greatest efficiency in the process. While it is possible to raise the temperature of the mash in the mash tun to different set temperatures, most uses of a mash tun involve a single temperature, known as a single infusion mash. The brewer selects the temperature of the single infusion mash to represent the flavor profile and alcohol content that they desire in the finished beer.

Most mash tuns are fitted with a false bottom, though the most basic vessels may have a network of perforated pipes that are placed in the bottom under the grist. The false bottom, or perforated pipe, allows the vessel to also serve as a lauter tun. The brewer could decide to pump the entire mash into a separate lauter tun, or when a false bottom is present, the wort can be drained from the spent grains.

6.3.4   Processes in Mashing

No matter which mashing vessel is used, the brewer takes great care to make sure that the temperature of the contents is monitored. The temperature is quite important in mashing as it determines the result of the wort. How does it do this? While hotter water can extract different components from the grain that cold water can't, the temperatures are chosen to allow enzymes in the malt to do their job and convert starches into sugars.

If the mash is too hot, the resulting beer would be very thick in its mouthfeel, very sweet in taste, and have a very low alcohol content. If the mash is too cold, the opposite would happen; the resulting beer would be very thin and have a significant amount of alcohol, but very little sweetness. If the temperatures were even further away from the values that are needed in the mash, the resulting beer may have the flavor of tea, or taste very similar to the water that was initially added.

6.4   Enzymes and What They Are

When the 20 essential amino acids polymerize in a biological system, the arrangement and combinations are what scientists call proteins. The specific amino acids used, the order in which they are arranged, and the specific three-dimensional structure that the resulting polymer adopts determines the function of the protein. In some cases, the protein adopts a fibrous shape and may find its use in the structure of a biological system. Hair, for example, is made of a protein known as keratin. Keratin is a polymer of amino acids bonded together in a chain (the amino acid cysteine makes up about 14% of the amino acids in human hairs). In other cases, the protein is more globular in nature. Myoglobin is an example of a globular protein containing over 150 amino acids. This polymer of amino acids holds a single iron ion that is used to bind to oxygen and store it near muscle cells that might need it later.

If the result is a globular protein, the molecule can have very specific properties. If one of those properties is to perform a reaction, we say that the arrangement of amino acids is actually an enzyme. Enzymes are biological polymers of amino acids that have a globular shape and cause specific reactions to take place in the biological system. For example, lactase is an enzyme that can catalyze the rate of reaction of lactose into glucose and galactose. In mashing, the enzymes come from the malt used in the process. And like all enzymes, they conduct very specific reactions.

Unfortunately, enzymes are very fragile. They are easily degraded, first by unfolding and untwisting in a process called *denaturation*, and then by breaking back into their individual amino acids. Once denatured, an enzyme is unable to be "fixed." Heat is one of the culprits that can denature an enzyme. If the temperature is low, the enzyme has very limited ability to conduct its reactions. As the temperature increases, the enzyme's activity increases until it reaches the maximum activity at a particular temperature. Above that temperature, the enzyme begins to denature and the overall activity decreases. If the temperature is reduced after the enzyme is denatured, the enzyme is still unable to be repaired and the reactivity of the molecule is lost (see Fig. 6.5).

The alkalinity or acidity (the pH) of the solution with the enzyme is also important in the reactivity of the enzyme. If the pH is too low (the solution too acidic), the activity of the enzyme is lowered. If the pH is too high (the solution too alkaline), the activity of the enzyme is lowered. Only within a specific pH range is the enzyme active. While it is possible for the enzyme to denature when the solution is outside of the pH range, as long as the deviation from the pH range isn't too great, the enzyme will not be denatured (Fig. 6.5).

Shear forces are the last of the damaging factors that can affect an enzyme's activity. A shear force occurs when one portion of the solution moves in an opposite direction from another portion of the solution. This can easily occur when a solution is stirred and mixed or, as we've seen, when the grist is dropped into the mash tun.

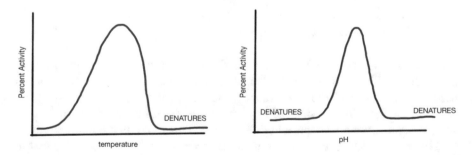

Fig. 6.5 Enzymes are active based on the pH and temperature of the solution. Note that enzymes do not denature below their optimum temperature, and that the pH must deviate a lot from the optimum before the enzyme denatures

The shear forces cause stress inside the molecules and can literally tear them apart. An enzyme can be torn apart into its original amino acids. Obviously, then, the enzyme is no longer capable of performing reactions.

When everything is right with the pH, temperature, and limited shear forces, the enzymes can do their reactions. Since every enzyme is a little different, each has a specific temperature and pH that describes its maximum activity. It is the job of the brewer to select the correct temperature and pH that allows the enzymes to do their job and to avoid high shear forces by stirring only as much as is absolutely necessary.

CHECKPOINT 6.4

In your own words, describe the term "shear."

Draw a picture of a mash tun and outline the differences it has with a mash mixer. What differences are there between a cereal cooker and a mash mixer?

6.5 Chemistry While Resting

As we've noted, the brewer selects the temperature of the mash to allow a particular enzyme to "turn on." The temperature is then held for a given amount of time to allow the enzyme to work. These hold times at specific temperatures are known as "rests." The enzymes involved at these particular rests perform specific chemical reactions. Because the enzymes are active over ranges of pH and temperature, many different enzymes might be active at any given pH and temperature.

In this section, we'll explore some of the more important enzymes that are active during the mashing process. In order to get a handle on how some of these enzymes work, we'll first uncover some of the chemistry associated with starch and related polymers of glucose. You may wish to refer back to Chap. 3 along the way.

6.5.1 Starch

As we learned in Chap. 3, glucose is an example of one of the carbohydrates. The carbohydrates themselves got their name because their chemical formula appears to be "hydrates" of carbon. For example, the formula for glucose is $C_6H_{12}O_6$ or $C_6(H_2O)_6$. Glucose is likely one of the most important of the carbohydrate because it is so prevalent in the world. For example, glucose is the monomer that makes up starch and cellulose. It is also found in barley as the monomer that makes up glucan (a structural component of the seeds). How does the same monomer make so many different polymers? If there were additional monomers included, we could assume that the main differences in the polymers were due to the number and order in which the monomers were attached. Interestingly though, starch and cellulose do not have other monomers in their structure. They differ only in the way in which the glucose molecule is attached to make the polymers.

Glucose itself is an interesting molecule. It contains a set of five different alcohol functional groups and one aldehyde functional group. In water, the molecule wraps around itself and one of the alcohol functional groups reacts with the aldehyde. The aldehyde functional group becomes known as the anomeric center. This gives rise to either a five-membered ring (known as glucofuranose) or a six-membered ring (known as glucopyranose) as shown in Fig. 6.6. To make things a little more complicated, as the alcohol group approaches the aldehyde group, it can either attack from the top or the bottom. The aldehyde functional group becomes known as the anomeric center. If the alcohol group attacks the aldehyde from the top, the resulting anomeric center ends up with an OH attached to it points down (known as the α-anomer). If it attacks from the bottom, the result is the β-anomer. Thus, in water, there are actually five different forms of glucose: open-chain glucose, α-glucofuranose, β-glucofuranose, α-glucopyranose, and β-glucopyranose.

Note that when we draw carbohydrates we often use a combination of the line drawing process and simply just writing all of the atoms. If the molecule exists in a ring, the carbon atoms in the ring are omitted from the drawing to make it easier to see the entire structure. When we draw the molecule in open chain format, this can also be done, but for our purposes, the molecules are drawn showing all of the carbon atoms.

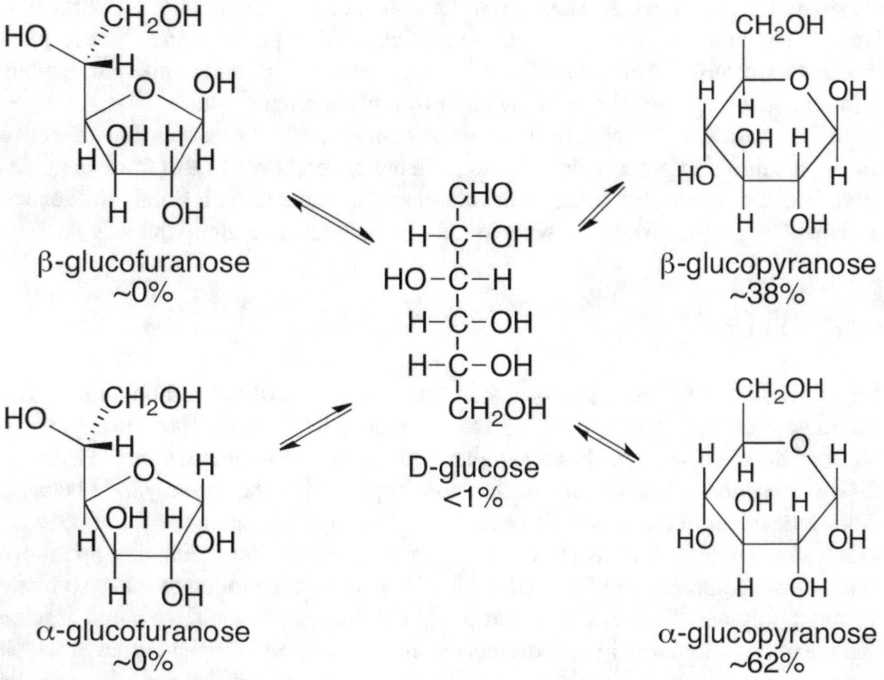

Fig. 6.6 The five forms of glucose found in water. Their relative percentages are <1% open-chain glucose, trace amounts of α-glucofuranose and β-glucofuranose, ~62% α-glucopyranose, and ~38% β-glucopyranose

The presence of an OH group on the anomeric center signals to the brewer that the ring can open and close on its own. In other words, when the glucose molecule bends over and closes to make a ring, it can also reverse the action and open back up to form the straight-chain form of glucose. As long as the OH group exists when the ring closes, the ring can reopen. Thus, the aldehyde group becomes masked and hidden when the ring is closed, but is easily observed when the ring reopens. Reduction of glucose is possible because the aldehyde functional group exists whether the ring is closed or not. We say that glucose is a *reducing sugar*.

As we noted in Chap. 3, glucose has a handedness. The common natural glucose found in nature has a mirror image of itself that is also known as glucose. That mirror image is completely unlike the glucose that we know. A designator is often used to distinguish between the two based upon the location of the alcohol functional group that is on the last chiral center farthest from the aldehyde. If the alcohol is on the right-hand side, the designator is "D" (from the Latin word "dexter"). On the left, and we call the molecule "L" (from the Latin word "laevus") as shown in Fig. 6.7. By convention, the L and D are written as small caps. The naturally occurring form of all carbohydrates is the D-sugar; living things on this planet have evolved to exclusively prefer the use of the D-carbohydrates. While this doesn't mean that an occasional L-carbohydrate might be encountered from a natural source, it does mean that they are fairly rare. Therefore, the most common form of natural glucose in water is known as α-D-glucopyranose.

When two glucose molecules condense together to make a dimer, a new molecule is born. The dimer results from the reaction of an alcohol functional group on one of the glucoses reacting with the aldehyde position on another glucose molecule in exactly the same fashion as when glucose folds over onto itself to make the five- and six-membered rings in water. In Fig. 6.8, we note that there are two possible dimers that result. These two forms arise from the OH of one glucose monomer attacking either the top or the bottom of the aldehyde on the other monomer.

The two resulting dimers of glucose are NOT the same. Maltose is the sugar that we need for the yeast to consume during the fermentation step of the brewing

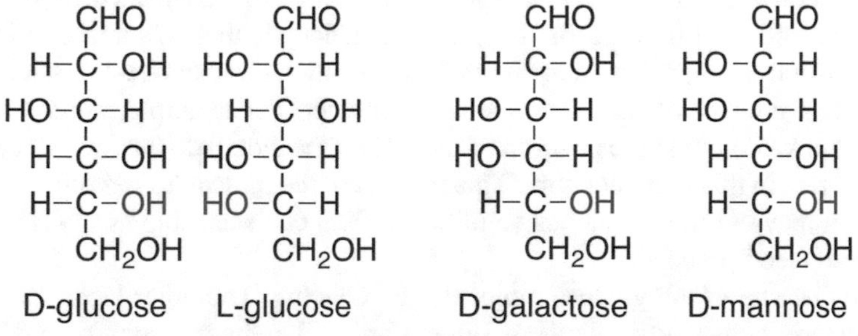

Fig. 6.7 D and L convention for carbohydrates. Note that the OH on the chiral center farthest from the aldehyde is on the right hand side for "D" and on the left hand side for "L." Also note that D-glucose and L-glucose are mirror images of each other that are not superimposable (they are enantiomers)

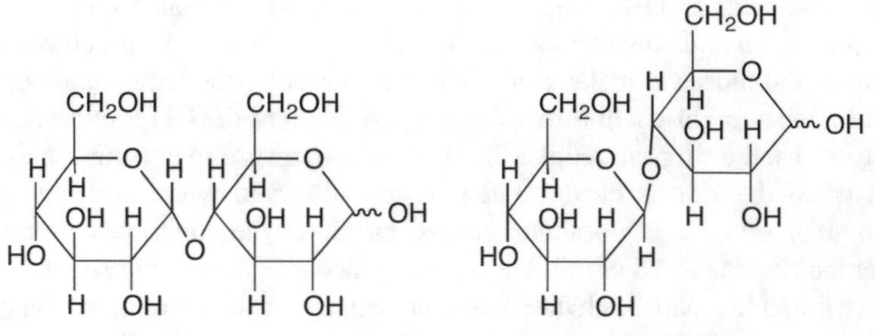

Fig. 6.8 Dimers of glucose. The major difference is the position of the oxygen attached to the anomeric carbon on the left-hand molecule. In maltose, that oxygen is down; in cellubiose, it is up. The wavy line on the anomeric carbon in the right-hand glucose molecule indicates that the OH can either be up or down

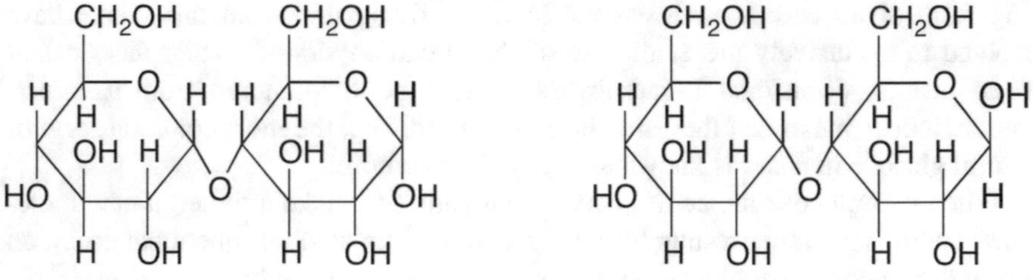

Fig. 6.9 Anomers of maltose. Note that the glucose ring on the right can open and close to give rise to a mixture of α-maltose and β-maltose

process. Cellubiose, on the other hand, is not digestible by yeast (and humans) and is not useful in brewing.

To distinguish between the possible dimers that can be made from the combination of carbohydrate monomers, we use a symbolism that represents the attachment. First, the position of the OH that is part of the aldehyde that was attacked is noted as either a or b. Then, in brackets, we write the position number of the first sugar and the position number of the second sugar monomer. For example, maltose has an α-[1→4] link. Note that the position of the OH on the aldehyde that wasn't attacked isn't reported in this nomenclature. This is because the molecules are often in water, and when they are in an aqueous solution, the OH can scramble as the ring opens and closes (see Fig. 6.9).

Note that one of the glucose molecules is reducing. The other lacks the OH on the anomeric center. Overall these dimers are still reducing sugars, and have a reducing end. But the other glucose ring in the molecule is unable to open and close and is not able to undergo reduction reactions.

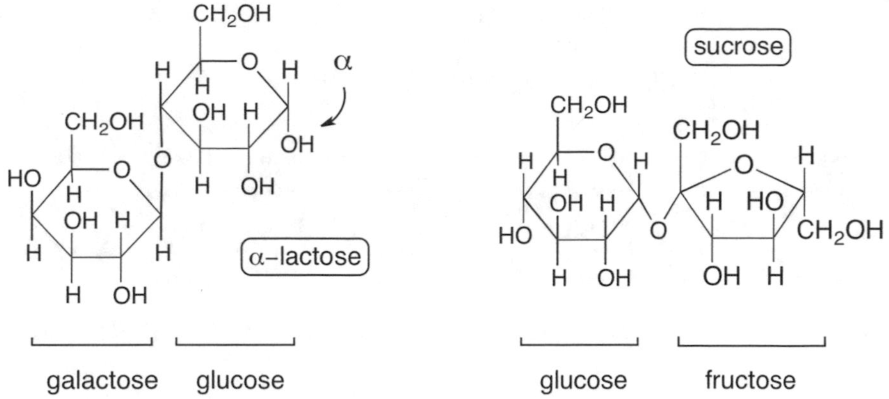

Fig. 6.10 Lactose and sucrose. Note that both an alpha and beta form of lactose exists in aqueous solution, but that sucrose does not have the capability of doing so

Other sugars can be linked together in the same fashion as two glucose molecules. Many of these dimeric carbohydrates are quite common in the brewery and used extensively in some recipes. For example, if glucose is condensed with galactose by a β-[1→4] linkage, the result is the formation of a molecule of lactose. Lactose intolerance results from this linkage. Some people lack the biological machinery to break a β-[1→4] linkage. Note that lactose is still a reducing sugar. In another example, glucose can be condensed with fructose (aka fruit sugar) to make a molecule known as sucrose. Sucrose, with an α-[1→2] linkage between the two sugars, is not a reducing sugar. We can tell this because it lacks an OH on the anomeric centers (Fig. 6.10).

If we carry the two dimers of glucose farther and create polymers, we end up with the two main forms of glucose that are found in nature. As maltose grows to maltotriose and maltotetrose (adding more glucose molecules), the chain starts to twist around itself into a telephone cord appearance. Every once in a while, a branch comes off of the main chain with an α-[1→6] linkage. Thus, the resulting polymer has a very tightly packed form of glucose where multiple chains are packaged in a twisting form that is a great way to store glucose in a system (Fig. 6.11).

If the cellubiose chain is continued, we get a much different form of the glucose polymer. Contrary to the starch molecule, the linkage between the glucose monomers is a β-[1→4] connection. This form of the linkage aligns the molecules in a linear zig-zag type arrangement (Fig. 6.12). The chain of glucose molecules just keeps growing and results in a long straight polymer. This polymer, known as cellulose, provides structure to cells in plants.

Another structural polymer of glucose can be found in the cell walls of grains, bacteria, and yeast. Known as β-glucan, the polymer is a slight modification of cellulose. The actual structure of β-glucan varies depending upon the source of the polymer. Yeast cell walls contain β-glucan with β-[1→3] linkages between the glucose subunits and occasionally along the chain there are β-[1→6] branches. In cereal

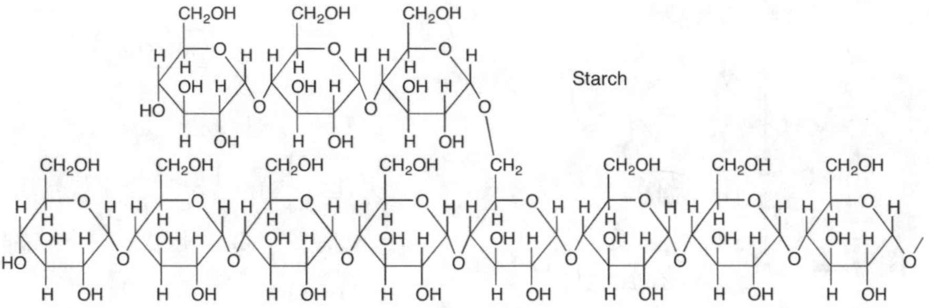

Fig. 6.11 Starch is a polymer of glucose molecules joined with α-[1→4] and α-[1→6] linkages

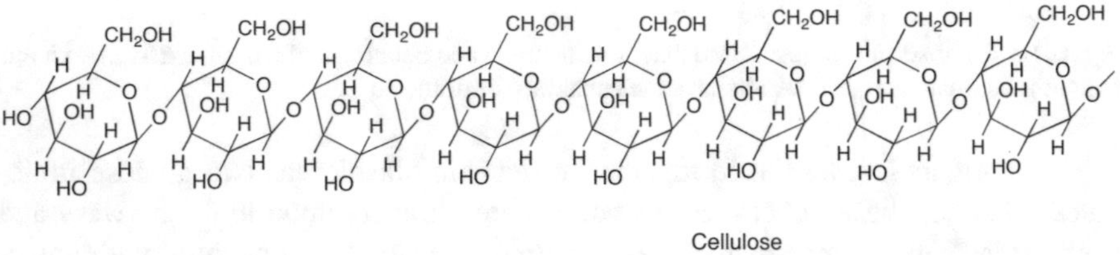

Cellulose

Fig. 6.12 Cellulose is a polymer of glucose molecules joined with a β-[1→4] linkage

grains, the β-glucan contains a mixture of glucose with β-[1→3] linkages and β-[1→4] linkages (Fig. 6.13). During mashing, the with β-glucans in the cereal grains need to be broken down to break the cells walls and free up the materials inside the cells (particularly the starch needs to be released from inside the starchy endosperm). Because of the differences in the linkages between β-glucan and starch, enzymes have evolved that are specific to breaking each of the bonds in these glucose polymers.

While the words sound similar, β-glucan and gluten are entirely different compounds. Gluten is a polymer of amino acids (i.e., a protein) whereas β-glucan is a polymer of glucose. Gluten does cause some problems with people that have celiac disease. β-Glucan, which is not related to gluten at all, does not have this issue.

CHECKPOINT 6.5

Identify the anomeric carbons in lactose and sucrose (see Fig. 6.10).

Galactose and glucose have very similar structures. What is the difference in the structures?

Can you draw L-glucose? Note that it is an enantiomer of D-glucose.

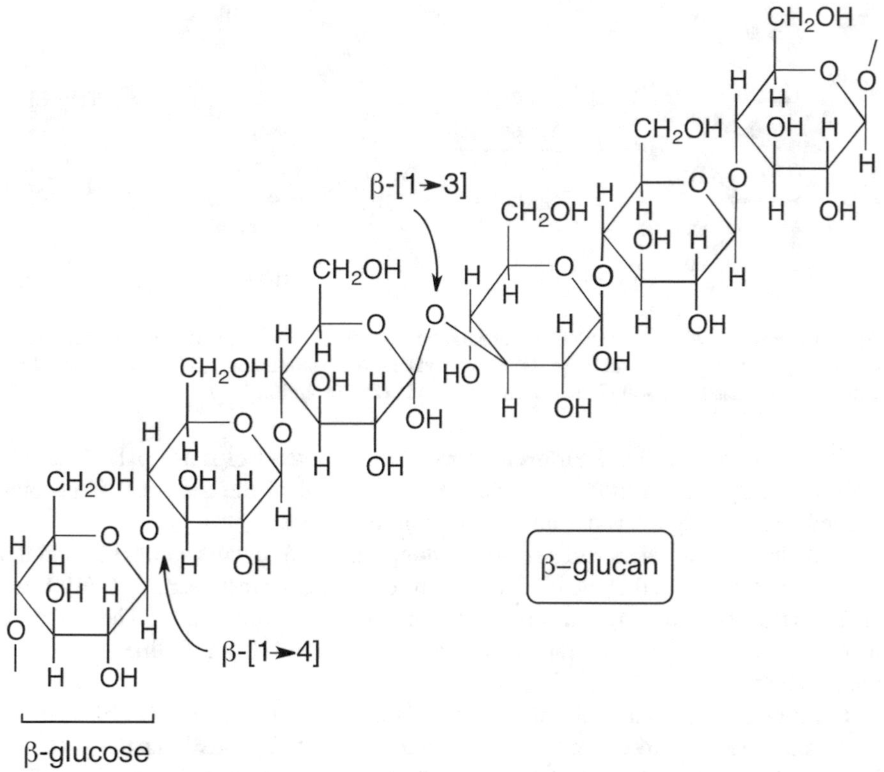

Fig. 6.13 β-Glucan is a polymer of glucose. The version shown here is found in cereal grains

6.5.2   Phytase

Phytase is active in the pH 3–5 range and is active at temperatures ranging from 30 to 52 °C (86–126 °F). This enzyme acts upon molecules that contain phosphate groups. In malt, phytase hydrolyses phytic acid. Phytic acid (see Fig. 6.14) is a molecule that looks very similar to glucose and contains six phosphate groups. This molecule is unusable by the growing barley plant, but when phytase becomes active, the molecule is hydrolyzed to release phosphorus in a usable form for the growing plant. That phosphorus is needed for the construction of DNA, other enzymes, and is involved in the energy production process.

In mashing, phytase removes up to five of the phosphate groups from phytic acid. This results in the primary formation of dihydrogen phosphate ($H_2PO_4^-$), although phosphoric acid (H_3PO_4) and hydrogen phosphate (HPO_4^{2-}) can also be formed based upon the current pH of the water. The result lowers the pH of the solution toward the 4.8–5.2 range. A rest at this temperature, known as an *acid rest*, was historically necessary for brewers that used soft water and malts that tended to be somewhat under modified. In those situations, the brewer often did not have the

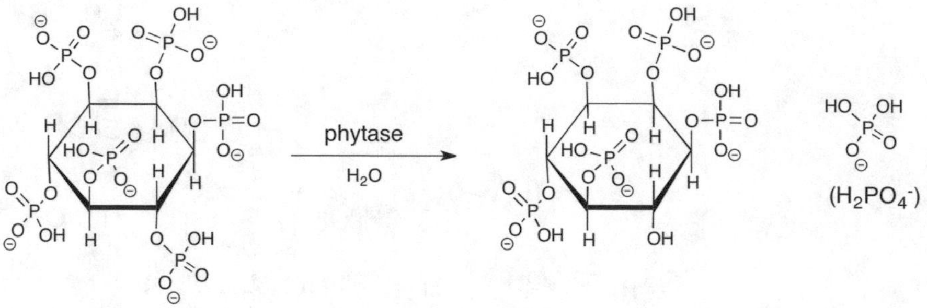

Fig. 6.14 Phytic acid is a source of phosphate. The first release of phosphate is shown here when phytase reacts with phytic acid (here shown as phytate – the penta-anion). Dihydrogen phosphate ($H_2PO_4^-$) is released, resulting in a lowering of the pH of the solution

ability to control mash pH (or even know about the effect of mash pH). This rest isn't necessary for the modern brewery where control of the mash pH is accomplished by adjusting the water chemistry prior to the mash.

With that said, brewers still perform a "dough-in." This occurs when they add the grist and hot liquor so that the final temperature results in a rest near 37 °C (100 °F). As the particles of the grist absorb the hot liquor, they swell in size. The rest here helps the particles get hydrate and limits small ones from falling through a false bottom.

Limit dextrinase is another enzyme that is active in the 32–49 °C (90–120 °F) temperature range and debranches the starch molecule by breaking the β-[1→6] linkages, but little of this enzyme survives the kilning process to be of utility. This isn't an issue as additional de-branching activity occurs later during the saccharification rest.

6.5.3 Proteases and Peptidases

Proteases and peptidases are enzymes that breakdown the proteins in the starchy endosperm. It is easy to tell what these enzymes do based on their names. A protease breaks down proteins. A peptidase breaks down peptides (short chains of polymerized amino acids). The result of the action of proteases and peptidases is not only the breakdown of the proteins and peptides, but the release of smaller amino acid chains and even individual amino acids themselves. This helps to free up the starch so that it can be solubilized. The proteases also break apart any other proteins that exist in the mash and help to thin the resulting beer. Unfortunately, in modern brewing using fully modified malts, an extended protein rests can significantly reduce the mouthfeel of the beer. As we'd expect, having some proteins in the beer is important to improve the mouthfeel.

These enzymes are most active in the 4.5–5.3 pH range and a temperature range of 47–54 °C (118–130 °F). A short rest at 50 °C (123 °F) activates these enzymes and lets them do their work. While not needed for a mash containing only malt barley, it can be very helpful for mashes that contain adjuncts such as oats and wheat. Keeping the rest short allows the enzymes to break apart the larger proteins that can lead to haze in the finished beer, but still allowing the mouthfeel of the added adjuncts to show through to the finished product.

6.5.4 Glucanase

There are many different *glucanases* that exist in malt barley. These enzymes act upon the β-glucans found in the malt. As we noted before, the β-glucans are associated with the structural components of the barley seed, and are the reason that bread is chewy. Breaking these down, helps free up the starch granules for the other enzymes to act upon and gets rid of the "stickiness" of the mash. This is particularly important when adjuncts such as oats or wheat are used in the mash, because these unmalted cereals contain a significant amount of β-glucans.

Specifically, the enzymes work to break both the β-$[1\rightarrow3]$ linkages and the β-$[1\rightarrow4]$ linkages in the cereal β-glucans. The result is that the mash gets a little thinner and less dough-like. In other words, the *viscosity* of the mash decreases. This improves the circulation of the hot liquor during the mash, increases the permeability of the mash when we remove the wort at the next step, and allows the starch to become more accessible during the mash.

The optimum activity of the glucanases in malted barley is in the range of 37–46 °C (100–115 °F). Resting at this temperature for 10–20 min allows those enzymes to do their work. Note that this is the same temperature as the one used for the dough-in. Remember, though, that it isn't absolutely necessary to rest at this temperature to force the glucanase to become active in a 100% malted barley recipe. While it can help thin the mash and break up the β-glucans, the malting process has already done much of the work.

6.5.5 Alpha-Amylase

By far, the most important enzymes in the mash are the amylases. Two important versions of this enzyme exist in malted barley. The first that we'll explore is α-amylase. α-Amylase exists in many different organisms; it is found in human saliva and in cereal grains. The structure of the enzyme involves a series of twists like a phone cord arranged into a circle, and known as a TIM barrel. The result is a central "corridor" along which the starch molecule can be threaded. The enzyme requires calcium to become activated, and once activated, it can do its job. Remember that starch is a long polymer of glucose where the linkages are α-$[1\rightarrow4]$.

The enzyme latches onto the long chain of starch and then holds it in place while it performs the reverse of the condensation reaction. In other words, it uses water to break the linkage between two adjacent glucose molecules. α-Amylase cleaves those linkages at random inside the starch molecule. The result is a large number of small glucose polymers. The smallest is glucose itself; maltose is also common, as is maltotriose, maltotetrose, etc.… When the enzyme comes close to a branch in the starch molecules (remember that starch has some α-[1→6] linkages), it runs into a problem. The cutting action of the enzyme cannot get close enough to the branch to eliminate it completely. Instead, the enzyme leaves a small twig on the main stem of the starch; the piece left behind is known as a *limit dextrin*.

Some researchers have found that the smallest limit dextrin possible in malted barley involves four glucose molecules bound to each other (Fig. 6.15). Because the structure of amylase differs in other organisms, the limit dextrin structure is also different. Limit dextrins are sweet to the taste and they add a thickness and richness to the flavor of the finished beer. In addition, limit dextrins are classified as unfermentable sugars. That is, limit dextrins cannot be taken up by yeast and converted into energy. Thus, a significant amount of α-amylase activity during the mash results in a heavier mouthfeel and a residual sweetness in the finished product.

α-Amylase activity is greatest in the temperatures from 145 to 158 °F (63–70 °C) and an optimum pH near 5.7. However, as we noted before, the enzyme is still active at temperatures lower than 145 °F (63 °C). The issue arises when the temperature is too warm and rises above 158 °F (70 °C). This causes the enzyme to denature and become permanently inactivated. As the pH deviates from the optimum 5.6–5.8, the activity of the enzyme also deviates from ideal. Does this mean that the activity drops to zero? No, instead, we find that the activity of the enzyme decreases as we move away from the optimal pH. If the pH of the mash moves significantly far away

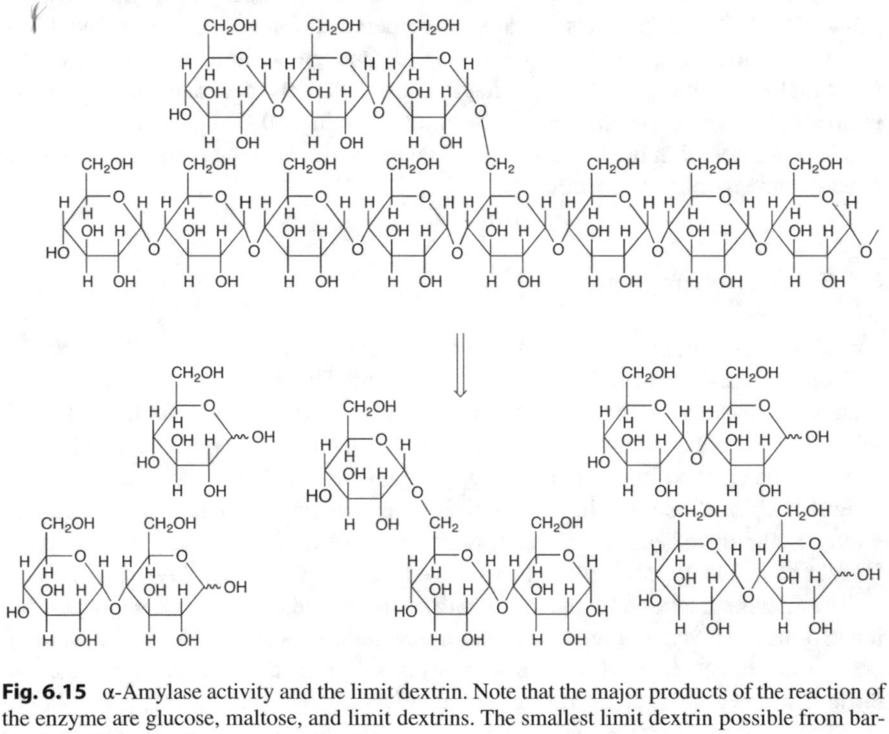

Fig. 6.15 α-Amylase activity and the limit dextrin. Note that the major products of the reaction of the enzyme are glucose, maltose, and limit dextrins. The smallest limit dextrin possible from barley malt amylase is shown

from the optimal pH, then, yes, a problem would exist. Too far away from pH 5.7 and the enzyme could be permanently denatured or degraded.

> **CHECKPOINT 6.6**
> What would occur in a mash that was operated at 60 °C (140 °F) and pH 4.9? What would you expect to happen if the pH of the hot liquor started out at pH 3.0 and 74 °C (165 °F)?

6.5.6   Beta-Amylase

β-Amylase, on the other hand, is a related amylase that differs in structure and function from α-amylase. This enzyme becomes activated during the germination portion of the malting process. Once malted, the β-amylase is ready to go to work on the starch molecule in the endosperm. This enzyme is most active between 131–149 °F (55–65 °C) and in the pH range of 5.4–5.6. It is, as is α-amylase, still active outside of the optimum pH range and at temperatures lower than the optimum range, but the rate of the reaction is much lower.

β-Amylase grabs onto the non-reducing end of the starch molecule and then begins hydrolyzing the α-[1→4] linkages between the glucose monomers. The second linkage is the one that is targeted, resulting in the formation of dimers of glucose from the starch molecule. In other words, α-amylase produces significant quantities of maltose from the starch molecule, but only as far as the branching points in the molecule (see Fig. 6.15). In addition, the speed of the conversion of starch into maltose is dependent upon the number of individual strands of starch. The more strands of starch, the faster the conversion of starch to maltose.

So, the best fermentability occurs with a compromise between the temperature and pH for the amylases. Too low and the increased activity of α-amylase becomes active, resulting in a wort that is highly fermentable (with lots of maltose), thinner in mouthfeel and a beer with high alcohol and very thin feel. Too high of a temperature and the β-amylase predominates, making a less fermentable wort that is thicker in mouthfeel and relatively sweet. The compromise is to find both a pH and temperature that lies in the middle of both amylase activities. Or, the compromise is to ramp the temperature through both temperatures in a stepped temperature mash. For the homebrewer, the best option for the single infusion mash with a cooler as the mash tun is to set the temperature at about 150–152 °F (65–67 °C) and hold it there for an hour.

The production brewery focuses on the saccharification temperatures and slowly ramps the temperature of the mash from 50 to 78 °C (120–172 °F). This ensures that each enzyme becomes activated at its optimum temperature as the process ensues. In addition, the ramping of the temperature allows the brewer to verify that each of the steps has been reached. If a reproducible temperature increase can be obtained from one mash to the next, the flavor profile can also be reproducible.

6.5.7   Mashout

At the end of the mash, the brewer raises the temperature of the mash to force the enzymes to denature and to denature the majority of the proteins that have become soluble in the liquid. Heat is added to the entire mash. This causes many of the enzymes to even begin degrading and breaking down into their amino acids. The same thing also happens to the other proteins in the mash. They are first denatured. Then, they are degraded and break apart into smaller pieces and even into individual amino acids.

The result of the large biological polymers (proteins and enzymes) breaking down into smaller pieces and individual amino acids is a reduction in the viscosity, or resistance to flow, of the solution. In other words, the liquid gets thinner and the enzymatic activity stops. This allows the liquid to be more easily separated from the grains during the lauter step.

As we'll explore in the next chapter, the heated and thinned mash can be separated in the lauter step. The liquid that results is known as sweet wort. The grains that remain, mostly husk material, are known as the *spent grains*, while the sweet wort will later be converted into beer. The destination of the spent grains is often a concern to the brewer. The sugars and most, if not all, of the starch in the grains has been converted into maltose, glucose, and limit dextrins (and a small amount of maltotriose and maltotretrose). Thus, the grains have very little usage left in them as a food source. However, since cattle and other ruminant animals can still digest cellulose rich materials, the spent grains are unique suitable as roughage (or silage) for those animals. Most brewers donate or sell their spent grains to feed-yards, ranchers, or other farmers for this purpose. Alternatively, once dried, the spent grains can be used as a supplement in breads, cakes, and other foods. Some unique uses for spent grains are also being explored (e.g., an abrasive in hand and face soap). Other entrepreneurs might think of additional uses for this material.

CHECKPOINT 6.7
Why is it important to mash-out?
Which enzyme is more likely to make the most maltose?

6.6   Efficiency of Extraction

The brewer is often concerned about obtaining as much of the sugars from the malt as possible. After all, the amount of sugar that can be extracted is directly related to the amount of alcohol that can be made during the fermentation step. And, malt costs money. It is not in the brewer's interest to extract less than the maximum amount of sugars because it can affect the bottom line.

6.6.1 Efficiency Calculations

The efficiency of a mash is related to the amount of carbohydrates in the malt's endosperm and how well they can be converted into sugars. The efficiency is so important to the brewer that it is often calculated after every mash. The reasons for such a dedication to the calculation of the mash efficiency are numerous.

Efficiency calculations are performed after every mash in order to evaluate the process. This allows the brewer the ability to adjust future mashes to improve the extraction of sugars from the malt. Because the efficiency determines the amount of sugars that are available for the fermentation process, the efficiency also dictates the alcohol content that will be available for the finished product.

The calculations are fairly straightforward to do. They require that the brewer knows the maximum amount of sugars that are possible in the malt. This amount can be measured in the laboratory using a standard procedure. The American Society of Brewing Chemists, for example, dictates a process where the malt is ground to exacting specifications. Then, the grist is added to a prescribed amount of warm water and stewed to allow the enzymes to act upon the endosperm. Then, it is slowly filtered and the amount of sugar measured. For most commercial malts, the approximate amount of extract sugars have been measured (Table 6.2).

The calculation for the efficiency of a mash is accomplished by first determining the expected amount of extract based on the given recipe. The best way to do this easily is to determine the number of "gravity points" for each of the ingredients based on the number of pounds of each material added to the mash. If a range of possible values is given, and the laboratory analysis for the specific malt or adjunct is not available, the best practice is to use a value in the middle of the range.

Table 6.2 Potential extract from malt, adjunct, and sugars

Ingredient	Potential extract (sp.gr./gal•lb)	Gravity points (points/gal•lb)	Gravity points (points/L•kg)
Pale malt	1.035–1.037	35–37	292–309
Barley, flaked	1.032	32	267
Barley, roasted	1.025	25	209
Sugar, cane or corn	1.046	46	384
Crystal malt	1.033–1.035	33–35	275–292
Chocolate malt	1.028–1.034	28–34	234–284
Corn, flaked	1.037	37	309
Honey	1.035	35	292
Dry malt extract	1.044	44	367
Oats, flaked	1.037	37	309
Pilsner malt	1.036	36	200
Wheat, flaked	1.035	35	292

The specific gravity in g/mL is based on the addition of 1.0 lb. of the ingredient dissolved in enough water to make up 1.0 gallon

Each malt/adjunct contribution to the total extract is determined by multiplying the mass of the material by the gravity points. After adding the gravity points for each of the malts or adjuncts, the total gravity points (or gravity units, GU) for the batch is then divided by the volume of the wort. The result is the calculated potential original gravity. For example, if 200 kg of pale malt and 20 kg of flaked wheat are used to make 9 hL (900 L) of wort, the calculations would be:

$$\text{pale malt}: 200\text{kg} \times 200\,\text{gravity points} = 40000\,\text{GU}$$
$$\text{flaked wheat}: 14\text{kg} \times 292\,\text{gravity points} = 4088\,\text{GU}$$

$$\text{total GU} = 44088\,\text{GU}$$

$$\text{Potential OG} = \frac{44088\,\text{GU}}{900\,\text{L}} = 48.987\,\text{GU} = 49\,\text{GU}$$

$$49\,\text{GU} = 1.049\,\text{OG}$$

The same calculation in imperial units (gallons and pounds) gives the same answer. The value of the potential original gravity is in g/mL. It can be converted to kg/m^3 by multiplying by 1000.

This is the potential original gravity. Thus, after the mashing step in the brewery, the specific gravity of the collected wort can be measured. The measured OG can be compared to the potential OG to determine the percent efficiency of the mashing process. The calculation is performed using the gravity units (GU) for the potential and measured systems:

$$\text{percent efficiency} = \frac{\text{measured GU}}{\text{potential GU}} \times 100\%$$

This efficiency relates the entire mashing process. It includes the efficiency of the enzymatic processes, the effects of the grist sizes, the effects of the pH, and the efficiency of the sparge and wort collection. For example, let's assume that the brewer obtained 9 hL of wort that had a measured original gravity of 1.042 g/mL (42 GU). If the potential original gravity was supposed to be 1.049 g/mL (49 GU), the efficiency of the mash can be calculated as follows:

$$\text{percent efficiency} = \frac{42}{49} \times 100\% = 86\%\text{efficiency}$$

6.6.2 Mash pH

The pH of the mash plays an important role in the efficiency of the process. As we've discovered, the enzymes in the malt operate most efficiently within a fairly narrow pH range. If the acidity of the mash is outside of this range, the saccharification and other rests within the mash profile will not work well. While the mash enzymes and the buffering capacity of the malt itself will adjust the pH somewhat, the initial strike water added to the grist must has a pH that can easily be adjusted to the appropriate range for the mash.

A mash pH within the appropriate range (pH 5.2–5.6) results in:

- Improved enzyme activity
- Improved hop isomerization during boiling (lower pH reduces hop conversion)
- Improved yeast health
- Inhibition of bacteria growth
- Improved precipitation of protein-tannin complexes during whirlpool
- Improved flavor and stability

The brewer is well aware of these benefits and takes steps to ensure that the hot liquor used in both the mash and sparge has a pH that allows the mash to hit a pH within this range. This is accomplished by a number of different methods, a common method is to add very small amounts of phosphoric acid or lactic acid to lower the pH of purified water. Alternatively, acid malt can be added in a small proportion to the grist to lower the pH during the mash.

6.6.3   Mash Thickness

The thickness of the mash also plays a role in the efficiency of the process. Some brewers report that the liquor to grist ratio is vital to maintaining wort quality, although in many evaluations, a small variability in the ratio does very little to affect the final beer quality, colloidal stability, or flavor. What is more important, however, is that the liquor to grist ratio adjusts the "buffering" capacity of the mash itself.

The liquor to grist ratio is typically expressed in liters of liquor to kilograms of grist. Values for the homebrewer, then, range from 2.6 to 5.2 L/kg.

As the ratio drops, the malt materials can help resist small changes to the temperature that can change the enzymatic efficiency of mashing. If the ratio drops too much (<2.6 L/kg), the wort that is produced becomes too concentrated to efficiently remove all of the sugars from the mixture. The amount of protein and tannin can also be slightly reduced in a thicker mash. While this can be a benefit, it is not often a significant improvement, and it is definitely not such an improvement that it would reduce the need to filter or further clarify the beer at a later stage in the brewing process.

At the higher end (>4.5 L/kg), the buffering of the grist becomes less effective, meaning that changes to the temperature become more damaging to the enzymatic activity. While the thinner mashes can absorb and extract the sugars much more efficiently, they can also increase the amount of protein and tannins that are extracted into the wort. This can increase the precipitate formed during whirlpool and increase the risk of chill haze. However, the temperature of the mash can become a significant problem. Because of this, thinner mashes with a high liquor to grist ratio need to be stirred periodically to ensure that the temperature is uniform throughout the mash.

Chapter Summary

Section 6.1

Milling involves cracking or crushing the malt to remove the endosperm from the husk. In some cases, it is advantageous to just crack the malt; in others it is best to convert the entire kernel into flour.

Section 6.2

Mashing is the process by which malt and hot liquor are added together. The process results in the formation of sweet wort containing fermentable and non-fermentable sugars.

Section 6.3

Pre-mash mixers allow the dough-in to start immediately upon mixing of the hot liquor and the grist.

Cereal cookers are required for mashes containing adjuncts that have not been malted.

Decoction and step-infusion mashes allow multiple rests to selectively activate enzymes to accomplish specific tasks during the mash.

Section 6.4

Enzymes are polymers of amino acids that catalyze reactions of molecules.

Enzymes are pH and temperature dependent. If the temperature of an enzyme rises hotter than their optimum temperature, the enzyme will denature and begin to decompose.

Section 6.5

Starch, cellulose, and β-glucan are polymers of glucose.

The linkages between the glucose monomers indicate the type of compound that exists.

Phytase, protease, glucanase, and the amylases are the major enzymes that are active in the mash. Each performs a very specific function, and each is activated at specific temperatures.

Section 6.6

The efficiency of a mash can be used to evaluate the process and suggest changes that can improve the recovery of sugars.

Mash efficiency depends highly upon the pH of the mash and the ratio of water to grist.

Questions to Consider

1. What are the reasons for milling and for mashing the malt?
2. Draw the structure of maltotriose.
3. Identify the fermentable sugars that exist in sweet wort.
4. Why does the brewer cook corn before adding it to the mash? Would this be needed if the corn was malted?
5. Describe a step-infusion mash that allowed the brewer to select the glucanase and β-amylase enzymes only.

6. Draw the temperature profile for a triple-decoction mash.
7. Draw the structure of D-galactose as a six-membered ring. ...as a five-membered ring.
8. The structure of β-glucan was presented in Fig. 6.13. What would the structure look like if it were part of a yeast cell?
9. Why must a brewer be careful in the amount of time spent during a rest at 50 °C (123 °F)?
10. Determine the amount of mash that should be decocted to raise the temperature of a mash from 50 to 65 °C. Assume the boiling point of water in the Colorado brewery is 95 °C. Would the brewer need to include more or less mash in the decoction if they were at sea level?
11. Compare lactose and sucrose. Which is a reducing sugar and why?
12. What potential problem may arise if the brewer doesn't perform the "dough-in"?
13. What benefit exists for the use of a pre-mash mixer versus just adding the grist to hot liquor already in the mash tun?
14. Knowing that enzymes denature at high temperatures, why do decoction mashes still make fermentable sugars from starch?
15. Why would a brewer want to do a decoction mash?
16. Many decoction mashes raise the temperature of the withdrawn mash to 72 °C and hold it there for 10 min before boiling it. Why?
17. From Fig. 6.6, what inference can you make about the relative stability of a five-membered ring versus a six-membered ring?
18. Describe the linkage in the di-saccharide shown below. Indicate the two monomers that make up the compound and then indicate the linkage.

19. Describe the differences between α- and b-amylase. Which one is responsible for a thinner, more alcoholic beer?
20. Assume a brewer wants to make a very alcoholic beer with a very thick mouthfeel. What should they do during the mash to make this happen?

21. List the major enzymes that are active in the mash and the temperatures at which they are active.
22. Why is rapidly stirring a mash not advised?
23. Consult Table 6.1. Could a mash be performed with ground oats and malted barley (without adding the oats to a cereal cooker before use)?
24. What differences would exist between a mash performed with water that had a pH of 7.0 and one performed using water with a pH of 6.0?
25. List two hazards associated with the use of a hammer mill.

Laboratory Exercises

The Effect of Temperature and pH on Mashing Efficiency

This experiment is designed to illustrate the differences that arise by adjusting the different parameters used in mashing. It further assists with the technical skills used in measurement of the density of wort.

Equipment Needed

600 mL beaker and 1000 mL beaker per group
2-row malted barley – ground into grist
Hydrometer and cylinder
100 mL graduated cylinder
10 mL volumetric pipet and pipet bulb
25 mL beaker
Thermometer
Hot plate or Bunsen burner setup for heating water with 600 mL beaker
8" × 8" baking pan
Funnel and filter paper

Experiment

Each group in the laboratory should take approximately 100 grams of the same barley malt (brewers 2-row will suffice) and place it into the inner beaker of a nested pair of beakers. Water is placed in between the two beakers and will act as an insulator for the mash. The water used in between the two beakers should be as close to the temperature of the mash as possible. The entire apparatus is then placed into a pan containing water at the same temperature as the mash. Then 275–300 mL of pH adjusted water at the specific temperature is poured onto the crushed grains and allowed to stand for 30 min. Additions of hot water to the pan should be made periodically to maintain the temperature of the mash. The mash is periodically (every 5 min) stirred with a spatula or spoon.

Table 6.3 Temperatures and pHs to explore in a class of 20

Trial	Temperature (°F)	pH
1	100	5.4
2	125	5.4
3	150	5.4
4	175	5.4
5	125	6.0
6	125	4.0
7	150	6.0
8	150	4.0
9	175	6.0
10	175	4.0

Additional students can duplicate trials 1–4

After the 30 min have passed (use a timer), approximately 75–100 mL the mash is gravity filtered into the hydrometer cylinder and then cooled to room temperature either by placing the cylinder in a bath of ice water or by leaving it sit at room temperature on the bench. The density is then determined using the hydrometer. Additionally, the density is measured by pouring ~50 mL into the graduated cylinder and then weighing the liquid. Finally, the density is confirmed by pipetting 10.00 mL of the wort into a clean tared beaker and obtaining its mass. The entire process should be repeated in triplicate and the average density reported for the sample.

Different groups of students should use different temperatures for their mashes. The table below indicates a set of temperatures that can be used to provide information about the activity of the saccharification enzymes. In addition, the pH of each group's water used in the mash (no adjustment needed to the water between the nested beakers or in the pan is needed) is set to a specific pH. The pH can be set by using phosphate buffers at the specific pH or by simply adjusting the pH of tap water using dilute phosphoric acid solutions (Table 6.3).

After the entire class has determined the density of their specific trial, the results are compared and then discussed. If multiple groups perform the same trial, those results are averaged before the discussion.

Lautering and Sparging

7.1 Introduction

Sparging is the process of rinsing the sugars away from the grain in the mash. After the mash is completed, we want to separate the dissolved sugar-water (sweet wort) from the grain. It isn't as simple as just letting the mash liquid drain. There are still a considerable amount of fermentable sugars left in the grain matrix. So, we rinse with hot water. The hotter the water is, the more soluble the sugar is in water, but excessive heat and/or an incorrect pH might also remove tannins from the grain husks. This is a condition that must be avoided. (The best way to illustrate the flavor of tannins in your beer is to take a tea bag and put it in your mouth. Not pleasant!) So, we sparge with enough water at the correct pH to extract as much sweet wort as possible, but not to the point where we extract tannins. Also, excessive sparge water will lead to excessively large volumes of more dilute sweet wort.

Sparging is the act of rinsing the grain with hot water. And, a lauter tun is the traditional vessel used for separating the wort from the grain. In this tradition, the brewer would pump the slurry of grist and hot liquor (after it had been mashed and while it was still hot) mash tun (or mash mixer) to the lauter tun. Here the mixture would be allowed to rest and settle, forcing the solids in the mixture to slowly settle to the bottom of the lauter tun. The process was referred to as lautering. Some breweries use a combination mash/lauter tun where the lautering and sparging occur in the same vessel where the mash takes place. We'll explore the modern version of the lauter tun later in this chapter.

Regarding sparging, there are several different approaches to how this accomplished. For example, the sweet wort could be drained from the lauter tun; the brewer could refill the tun with sparge water, and drain again, and again. This is the basic method behind *batch sparging*, a method attributed to the English brewing process. Each of the drainings (or *gyles*) could be used separately to make different beers, or combined in ratios to ensure a certain specific gravity for a particular beer. This *parti-gyle* method of brewing was the "technology" of the day when it was first

M. Mosher, K. Trantham, *Brewing Science: A Multidisciplinary Approach*, https://doi.org/10.1007/978-3-030-73419-0_7

used industrially. Because of its utility in ensuring a consistent pre-boil gravity, it is slowly seeing resurgence in some brewhouses today.

Alternatively, the brewer could slowly drain the lauter tun and continuously supply fresh sparge water to the top of the grain bed at the same rate that is being grain. This is *fly sparging*, a method attributed to the German brewing process.

In both cases, the first runnings were often returned to the top of the lauter tun to be refiltered through the grain. Initially, the runnings would drip out of the grain bed loaded with insoluble proteins, cellulose, small amounts of insoluble starch that didn't get mashed, and even small grain particles. By passing these again through the system, the grain bed would act like a filter and help remove the insoluble material. This process is known as the *vorlauf*, from the German word meaning "the first amount." Brewers often use the word as both a noun (its intended use) and as a verb.

In this chapter, we will explore a lot of details about the mechanical process of moving liquids around the brewery. This is very important to the brewer, because large quantities of water must be moved from place to place. In home brewing, it is relatively easy to lift 5 gallons of hot water and move it by hand – well, it's "relatively" easy. But brewing on a larger scale requires larger storage vessels, pipes, hoses, and pumps. We certainly don't want to move 50 barrels (1550 gallons, 5800 L) of sweet wort by hand!

Because of that issue alone, fluid engineers design modern breweries. Significant thought goes into the design of pipe size or pumps for the job. This chapter is not meant to train a brewer to design modern piping systems, but it will at least help the brewer understand why certain choices are made and understand the limitations of their installed pumping systems. And understanding all of this requires some basic understanding of the physics of fluid transfer.

7.2 Fluid Physics: Static Case

Before we launch into the more complicated case of moving fluids around the brewery, we will first consider stationary, or "static," fluids to set some basic definitions and build a knowledge base. Once we have considered the simple static case, we will look at moving fluids, and finally we will add extra complications such as moving fluids through a grain bed.

7.2.1 Pressure

When considering movement fluids in pipes, we often want to know the force, or the pressure, that a fluid exerts on a container. The words pressure and force are frequently confused and used interchangeably. However, these two terms are very different. By definition, pressure is defined as the force that is exerted on a surface divided by the area that the force is distributed across:

$$P = \frac{F}{A}. \tag{7.1}$$

In the SI system, with forces measured in Newtons (N) and area in square meters (m²), the pressure is measured in N/m². This combination of units is known as the Pascal (Pa), named after the French scientist Blaise Pascal (1623–1662). A unit that is sometimes used is known as the bar (bar). One bar is equal to 100,000 Pa. Another common unit for pressure, particularly in fluids engineering in the United States, is pounds per square inch (commonly abbreviated psi, or sometimes as lbs/in²).

Let's consider the difference between force and pressure using a simple experiment: Place a thumbtack between your index finger and thumb such that the pointy-end is pressing on your thumb. Press together gently. You will notice that (ouch!) the pointy end of the tack is digging in and hurting compared to the other, flat end. It is important to note that the force is the same at either end of the tack and thus the same on either finger/thumb. But since each end of the tack has considerably different areas, the pressure is significantly different. You should see that for the same force, a smaller area gives a larger pressure. Likewise, a larger area will give a smaller pressure – again with the same force.

7.2.2   Pascal's Law

Pascal's law, (also Pascal's principle) is a principle in fluid physics that states that pressure exerted anywhere in a confined and incompressible fluid (such as water) is transmitted equally in all directions throughout the fluid. This means that, at any given depth of the fluid, the pressure is the same.

Another example here will highlight once more the difference between force and pressure, and use Pascal's principle as well. Consider a hydraulic car lift as shown in Fig. 7.1. A car, weighting 26.7 kN (about 6000 lbs), is placed on one piston with a radius of 45 cm. The hydraulic fluid is connected to another piston, but with a radius of 1 cm. Since the pressure must be the same in the fluid at identical depths, the required force at the smaller piston to lift the car can be determined from

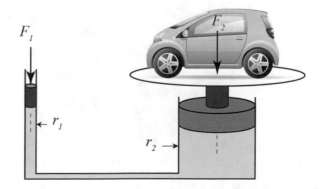

Fig. 7.1 Example illustrating Pascal's principle. The pressure at each piston must be the same

$$P_1 = P_2$$

$$\frac{F_1}{\pi r_1^2} = \frac{F_2}{\pi r_2^2} \tag{7.2}$$

or,

$$F_1 = F_2 \frac{r_1^2}{r_2^2} = 26,700\text{N} \frac{1^2}{45^2} \tag{7.3}$$

$$F_1 = 13.2\text{N} \left(\text{about 3lbs}\right)$$

Of course, for a given amount of travel in the large piston, the smaller piston must travel further. The amount, or volume, of the fluid moved must be the same at either piston:

$$V_1 = V_2. \tag{7.4}$$

So for every centimeter that the large piston moves, the smaller one must move:

$$\pi r_1^2 x = \pi r_2^2 \left(1\text{cm}\right)$$

$$x = \left(1\text{cm}\right)\frac{45^2}{1^2} \tag{7.5}$$

$$x = 2025\,\text{cm}$$

Pascal's law is actually a statement of the weight of the fluid above a certain point. Looking at Fig. 7.2, we now consider a liquid and draw an imaginary cube somewhere in the liquid. The difference in pressure between the top and the bottom of the cube can be determined by finding the weight of this cube of fluid. The weight (a force) of this cube is

$$W = mg = \rho V g \tag{7.6}$$

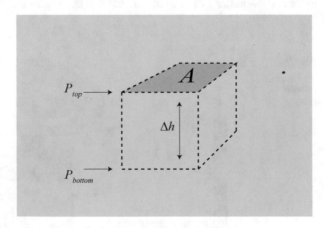

Fig. 7.2 An imaginary cube is drawn around a parcel of fluid to illustrate Pascal's law

where ρ is the density of the fluid and V is the volume of the cube. So, the pressure at the bottom of the cube is simply the pressure at the top of the cube *plus* the pressure-"force" due to the weight of the cube:

$$P_{\text{bottom}} = P_{\text{top}} + \frac{\rho V g}{A}. \tag{7.7}$$

Rearranging, Pascal's law is more generally stated as

$$\Delta P = \rho g \Delta h \tag{7.8}$$

with ΔP being the difference between the top and bottom pressures. It just states that the pressure difference between two points depends only on the net difference in height between the two points. So, if two points in a fluid are at the same depth, they will have the same pressure.

Consider another example: A tube, bent into a U-shape, is partially filled with water ($\rho = 1$ g/cm^3). We add enough water such the bottom part of the U is completely covered. At this point we expect after equilibrium has been reached, that the water will be at the same level on either side of the U. Then, on the left side of tube we add oil ($\rho = 0.85$ g/cm^3) so that the total height of the oil in the tube is 3 cm as shown in Fig. 7.3. Then, the question is: what is the difference in the water level, x, on either side of the tube?

The weight of the oil on the left side will push down on the water below it until the pressures at positions a and b are equal. Since the pressures at the top of the fluids due to the atmosphere are the same, we start with Pascal's law:

$$\begin{aligned} \Delta P_{\text{oil}} &= \Delta P_{\text{water}} \\ \rho_{\text{oil}} g \Delta h_{\text{oil}} &= \rho_{\text{water}} g \Delta h_{\text{water}} \end{aligned} \tag{7.9}$$

The densities are given for both fluids, and the gravitational constant g cancels on both sides. Then substituting,

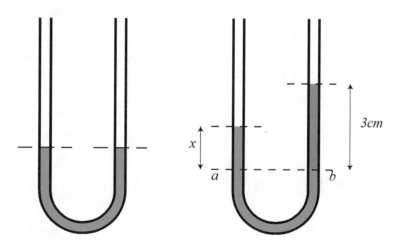

Fig. 7.3 A U-tube is first filled with water, and then oil

$$\frac{0.85\,\dfrac{\mathrm{g}}{\mathrm{cm}^3}}{1.0\,\dfrac{\mathrm{g}}{\mathrm{cm}^3}}\,3\mathrm{cm}=x \tag{7.10}$$

$$x = 2.55\mathrm{cm}$$

We can use Pascal's law to illustrate another idea. Let's consider an elevated container of water, such as a hot liquor tank, as shown in Fig. 7.4. The container has 1.5 m of water, and it is elevated by 3 m. The container has a pipe leading down to the floor where we have a closed valve. It is important to note that the valve is shut, and the fluid is not moving. We can use Pascal's law to calculate the pressure difference between the top of the water and the location of the valve. Since, the density of water is 1000 kg/m³, then

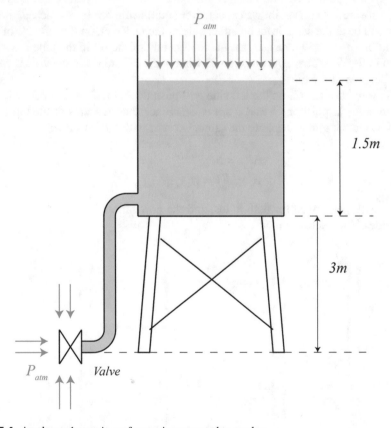

Fig. 7.4 An elevated container of water is connected to a valve

$$\Delta P = \rho g \Delta h$$

$$\Delta P = 1000 \frac{\text{kg}}{\text{m}^3} 9.8 \frac{\text{m}}{\text{s}^2} 4.5\text{m} \qquad (7.11)$$

$$\Delta P = 44100 \frac{\text{N}}{\text{m}^2} = 44.1\text{kPa}$$

where we have used the total height in meters.

In the above example, note that we ignored the atmospheric pressure pressing on the top of the water in the container. Pascal's law only gives the pressure change due to the weight of the fluid. Since there is about 101.3 *kPa* (14.7 psi) of atmospheric pressure pressing at the top of the water the total, or *absolute pressure*, is 44.1 + 101.3 = 145.4 kPa. We will write this as $P = 145.4$ kPa$_{\text{absolute}}$. In the past this would have been expressed as 14.7 psia where the "*a*" means absolute. Absolute pressure is measured against a reference pressure of zero, a perfect vacuum.

On the other hand, there is also atmospheric pressure pressing on the outside of the valve. Since it is pressure *differences* that will tend to cause fluids to move (or vessels and pipes to expand), we are frequently only interested in the pressure relative to the outside, or the atmosphere. Thus the relative pressure, or *gauge pressure* is always measured relative to the local atmospheric pressure. The pressure measured at the valve in the example will be written as $P = 44.1$ kPgauge. If a pressure measurement does not specify absolute or gauge, then gauge is assumed.

CHECKPOINT 7.1

A famous experiment is attributed to Pascal in which it is said that he inserted a long tube into a wooden barrel. It is reported that the barrel busted after he filled the tube with water. While the accuracy of this tale has not been verified, it is educational to calculate the **force** of the water on the bottom of the barrel. Why calculate force and not just pressure? Because the staves of the barrel must resist this total force to hold the bottom in place – in the same way the piston held up the total weight of the car in Fig. 7.1. Consider a standard-sized bourbon barrel; the ends have diameters of about 50 cm and are 1 m tall. It is said that Pascal used a 10 m tube. Calculate the total force on the bottom of the barrel once the tube is filled with water.

7.3   Fluid Physics: Dynamic Case

Fluid dynamics deals with fluids in motion. The motion of fluids can change the apparent pressure exerted by the fluid. If this were not the case, airplanes would not fly. The physics behind fluid dynamics sets limits on pumping speeds in the brewery, so it is important to understand the underlying physical laws.

In fluid dynamics, there are four basic assumptions made from physics: (*i*) that mass is conserved when an incompressible fluid encounters a junction or a change in pipe diameter, (*ii*) that both potential energy and kinetic energy are conserved, (*iii*) that momentum is conserved, and (*iv*) that the fluid is a continuum. This last condition basically ignores the fact that a fluid is made of discrete atoms and molecules. For example, liquid water is actually composed of discrete molecules of H_2O separated by empty space. Stating that water is a continuous fluid ignores this extreme microscopic view. Said in yet another way, the mean free path between collisions of neighboring particles in the fluid is very small compared to the size of the pipe or container. Liquids can almost always be treated as continuous; however, this condition fails under certain situations with vapors.

7.3.1 Conservation of Mass: The Continuity Equation

The first assumption from physics is the conservation of mass. It is as simple as stating for a given mass entering a region, that we must have a commensurate mass exit the same region, assuming that the fluid is incompressible. The vernacular "conservation" is borrowed from physics and basically means that it does not change. The flow of a fluid is usually expressed as mass per unit of time, so conservation of mass takes the form algebraically as

$$\frac{\Delta m_{in}}{t} = \frac{\Delta m_{out}}{t}. \tag{7.12}$$

As a simple example, consider a fluid flowing through a pipe which narrows as shown in Fig. 7.5. Since the mass flowing in must equal the mass flowing out, and since mass is density times volume $m = \rho(Ax)$, we can re-write Eq. 7.12 as

$$\frac{\rho A_{in} \Delta x_{in}}{t} = \frac{\rho A_{out} \Delta x_{out}}{t}. \tag{7.13}$$

Recognizing that $\Delta x/t$ is a velocity, the mass flow rate is then $\Delta m/t = \rho A v$. If the density of the fluid does not change, conservation of mass implies

$$A_{in} v_{in} = A_{out} v_{out} = Q \tag{7.14}$$

where v is the velocity of the fluid through the appropriate cross-sectional area of pipe. The product of an area with velocity gives a volumetric flow rate, customarily symbolized as Q.

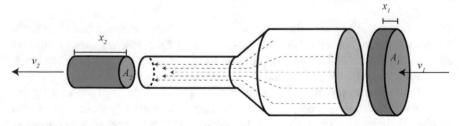

Fig. 7.5 A horizontal section of pipe that changes diameter. The mass flowing into the pipe must equal the mass flowing out of the pipe

For example, consider a level section of 15 cm diameter pipe that carries water at a rate of 45 liters per minute. The pipe tapers to another pipe, 10 cm in diameter. (A) What is the flow rate out of the pipe? (B) What is the velocity of the water at the entrance? (C) What is velocity of the water at the exit?

(A) Since there are 45 liters per minute (L/min) flowing into the pipe, there must also be 45 liters per minute flowing out – assuming that the density of the water did not change. It really is as simple as that: *amount in* must equal *amount out*. Note that we expressed this flow rate as a volume per unit of time. The most accurate flow rate should be mass per unit of time (kg/s in SI units), but sometimes weight per unit of time is used in imperial units (such as pounds/minute). If the density does not change, then volumetric flow rate (as in L/min) is acceptable.

(B) Recall the volumetric flow rate is $Q = Av$. We are given this (45 L/min), but we need to be careful with units:

$$Q = 45\frac{\text{liters}}{\text{min}} \cdot \frac{1\,\text{m}^3}{1000\,\text{liters}} \cdot \frac{1\,\text{min}}{60\,\text{s}} = 0.00075\frac{\text{m}^3}{\text{s}}. \qquad (7.15)$$

We need to find the cross-sectional area of the larger pipe before we find velocity. So, since the area of a circle is $\pi\,r^2$, we can determine the area (A) if we know the radius:

$$A = \pi r^2 = \pi(0.075\text{m})^2 = 0.0177\text{m}^2. \qquad (7.16)$$

Then,

$$Q = Av$$

$$0.00075\frac{\text{m}^3}{\text{s}} = 0.0177\text{m}^2 \cdot v \qquad (7.17)$$

$$v = 0.042\frac{\text{m}}{\text{s}}$$

(C) The area of the smaller pipe,

$$A = \pi r^2 = \pi(0.05\text{m})^2 = 0.00785\text{m}^2. \qquad (7.18)$$

Thus,

$$A_{in}v_{in} = A_{out}v_{out}$$

$$0.00075\frac{\text{m}^3}{\text{s}} = 0.00785\text{m}^2 \cdot v_{out} \qquad (7.19)$$

$$0.096\frac{\text{m}}{\text{s}} = v_{out}$$

Notice that the velocity in the smaller pipe is $(15/10)^2 = 2.25$ times the velocity in the larger pipe (without rounding errors).

7.3.2 Bernoulli's Principle and Laminar Flow

Bernoulli's principle is a statement of conservation of energy applied to fluid flow. It considers both potential energy of an elevated static fluid, and the kinetic energy of a moving fluid. Bernoulli's principle, named after Daniel Bernoulli, a Swiss mathematician and physicist that lived in the 1700s, states that for an increase in speed of a fluid there will be a simultaneous decrease in pressure.

Bernoulli's principle can be derived by applying conservation of energy to a streamline of fluid moving through a system. A streamline is the path that a mass element of the fluid will follow as it moves through the piping system. In this analysis we make two very important assumptions. First, we assume that friction caused by viscous forces is small. This is an important consideration to keep in mind. As piping systems neck-down to smaller sizes or we are moving viscous fluids such as sweet-wort, this assumption may not hold. Secondly, we assume that the density does not change along a streamline. This assumption is more likely to hold for water-based fluids in the brewery since the only thing that will cause a density change is a significant change in temperature.

From physics, the conservation of energy states that work done must equal the change in kinetic and potential energies:

$$W_{NET} = \Delta K + \Delta U. \tag{7.20}$$

Work done is defined as force times a distance that a certain mass is moved. We'll consider a small bolus of mass entering a streamline at point 1 of Fig. 7.6. Thus, the work done moving this mass from point 1 to point 2 can be expressed in terms of the difference in pressure:

$$W = \left(F_1 \Delta x_1 - F_1 \Delta x_1 \right) = \left(P_1 A_1 \Delta x_1 - P_1 A_1 \Delta x_1 \right) = \left(P_1 - P_2 \right) V. \tag{7.21}$$

The change in kinetic energy is related to the change in speed of the fluid. Recall that kinetic energy is a measure of "energy associated with motion." So,

$$\Delta K = \frac{1}{2} m v_2^2 - \frac{1}{2} m v_1^2 = \frac{1}{2} \rho V \left(v_2^2 - v_1^2 \right). \tag{7.22}$$

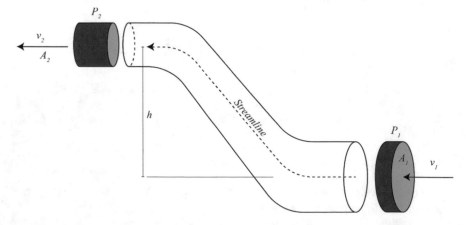

Fig. 7.6 General configuration for Bernoull's principle

And, the change in potential energy is

$$\Delta U = mgy_2 - mgy_1. \tag{7.23}$$

Putting these together and re-arranging, Bernoulli's principal can be summarized in equation form by

$$P_1 + \rho g y_1 + \frac{1}{2}\rho v_1^2 = P_2 + \rho g y_2 + \frac{1}{2}\rho v_2^2 = \text{constant}. \tag{7.24}$$

First notice that each term has units of pressure. Also, since it is only the differences in elevation that matter, we frequently set $h = (y_2 - y_1)$, or alternately let $y_1 = 0$ and $y_2 = h$, the height above the starting point. Now, let's identify different terms in Bernoulli's equation. If we set the velocity to zero at point 1 and point 2, Bernoulli's equation simply reduces to Pascal's law. So the terms $P + \rho g h$ can be called "static pressure." The other term involving velocity, $\frac{1}{2}\rho v^2$, is called the "dynamic pressure." So in words, Bernoulli's principle simply states that the total pressure is the static pressure, caused by an applied pressure and the weight of the fluid, plus the dynamic pressure caused by a moving fluid.

As an example consider a Venturi tube in which water (density $\rho = 1000$ kg/m³) is flowing through a pipe that gradually tapers to a smaller size. Let's say that the larger pipe has a radius of 3 cm and the smaller pipe has a radius of 1 cm and that the fluid is forced through the larger pipe at a velocity $v_1 = 10$ m/s (this is quite fast, ~22 mph). We wish to calculate the pressure difference between the two regions. Given the statement of conservation of mass, the fluid *must* travel faster through the smaller region. Since we will need it, we first calculate the speed in the smaller pipe. Following the example above, the speed can be found from

$$A_1 v_1 = A_2 v_2$$

$$\pi\, 3^2 \cdot 10\,\frac{\text{m}}{\text{s}} = \pi\, 1^2\, v_2. \tag{7.25}$$

$$90\,\frac{\text{m}}{\text{s}} = v_2$$

Since the two sections are at the same level, $h = (y_2 - y_1) = 0$, we don't need to worry about those terms in Bernoulli's equation. We start with

$$P_1 + \frac{1}{2}\rho v_1^2 = P_2 + \frac{1}{2}\rho v_2^2. \tag{7.26}$$

Rearranging,

$$P_1 - P_2 = \frac{1}{2}\rho\left(v_2^2 - v_1^2\right)$$

$$P_1 - P_2 = \frac{1}{2}1000\,\frac{\text{kg}}{\text{m}^3}\left(90^2 - 10^2\right)\left(\frac{\text{m}}{\text{s}}\right)^2 \tag{7.27}$$

$$P_1 - P_2 = 4{,}000{,}000\,\text{Pa} \approx 580\,\text{psi}$$

This calculation means that the pressure at the smaller pipe is about 4000 kPa *less* than the pressure at the entrance. Said another way, the kinetic energy of the fluid increases at the expense of the pressure. Note that if the working fluid was air

($\rho = 1$ kg/m³) instead of liquid, the pressure difference would be a thousand times less than that for water, but P_1 is still greater than P_2.

CHECKPOINT 7.2

A *pitot tube* (named after Henri Pitot, pronounced "pee-toe") is a device for measuring velocity of a fluid. These devices are generally used in aircraft to measure air-speed but can be adapted to measure liquid flow in the brewery. Air, with density $\rho_{air} = 1$ kg/m³ is made to flow past two points of a tube. Often, these points on the tube are housed in a larger, main tube as shown in Fig. 7.7. One point of the tube is directed into the air stream; thus, the air stagnates at this point and its velocity is zero. This point registers pressure P_2. At the other point, the air stream is allow to flow past the tube opening unobstructed. This point registers pressure P_1.

(A) If the airspeed is 35 m/s, find the pressure difference P_2-P_1.

(B) These two points are connected to a U-tube filled with water, $\rho_{water} = 1000$ kg/m³. Based on the pressure difference found in part A, find the height difference, h, of the water.

(C) Now let's work this backward. If the height difference is $h = 5$ cm, determine the air velocity in the main tube.

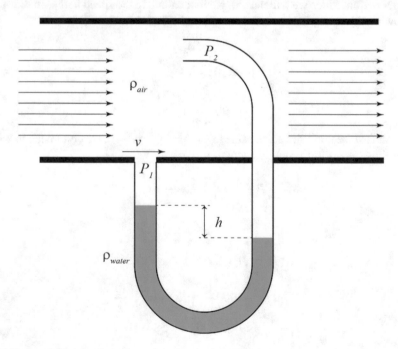

Fig. 7.7 A typical pitot tube

7.3.3 Pressure and Hydraulic Head

Bernoulli's equations suggest that there is a flow speed for which pressure could be zero or even negative. In normal situations, it is not possible for fluids to have negative pressure so it is apparent that Bernoulli's equations are not valid in this regime. Yet, it is instructive to think about this situation. Recall the water tower in Fig. 7.4. With the water level used in that example, we found that the static pressure at the valve is about $P = 44.1$ kPa$_{gauge}$, or $P = 145.4$ kPa$_{absolute}$ given an approximate atmospheric pressure of 101.3 kPa$_{absolute}$. Now let's open the valve and consider the flow velocity with different back-pressures on the other side of the valve. To do this using Bernoulli's equation, we will assume that the volume of water in the tank is so large that the water level does not change – the velocity of the water at the top of the tank is zero. Inserting this situation into Bernoulli's equation,

$$P_{top} + \rho g h = P_{valve} + \frac{1}{2} \rho v_{valve}^2$$

$$101.3\,\text{kPa} + 44.1\,\text{kPa} = P_{valve} + \frac{1}{2} \rho v_{valve}^2. \tag{7.28}$$

So, if the back-pressure at the valve is $P_{valve} = 145.4$ kPa$_{absolute}$, then the flow velocity at the valve is zero. If we lower the back-pressure to, say, $P_{valve} = 80$ kPa$_{absolute}$, then the velocity is

$$101.3\,\text{kPa} + 44.1\,\text{kPa} = 80\,\text{kPa} + \frac{1}{2} 1000 \frac{\text{kg}}{\text{m}^3} v^2$$

$$65.4\,\text{kPa} = 500 \frac{\text{kg}}{\text{m}^3} v^2 \tag{7.29}$$

$$11.44 \frac{\text{m}}{\text{s}} = v$$

This relationship also suggests that there is an upper limit to the velocity through the valve, since there is a lower limit on the pressure at this point. Setting $P_{valve} = 0$, we get an upper limit for a fluid velocity of $v_{valve} = 17.1$ m/s. Installing a pump at the valve and attempting to pump faster than this will lead to cavitation at the pump. *Cavitation* and its deleterious effects are discussed later, but note in this example that pipe sizes and restrictions were not considered. Considering real pipes with real restrictions will only serve to lower the maximum velocity and exacerbate the problem.

So, how could we increase the *speed* (not to be confused with volumetric flow rate) of the fluid in this example? The only possible way is to increase *hydraulic head* at the valve by either increasing the height of the water level, and/or increasing the pressure on top of the water. In fluid dynamics, it is customary to re-write Bernoulli's equation as

$$q + \rho g h' = \text{constant} \tag{7.30}$$

where $q = \frac{1}{2} \rho v^2$ is the dynamic pressure, and

$$h' = h + \frac{P}{\rho g} \tag{7.31}$$

is the hydraulic head which is due to the height, h, of the fluid above a given point plus the pressure-head on the fluid. In this example, it is very important to distinguish between speed of the fluid (in m/s) and volumetric flow rate (m³/s). Hydraulic head will place an upper limit on the speed of a fluid through a pipe. But, if we want a greater volumetric flow rate, all we have to do is make the pipe larger.

7.3.4 Head and Pump Dynamics

Pumps are used to transfer fluids around the brewery. The example surrounding the water tank used gravity to transfer liquid to a lower level. But many times we are trying to lift fluids to a higher elevation, such as transferring hot water from the hot-liquor tank to a sparging vessel. Here, we will explore some of the relevant issues surrounding pumping system design. Much of what is presented here will draw on, and apply the fundamental fluid dynamics discussed above. We'll first consider the model system shown in Fig. 7.8 where we are drawing fluid from below the pump, and transferring to a higher tank. In this discussion, we'll introduce some of the vernacular associated with pump installations.

Suction head refers to the distance below the pump from where we are drawing fluid. This distance will have special meaning, which will be explained later. For the moment, let's apply Bernoulli's principle to this section of the pumping system and explore some of the physical constraints of this system.

The two points in our system that we consider in applying Bernoulli's equations are (1) the top of the liquid in the lower tank and (2) the pump inlet. As we have done before, we'll assume that the tank is large enough that the velocity of the fluid in the tank is very small. So, the only relevant term in Bernoulli's equation is the atmospheric pressure pressing on the top of the fluid. The pump inlet is some height h_1 about the liquid level and we'll assume some arbitrary pressure and fluid velocity here. So, setting up Bernoulli's equation,

$$101.3 \, \text{kPa} = P_{\text{inlet}} + \rho g h_1 + \frac{1}{2} \rho v_{\text{inlet}}^2. \tag{7.32}$$

Now we'll consider some extremes in this situation. First consider the condition with the lowest possible pump inlet pressure, $P_{\text{inlet}} = 0$ and a fluid velocity $v_{\text{inlet}} = 0$. Granted, this is not a very useful situation in practice. With zero velocity, we are not transferring any fluid; however, this example gives us the maximum height, h_1 (i.e., maximum suction head), that is possible with normal atmospheric pressure. Inserting the numbers,

$$101.3 \, \text{kPa} = 1000 \frac{\text{kg}}{\text{m}^3} 9.8 \frac{\text{m}}{\text{s}} \cdot h_1 \tag{7.33}$$

$$10.34 \, \text{m} = h_1$$

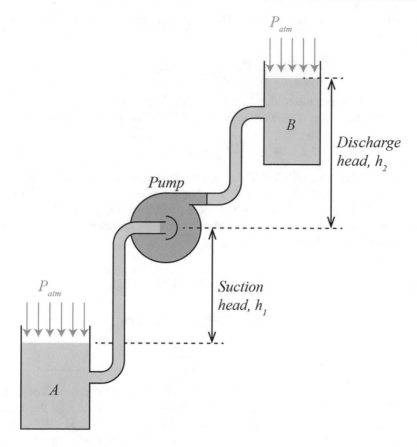

Fig. 7.8 Model pump system to illustrate pump dynamics

The above example is an important illustration of basic fluid physics. The movement of fluid is caused by a pressure difference. The pump lowers the pressure at its inlet, so that the atmospheric pressure can <u>push</u> it in. It is technically incorrect to say that the pumps sucks the fluid into the pump, although we may say this in informal discussion.

CHECKPOINT 7.3

Determine the suction head and pump inlet pressure in Fig. 7.8 that gives the maximum possible fluid velocity at the pump. What is this velocity?

Discharge head, h_2 in the figure, is the distance above the pump that we are discharging the fluid. Note that this distance is from the pump to the actual, upper fluid level – not just where the pipe ends. Again, we'll discuss discharge head in a greater context, but for now we'll use this distance to again explore what Bernoulli's equations imply. Here we look at the two points (1) pump outlet and (2) the top of the highest point of the fluid at discharge. Using arbitrary pressures and velocities at the pump as before,

$$P_{outlet} + \frac{1}{2}\rho v_{outlet}^2 = 101.3\,\text{kPa} + \rho g h_2.\ (7.34)$$

As we will see in a moment, it somewhat misleading to ask what minimum pressure at the pump outlet is required to move the fluid. Inserting a velocity $v_{outlet} = 0$,

$$P_{outlet} = 101.3\,\text{kPa} + \rho g h_2, \tag{7.35}$$

which is essentially Pascal's law. The minimum pressure depends on the final height of the elevated fluid. So, for example, if the height is $h_2 = 4.5$ m as in our previous examples, then $P_{outlet} = 145.4$ kPa. What is the pressure at the pump outlet that will give maximum velocity? We'll get our maximum velocity when the pressure at the pump outlet is zero! Inserting the numbers,

$$P_{outlet} + \frac{1}{2}\rho v_{outlet}^2 = 101.3\,\text{kPa} + \rho g h_2$$

$$0 + \frac{1}{2}\cdot 1000\,\frac{\text{kg}}{\text{m}^3}\cdot v_{outlet}^2 = 145.4\,\text{kPa} \tag{7.36}$$

$$v_{outlet} = 17.1\,\frac{\text{m}}{\text{s}}$$

This example illustrates what Bernoulli's equation really means. Recall that Bernoulli's equation was derived based on energy. In this example, the kinetic energy of the fluid is converted to potential energy at the final fluid height. When fluid is moving, the pressure must necessarily drop. This example **does not** imply that if we put a (static) zero pressure at the outlet that we will get maximum velocity. It is important to realize here that the pump is doing work and adding energy to the system. This work shows up as either kinetic energy of the fluid, or as an incremental work (related to pressure), or a mixture of the two. In the end, however, the final energy of the fluid is simply potential energy.

At the end of the day, the pump's job is to transfer fluid. In the simplest of terms using the smallest amount of energy, the fluid starts at rest, moved to a higher elevation, and ends at rest. The energy put into the system, or the work done by the pump, simply changes the potential energy of the fluid. The total change in potential energy depends only on the <u>net change in height</u>. So, we define the *total static head* as the sum of the suction head plus the discharge head:

$$\text{Total static head} = \text{suction head} + \text{discharge head}. \tag{7.37}$$

The total change in potential energy of the fluid can be expressed as

$$\Delta PE = mgh_1 + mgh_2 \tag{7.38}$$

where h_1 and h_2 are the suction and discharge head respectively. Note also that the term mg is the weight of the fluid we are moving. It is customary in fluid engineering to express this in terms of the specific weight,

$$\gamma = \frac{mg}{V} \tag{7.39}$$

the weight per unit volume; $\gamma_{water} = 9800 \frac{N}{m^3}$. So, the change in potential energy will occasionally be reorganized to look like

$$\Delta PE = \gamma V \left(h_1 + h_2 \right). \tag{7.40}$$

Also, if the level of the suction head is above the pump, we would subtract h_1; the only thing that matters is the net difference in elevation between the starting and ending points.

As an example, consider how much energy is required to lift one cubic meter of water. Let's assume we are using a pump with a suction head of 1.2 m and a discharge head of 3.2 m. So, the total static head is 4.4 m. Recalling the definition of work and potential energy, the work done is

$$\begin{aligned} W &= mgh \\ &= \rho Vgh \end{aligned} \tag{7.41}$$

Inserting numbers,

$$W = 1000 \frac{kg}{m^3} \cdot 1m^3 \cdot 9.8 \frac{m}{s^2} \cdot 4.4m$$
$$W = 43120 \frac{kg\,m}{s^2} = 43120 \,\text{Joules}. \tag{7.42}$$

where a kg•m/s^2 is the unit known as a Joule (J).

Now, if we use a pump that is rated at one horsepower, how much time will this take? The horsepower (h.p.) is a traditional measure of work per unit of time for pumps, and has the definition 1 h.p. = 745.7 Watts. To solve this, we use the definition of power (work divided by time):

$$P = \frac{W}{t} = \frac{43120J}{t} = 745.7W, \tag{7.43}$$

and then solve for time, $t = 57.8$ s. So, in conclusion, a one horsepower pump can transfer one cubic meter of water vertically up by 4.4 m (about 14 ft) in about a minute – this is the minimum energy required.

This example is somewhat misleading in that it assumes the fluid starts at rest and ends at rest. However, if we're actually trying to fill a container in a given amount of time, we need to consider giving the fluid a bit more energy in the form of kinetic energy. Kinetic energy is the energy associated with motion and expressed as

$$KE = \frac{1}{2} mv^2 \tag{7.44}$$

Rather than specifying the final velocity, or the energy put into the system, it is customary to express the kinetic energy as a "velocity-head," in units of height. Essentially, this means expressing the height that gives an equivalent amount of potential energy as the kinetic energy:

$$\frac{1}{2}mv^2 = mgh_v.$$ (7.45)

where h_v is the *velocity-head*. Expressing energy in this way simplifies calculations since changes in elevation are already expressed in height.

Another customary simplification is to approximate friction losses as equivalent losses in head. Friction losses are often tabulated this way for different types and sizes of fittings, and even for the pipes themselves. So, one might consult a table for a given set of fittings in the application. These tables are easily found on the internet by searching for "friction head table."

The total amount of energy required to move fluid from one point to another includes changes in potential energy due to elevation, the final kinetic energy of the fluid at the end point (assuming the fluid starts from rest), and any friction losses along the way. So, *Total Dynamic Head* captures all of this information:

Total Dynamic Head = suction head + discharge head + friction head + velocity head

$$= \left(mgh_1\right) + \left(mgh_2\right) + \left(mgh_f\right) + \left(mgh_v\right)$$ (7.46)

This can be viewed as the total energy required to move the fluid. And by rearranging the equation we get:

$$E_{\text{total}} = mg\left(h_1 + h_2 + h_f + h_v\right) = \lambda V\left(h_1 + h_2 + h_f + h_v\right).$$ (7.47)

As a final example, consider the model system shown in Fig. 7.9; note that the discharge arrangement is somewhat different than Fig. 7.8. Let's assume that we are moving water and that the suction head is $h_1 = 3.5$ m, the discharge head is $h_2 = 5$ m, and we want to move water at a rate of 3.2 L/s = 0.0032 m³/s (about 51 gallons per minute) into the upper tank. It a simple system using 5 cm (~2 in) diameter pipe, and both the suction piping and the discharge piping have one 90° elbow. Accounting for friction losses, the goal is to determine the minimum horsepower rating for a pump.

The first step in this analysis is to determine the velocity of the water. Again, the flow rate is equal to the product of the cross-sectional area of the pipe and the velocity, so since we know the flow rate ($Q = 0.0032$ m³/s) and the radius of the pipe (diameter = 5 cm; radius (r) = 2.5 cm = 0.025 m), we can solve for the velocity of the fluid in the pipe:

$$Q = Av$$

$$0.0032\frac{\text{m}^3}{\text{s}} = \left(\pi\left(0.025\,\text{m}\right)^2\right) \cdot v$$ (7.48)

$$1.63\,\text{m}/\text{s} = v$$

This means that we need a velocity head of

Fig. 7.9 Final example system that uses all concepts

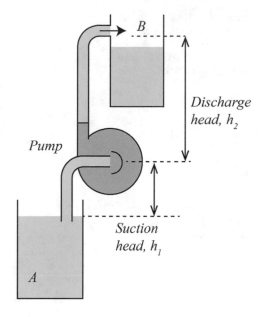

$$\frac{1}{2}mv^2 = mgh_v$$

$$\frac{\frac{1}{2}\cancel{m}v^2}{\cancel{m}g} = \frac{v^2}{2g} = h_v$$

$$\frac{\left(1.63\frac{m}{s}\right)^2}{2 \cdot 9.8\,m\big/s^2} = h_v$$

$$0.136\,m = h_v \qquad\qquad (7.49)$$

To calculate the friction losses, we consult Tables 7.1 and 7.2. Each elbow will add 2.7 m of equivalent pipe for friction purposes. So, adding the suction head and discharge head with this yields 13.9 m of pipe. Table 7.1 indicates that there will be a 0.394 m loss of head for every 10 m of pipe at 3 L/s (the closest entry to our problem). So we simply scale our result to find the friction head:

$$h_f = \frac{0.394\,m}{10\,m} \cdot 13.9\,m = 0.548\,m. \qquad\qquad (7.50)$$

The total dynamic head is then the sum of the suction, discharge, velocity, and friction head. Then, the total dynamic head for this problem is: Total Dynamic Head = 5 m + 3.5 m + 0.548 m + 0.136 m = 9.184 m. We can then find the power required by multiplying by the specific weight. Note that we also recognize that the volumetric flow rate is volume per unit of time, $Q = V/t$,

Table 7.1 Head loss (in meters) per 10 meters of 5 cm diameter (~2 in) high-density poly ethylene (HDPE) pipe moving clear water

Liters per second	Head loss (m)
0.1	0.002
0.5	0.016
1.0	0.051
1.5	0.107
2.0	0.182
3.0	0.394
5.0	1.056
10	4.112

Table 7.2 Equivalent length of pipe in meters caused by a 5 cm (~2 in) joint

90° Elbow	2.7
T branch	5.2

$$P = \frac{E}{t} = \frac{\lambda V\left(h_1 + h_2 + h_f + h_v\right)}{t}.$$

$$P = 9800 \, \frac{N}{m^3} \cdot 0.0032 \, \frac{m^3}{s} \cdot (9.184 \, m) = 288 \, \text{Watts}$$

$$P = 288 \, W \cdot \left(\frac{1 \, \text{h.p.}}{745.7 \, W}\right) = 0.386 \, \text{h.p.} \tag{7.51}$$

Earlier, we briefly touched on the fact that the pressure at the inlet of a pump cannot be less than zero; otherwise, a condition called *cavitation* will occur. Let's look at this again armed with more background knowledge. When the total pressure, or more correctly the total energy, of the fluid drops below a certain point it will flash into a vapor – usually little bubbles of vapor. This will most likely happen at the pump inlet where the pressure is the lowest. As the vapor travels through the pump, and then encounters higher pressures, the bubbles implode with enough force to pit the metal of the pump impeller. To prevent this, we want the pressure at the pump inlet plus the dynamic pressure (the kinetic energy, velocity term in Bernoulli's equation) to be greater than the vapor pressure of the liquid. The vapor pressure of a liquid is the minimum pressure to keep it in a liquid state. So, we want

$$P_{inlet} + \frac{1}{2}\rho v_{inlet}^2 \geq P_{vp}. \tag{7.52}$$

We can rearrange this to set the variables on the left hand side of the greater than or equal to sign,

$$P_{inlet} + \frac{1}{2}\rho v_{inlet}^2 - P_{vp} \geq 0. \tag{7.53}$$

The *Net Positive Suction Head* (NPSH) is the equivalent height (i.e., head) of liquid above the inlet at the pump required to avoid cavitation. It is defined as

$$\text{NPSH} = \frac{P_{inlet} + \frac{1}{2}\rho v_{inlet}^2 - P_{vp}}{\rho g}. \tag{7.54}$$

NPSH is a characteristic of a particular pump – each pump has requirements for the amount of liquid that must be fed into it in order to run without cavitating. If the liquid comes into the pump too slowly or without enough pressure, the pump will not work. Since the pressure at the inlet depends on the suction head, the atmospheric pressure pushing the liquid from the suction tank, and friction losses for the pipe leading to the pump, NPSH can be expressed as

$$\text{NPSH} = h_{atm} \pm h_s + h_v - h_f - h_{vp}, \tag{7.55}$$

where

h_{atm} is the equivalent liquid height for the atmospheric pressure (10.34 m of water in the earlier example)
h_s is the suction head above (use the +) or below (use the -) the pump inlet
h_v is the velocity head
h_f is the friction head
h_{vp} is the equivalent liquid height for the vapor pressure

It is important to realize that the vapor pressure is dependent on temperature of the fluid. For example, water at 10 °C (50 °F), h_{vp} is just 0.12 m. But at 70 °C (160 °F) a head of $h_{vp} = 3.4$ m is required to prevent the water from flashing into vapor.

Cavitation leads to excessive noise and vibration in the pump. If the pump operates in this condition for an extended length of time, the pump impeller will begin to degrade and pits will form in the metal parts of the impeller. The violent collapse of vapor bubbles forces liquid at high velocity into the pores of the metal. The forces due to the imploding bubble can exceed the tensile strength of the metal, and actually blasts out bits of metal giving the impeller a pitted appearance. This looks like corrosion to the naked eye and is often talked about at the same time as corrosion, but the pits in the metal result from a very different process. Cavitation also causes excessive vibration leading to bearing failure, shaft breakage and other fatigue related failures in the pump. And let's not forget the key product of cavitation, the liquid itself. Excessive cavitation can heat up the liquid and potentially cause wort to brown via Maillard reactions, or result in degassing beer entirely.

CHECKPOINT 7.4
Calculate the NPSH for the example shown in Fig. 7.9 assuming 70 °C water.

7.3.5   Darcy's Law and Laminar Flow in Porous Media

Darcy's law was originally developed for hydrology problems, for example, ground water moving through sand. This law addresses the resistance that a sand-bed offers to the flow of water through it. It can also be loosely applied to the flow of sparge water through a grain bed. In other words, Darcy's law gives a very good approximation of the sparging process. The law basically relates the flowrate Q to the physical *dimensions* of the bed, the physical *properties of the bed*, and the physical *properties of the fluid* moving through it. The main idea behind Darcy's law is virtually identical to the idea behind resistance in electricity (Ohm's law) and heat conduction through materials (Fourier's law).

Before we get into the details of Darcy's law, it is useful to consider a mechanical analogy to the problem: the *Plinko* game. In the game of *Plinko*, users drop "chips" onto a board filled with regularly space pegs. Gravity provides the driving force pulling the chips down, but regularly spaced pegs momentarily stop the chips. Through collisions with the pegs, the chips will take a more-or-less random path through the array of pegs. The angle of board will vary the gravitational force pulling the chips (a *Plinko* game with a horizontal board would not work). The maximum possible "flow rate," i.e., number of chips per second through the board, will depend on how wide the board is, how long (how many total pegs) and the density of pegs.

The flow of the *Plinko* chips in this analogy accurately models: (1) flow of water through a permeable substance such as sand, (2) flow of charges through a conductor, and (3) flow of energy through a heat-conductor like metals. In each case, there is a driving force trying to move something. But due to constant collisions with the bulk material, the flow is restricted. These three examples also have something else common: they can be modeled as a diffusive system. If we put a *source* of water, or a *source* of charges, or a *source* of heat in a system, then the water, or charges, or energy will tend to diffuse away from this source. This is one aspect that's a bit more difficult to see with the *Plinko* analogy. Image dumping a large number of *Plinko* chips at the top of the board in one single location. Due to random collisions with the pegs and other chips, the chips will tend to spread out as they make their way through the board: diffusion.

In diffusion problems such as these, it is customary to discuss the dynamics of the system in terms of a *potential*, rather than a driving force. A "potential" is closely related to potential energy, but certain material properties of the system under study have been separated. In electrostatics, the potential energy of a charge in an electric field is given by

$$PE = qV, \tag{7.56}$$

where q is the charge, and V is the potential in volts. We say that charges tend to "flow" from high potential to low potential. In hydrodynamics, the potential energy of a bolus of liquid is given by

$$PE = mgh, \tag{7.57}$$

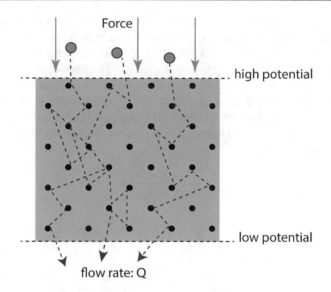

Fig. 7.10 "Flow" of *Plinko* disks through a peg board

where m is the mass of the liquid, g the gravitation constant, and h is now called the "potential". We have already been introduced to this earlier – this is simply the head. So, using the "potential" vernacular, water flows from higher potential (i.e. head) to lower potential (lower head) (Fig. 7.10).

Ohm's law relates the flow of charges through a conductor and is the electrical analog to Darcy's law. The conduction of charges through a conductor is virtually identical to the *Plinko* analogy. The nuclei of the bulk material represent the pegs that the charges (chips) continuously collide. The average velocity, "drift velocity," of the charges is very small, and flow rate is largely determined by the "chip-peg" collisions. An applied electric field will try to move the charges in the same way as gravity tries to move the chips, but collisions with nuclei limits the velocity to a very small value. Overall, and from a macroscopic point of view, the flow *rate* of the charges can be expressed, in words, as

$$\frac{\text{charge}}{\text{time} \cdot \text{area}} = \text{constant}\left(-\frac{\text{difference in potential}}{\text{difference in distance}} \right) \qquad (7.58)$$

The minus sign is there indicating that flow is from higher potential (volts) to lower potential. For those that are interested, the full form of Ohm's law is

$$\frac{I}{A} = -\sigma \frac{\Delta V}{\Delta x}. \qquad (7.59)$$

Here, I is the current in Amps, A is the cross-sectional area that the current is going through, σ is the conductivity of the material, and $\Delta V/\Delta x$ is the change in potential (in volts) per distance applied across the conductor.

Darcy's law is conceptually identical to Ohm's law. We can use the *Plinko* game, and lessons learned from electrostatics to model fluid flow through a resistive media, such as water through a sand bed or grain bed. The driving force for fluid through a bed will either be gravity, or a pressure difference. In either case, we can say that fluid flows from higher potential to a lower potential – i.e., head. Furthermore we can say, like Ohm's law, that the fluid flow rate is proportional to material specific properties, pressure differences, and geometry of object. The only difference in concept between Darcy's law and Ohm's law is that we are moving volumes of liquid instead of charges. Thus,

$$\frac{Q}{A} = -\frac{\kappa}{\mu}\frac{\Delta h}{\Delta x}.$$

(7.60)

where

Q is the flow rate (volumes per second)
A is the cross-sectional area
$\Delta h/\Delta x$ is the change in potential (head) applied across a sand or grain bed
κ is the intrinsic permeability of the bed (similar to σ in Ohm's law)
μ is the viscosity of the fluid through the bed

It can be argued that modeling the flow of water through a sand bed, as in an aquifer, is not the same thing as modeling flow of sparge water through a grain bed. First, Darcy's law, and the hydrodynamics behind it as applied to permeable materials, assumes that the bulk material is not compressible. It assumes that the permeability of the bed does not change. This is certainly not absolutely true for a grain bed during sparging. Secondly, Darcy's law assumes that the viscosity is also a constant throughout the flow – also not true in a sparging situation. As water rinses away sugars from the grain bed, the viscosity of the fluid decreases.

With all of this in mind, this does not mean that Darcy's law is useless in the sparging problem. As with virtually every scientific law or principle, there are limits of applicability. The laws and principles, however, are very useful in determining overall behavior and approximating the problem at hand. In the case of sparging, the changing bed permeability and wort viscosity would make this problem virtually impossible to solve accurately without a huge number of sensors in the grain bed and a computer program. But, we can still learn about the overall, "big-picture" behavior by exploring Darcy's law and the associated hydrodynamics.

Let's explore Darcy's law with the use of a spreadsheet on a computer. Doing so will allow us to approximate the flow of liquid through a grain bed to explore changes to the bed and to the flow. The setup will require a little bit of math to show how to build the spreadsheet. But, once we've set it up, we can play with different situations and see how the sparging system behaves quite easily.

To start, we introduce two new variables as a form of shorthand. First, the specific flow rate (q),

$$q = \frac{Q}{A}, \tag{7.61}$$

is the volumetric flow rate normalized to the area that it goes through. Second, we replace the ratio of permeability to viscosity with the variable k,

$$\frac{\kappa}{\mu} = k \tag{7.62}$$

Finally, we extend Darcy's law to two dimensions,

$$q_x = -k\frac{\Delta h}{\Delta x} \qquad q_y = -k\frac{\Delta h}{\Delta y}. \tag{7.63}$$

In words, q_x is the specific flow rate in the x dimension and q_y is the specific flow rate in the y dimension. These flow rates are related to the degree of the potential (head) in different directions. In other words, if the flow goes through the bed at an angle, we break the flow down into the movement in the x and y dimensions (Fig. 7.11). Extending into the third dimension is straightforward, but we will limit our discussion to two dimensions.

Now let's consider a small "square" of bulk material; the grains in our lauter tun. Darcy's law will tell us the flow rate through this square, both in the x and y direction and how it depends on some arbitrarily applied potential difference. Applying the principles of the conservation of volume flow rate implies

$$\frac{\Delta q_x}{\Delta x} + \frac{\Delta q_y}{\Delta y} = 0. \tag{7.64}$$

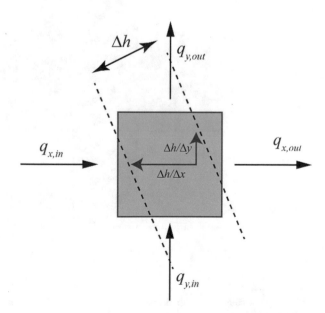

Fig. 7.11 Model square "volume" of the bulk material with an applied potential difference

where $\Delta q_x = q_{x,\,out} - q_{x,\,in}$ is the difference between what flows in and out of the square in the x-direction; and a similar definition for Δq_y. In other words, if there is no source of q in the square (if no liquid is being added at that point), then what flows in must flow out. Inserting Darcy's law into Eq. 7.64 gives

$$\frac{\Delta}{\Delta x}\left(k\frac{\Delta h}{\Delta x}\right) + \frac{\Delta}{\Delta y}\left(k\frac{\Delta h}{\Delta y}\right) = 0 \qquad (7.65)$$

If there is a source S of specific flow into our little square, then we can modify this to

$$-\frac{\Delta}{\Delta x}\left(k\frac{\Delta h}{\Delta x}\right) - \frac{\Delta}{\Delta y}\left(k\frac{\Delta h}{\Delta y}\right) = S \qquad (7.66)$$

While that is a LOT of Δ's, we are now in a position to build a spreadsheet that will calculate the flow through a bed of grain.[1] Consider the four nearest neighbors to our little square, to the North, South, East, and West as shown in Fig. 7.12. Each nearest neighbor square will be allowed to have its own potential (head), and constant k. This second allowance will let us play with the possibility of "channeling" through the grain bed. What we need to do now is approximate the differences as differences in values in neighboring cells.

Fig. 7.12 A system of discrete squares to approximate fluid dynamics in a permeable media

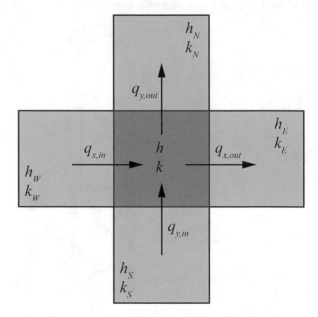

---

[1] For those who enjoy a good calculus problem, the full, time independent expression in three dimensions will look like $-k\nabla^2 h = S$

CHECKPOINT 7.5

Approximate the difference in potential on the right-hand side of our square as

$$\frac{\Delta h}{\Delta x}\bigg)_{\text{right}} = \frac{\left(h_E - h\right)}{\Delta x},$$

with h_E the potential in the East cell and similar definitions for the West, North, and South cell. Insert these into

$$\frac{\Delta}{\Delta x}\left(k\frac{\Delta h}{\Delta x}\right) + \frac{\Delta}{\Delta y}\left(k\frac{\Delta h}{\Delta y}\right) = 0,$$

and solve for the potential h in our square. You should show that the potential in our square is simply the weighted average

$$h = \frac{k_N h_N + k_S h_S + k_E h_E + k_W h_W}{k_N + k_S + k_E + k_W}$$

What assumptions did you have to make? Repeat the derivation, but now consider the possibility of a source S in our square. What happens if all of the k's are the same everywhere?

After working the above checkpoint, we see that the head at a given location is just the weighted average of the heads at the nearest neighboring cells. This will be easy to build a spreadsheet to make these calculations. The only issue is deciding on what to do at the boundaries of the model system.

To illustrate the process, we start by modeling a very simple system where a constant head of 10 (arbitrary units) is applied to the top of a container of permeable material. This container has one exit, and we assume that a constant head of 0 is applied here. This will model our lauter-tun, filled with a constant layer of liquid above the grain bed with a single pipe exiting to the atmosphere. An example spreadsheet, with a very coarse grid, is shown in Fig. 7.13. The positions of constant head are identified in yellow and these cell values are fixed. In the bulk of the permeable bed, the head calculation is just the average of heads around each cell – we are assuming that the k value (permeability and viscosity) is the same everywhere. The container walls are boundaries where there should be zero flow out of the

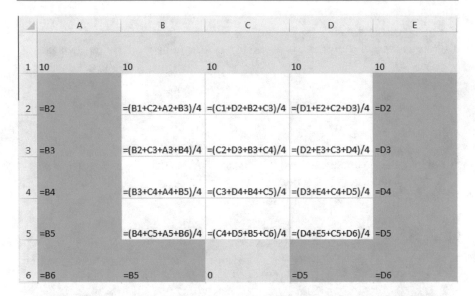

Fig. 7.13 Spreadsheet formulas to model a simple lauter-tun with (i) constant head applied at the top – row 1 contains the same values, (ii) a single exit to the atmosphere at the bottom – cell C6 with a value of 0, (iii) constant permeability and viscosity everywhere, and (iv) no flow out of the boundaries of the container

system. Setting the head at these cells (green) equal to their interior, neighboring cells will ensure flow *out* of the container boundaries is zero.

One final note is in order. Since each cell in the spreadsheet references another, this is by definition a circular reference. Most spreadsheet software can be set to ignore this "error." Also, the spreadsheet must "recalculate" until there is very little change observed in the cell values. Again, modern spreadsheet packages can automatically iterate until the changes are small.

Using a finer grid, that is to say, using more cells will give us a reasonable picture of what is happening in our simple lauter-tun. Extending the formulas to a larger array of spreadsheet cells is straightforward; however, excessively large arrays are not necessary to give a reasonable approximation. Figure 7.14 shows the resulting potential (head) map inside our simple lauter-tun using a 14x15 spreadsheet grid. You can imagine a little ball-bearing rolling down the surface of this map to mimic the flow of fluid – with one important difference. A ball-bearing will roll down the surface and *gain* speed. Due to the restrictive nature of the grain bed, the flow is limited; this is the *Plinko* game again. So, in this type of map the slope indicates the flow rate – the steeper the surface the greater the (constant) rate. If the map is nearly 'flat' then the flow rate is nearly zero. We use this potential map, and Eq. 7.63 to produce a velocity-vector map as shown in Fig. 7.15. This map is like the weather forecaster's map showing wind speed and direction.

Our model shows that the flow is uniform at the top of our simple lauter-tun, but near the exit the flow is greater and not uniform. We are now able to reach a very

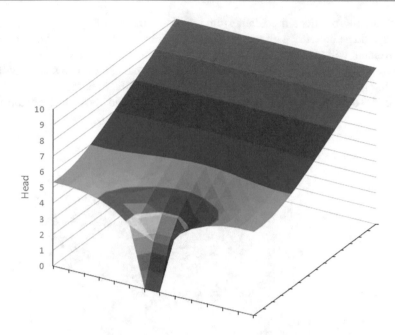
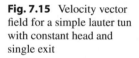

Fig. 7.14 Potential map for simple lauter tun with constant head and single exit

Fig. 7.15 Velocity vector field for a simple lauter tun with constant head and single exit

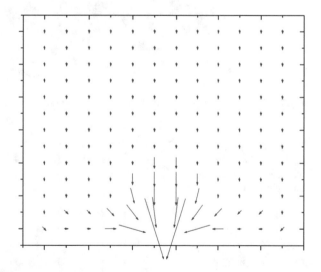

important conclusion. With the greatly increased flow rate near the single exit, we might expect this area of the grain bed to be over-sparged, and regions far from the exit under-sparged. This would give us two issues in the brewery. First, we run the risk of extracting tannins from the portion of the grain bed that is over sparged, and secondly we are compromising our brewhouse efficiency.

We can improve our simple lauter-tun by adding more exits, perhaps making the bottom a "false bottom" with many small exits. Figs. 7.16 and 7.17 show that this improvement will lead to more uniform flow over the grain bed near the exit.

We now want to define a new boundary condition that allows for a source of flow. Note that before we just fixed a constant head and let the flow rate be determined from this. To do this, we will eliminate the "constant head" boundary condition at

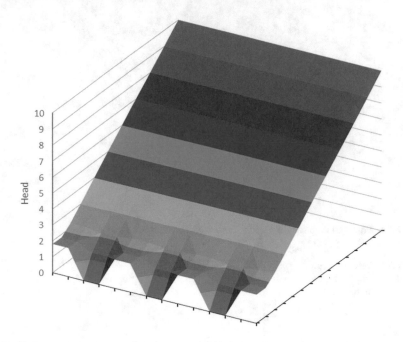

Fig. 7.16 Potential map for a simple lauter tun with constant head and multiple exits

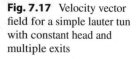

Fig. 7.17 Velocity vector field for a simple lauter tun with constant head and multiple exits

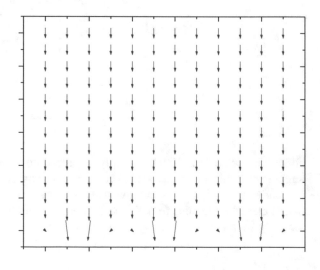

	A	B	C	D	E
1	=B1	=(C1+A1+2*B2)/4	=(D1+B1+2*C2)/4+2.3	=(E1+C1+2*D2)/4	=D1
2	=B2	=(B1+C2+A2+B3)/4	=(C1+D2+B2+C3)/4	=(D1+E2+C2+D3)/4	=D2
3	=B3	=(B2+C3+A3+B4)/4	=(C2+D3+B3+C4)/4	=(D2+E3+C3+D4)/4	=D3
4	=B4	=(B3+C4+A4+B5)/4	=(C3+D4+B4+C5)/4	=(D3+E4+C4+D5)/4	=D4
5	=B5	=B4	0	=D4	=D5

Fig. 7.18   Spreadsheet formulas to model a simple lauter tun with a constant source (2.3 arbitrary units) at the top

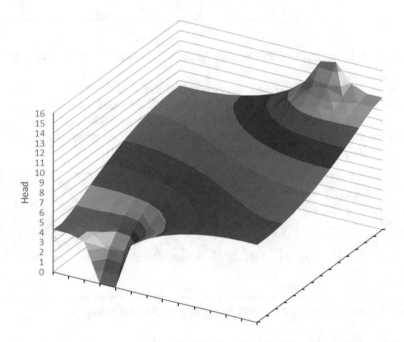

Fig. 7.19   Potential map for a lauter tun with a single constant fluid source and a single drain

the top of the lauter-tun by setting each cell to the average of those around it like the interior of the vessel, but with one small change. Since there isn't a cell above the top layer of cells, we count the cell below a given cell twice, as illustrated in the blue cells of Fig. 7.18. Finally, we simply add a constant term to the cells that we wish to have a constant source of flow into the system: the single orange cell in the figure.

By adding a single source term into the spreadsheet, we can approximate a stuck sparge arm where the source of rinse water is stationary. We further exacerbate the problem in our model by considering a single exit which is offset from the source. A contour and flow map for this model is shown in Figs. 7.19 and 7.20. Again notice

Fig. 7.20 Velocity vector field for a single source and single exit

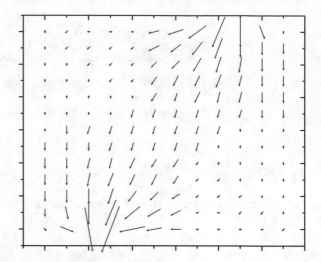

that the flow is not uniform through the grain bed, which will lead to very uneven sparging.

Uneven sparging such as this, or a grain bed that is not constantly stirred, will lead to excessive rinsing along certain "channels." The excessive rinsing will tend to reduce the viscosity of the liquid in this area, making the problem worse. This will make recovery of sugars from the other parts of the grain-bed more difficult. Finally, this will lead to poor brew house efficiency or weaker worts.

CHECKPOINT 7.6

Given the information that has just been discussed about sparging, what inferences can you draw about the flow of water at the top and bottom of the bed of grain?

7.4   Equipment Used in Sparging and Lautering

As we've discussed in this chapter, after completion of the mash, the liquid (also known as the sweet wort) is separated from the spent grains by filtration. Traditionally, draining the liquid from the grains using a process known as batch sparging did this. While the batch sparge process is not used as much anymore, some brewers recognize the usefulness of the parti-gyle method for beer production. Modern techniques include the use of fly sparging and the *mash filter*.

The purpose of these different techniques is to completely remove the fermentable sugars from the residual grains. The efficiency of the process is of paramount concern to the brewer, because any loss of fermentable sugars results in a decrease in the potential quality of the finished beer (and a concomitant decrease in profit.)

7.4.1 Batch Sparging

Batch sparging is typically accomplished by simply draining the liquid from the grains. Some brewers continue to use this process today, but it was much more common in the early days of brewing. In this process, the first batch of liquid drains slowly from the mash, and is then returned to the top of the mash. This process is continued until the wort is clear (known as the vorlauf). After the liquid draining from the mash is clear, the liquid is collected in the underback (also known as the grant). This collection vessel is typically located underneath the tun and has multiple drain holes from the mash directly into the underback. The grant is a specialized version of this piece of equipment that fills with the drained liquid. The volume of the liquid in the grant is maintained to ensure that the pressure of the filtration can be monitored. If the difference in the height of the liquid in the grant and the height of the liquid in the mash tun gets too great, the brewer can slow down the removal of the liquid from the grant or turn on the mash rakes to increase the flow rate.

At this point, after all of the liquid has been removed (the first runnings), the flow of liquid across the grain bed is stopped and another batch of hot liquor is added to the mash tun. Often the entire bed is mixed, stirred, and then allowed to resettle. The mixing is important because it allows the new hot liquor to rinse the grains and extract any sugars from the grains. Then, the process of running the liquid through the underback or grant is repeated. This second batch of wort (the second runnings) can either be added to the first or treated separately as a different beer. The process is often repeated a third time. By the end of the third runnings, there is very little sugar left in the grains.

The equipment used for the batch sparge is often the same as that used for the mash. In other words, the mash tun can double as the lauter tun when the batch sparge is performed. Alternatively, the entire mash can be pumped (grain and liquor) into a lauter tun to perform the batch sparge. The benefit is the same as that which is described in the subsequent section on fly sparging.

Each of the drainings (or runnings) in a batch sparge operation is known as a *gyle*. These worts were evaluated for the concentration of sugars and then mixed to provide different beers. The mixing and the overall process of doing so was referred to as *parti-gyle* brewing. The strongest (and first draining of the wort) has the greatest maltose content and can be used to make export beer, barleywine, or strong ale. The weakest and last drainings of the wort can be used (historically this was the case) to make *small beer* – a lower alcohol beer that was consumed regularly by everyone in the family.

Parti-gyle brewing is not performed much anymore, but there are some commercial brewers who have reconsidered this process as a viable method for creating a wide variety of beers from a single wort stream.

7.4.2 Fly Sparging

In this process, which we discussed at length in the bulk of this chapter, hot liquor is added to the mash as the wort is removed from the mash. Again, the goal is to remove all of the fermentable sugars. A special sparge arm is employed to add the additional hot liquor without disturbing the bed of grains. Again, the first runnings are returned to the top of the mash (the vorlauf) in order to use the grain as a filter and clarify the wort.

There are two forms of fly sparging that are used commercially. In the simplest system, the mash tun serves as the location for the process. If the mash tun is equipped with stir paddles, these are stopped and held in place during the sparge. A layer of water at least 2–5 cm (1–2 inches) thick is placed on top of the spent grains and the liquid slowly drained out of the system. Additional hot liquor is added at the same rate that it is removed so that the pressure differential across the bed of grain is kept small. The first of the sparge water that is added during the sparge must be at about the same temperature as the mash during mash-out. After about ½ of the wort has been removed, the sparge water can be traded out for room temperature water. Studies have shown that no benefit is obtained by sparging with hot liquor during the entire sparge.

In more efficient processes, the entire mash is pumped into a much wider tun that resembles the mash tun, but is often fitted with rakes rather than stir paddles (see Fig. 7.21). The wider lauter tun allows the thickness of the grain bed to be significantly reduced. This allows the sparging process to take place faster (with a higher flow rate) while keeping the pressure differential across the bed the same as it would be in the mash tun. Design of the mash rakes often used in the lauter tun indicates that their shape and dimensions can seriously impact the extraction efficiency during the sparge. Often the rakes have metal triangles placed at angles along the rake. The rakes are also not simply straight; they are bent into a zig-zag pattern from top to bottom. The design of the rakes allows maximum effect in increasing the bed permeability. These mash rakes are employed if the pressure differential begins to increase at the flow rates employed. This is simply done by rotating the rakes around the grain bed. In some systems, the rakes can be raised and lowered, allowing even more control over the effect of the rakes on the process of sparging.

The sweet wort that exits the fly sparging setup is usually emptied immediately into the underback or grant. As we noted before, the grant collects all of the wort

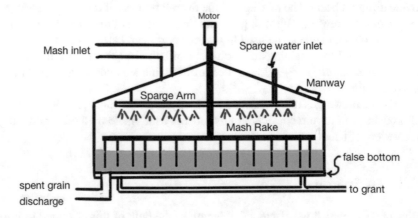

Fig. 7.21 The lauter tun versus the mash tun for sparging

Fig. 7.22 Visual comparison of the wort levels in the lauter tun and grant can be used to determine the pressure of the system. If the height is measured, the head (in m) or the pressure (in kPa) can be determined

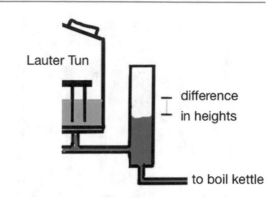

across the entire bottom of the lauter tun or mash tun and fills slowly. Visually, the difference in the heights of the liquid in the grant and the level of the liquid in the mash/lauter tun can be compared (see Fig. 7.22). In doing so, the pressure across the grain bed can be either estimated or calculated (if the difference in heights is measured.) This is done to verify that the pressure doesn't exceed the capabilities of the false bottom. Too high and the false bottom can buckle under the pressure. In addition, the difference can be used to verify the flow rate of the wort being removed from the vessel.

7.4.3   Mash Filter

The most efficient extraction of fermentable sugars occurs with the use of the mash filter. This process also requires that the grist is much more finely ground. In fact, it is often ground to the point where it resembles flour. In these systems, the mash is much thicker and first runnings of the wort after removal from the mash can be greater than 24°P. The efficiency of the mash filter drives its usage in the brewing industry.

When a mash filter is used, the entire mash resembles a thick porridge. Malt flour is added to the hot liquor in a mash mixer. The flour can be added directly to the hot liquor, or as is done with the ground grist using a lauter tun, a pre-mash mixer is often employed. Constant mixing during the mash aids in the conversion of starches into fermentable sugars. Just as with the other systems, care must be taken to avoid high shear forces that can destroy the proteins and other compounds in the mixture. Once the mash is complete, the entire mixture is pumped into the mash filter apparatus and then squeezed with bladders to push the wort away from the flour (Fig. 7.23). Hot liquor is the added to the pressed flour and the mash resqueezed. The wort is collected and passed along to the boil kettle.

Use of the mash filter allows extraction efficiencies to approach or even be better than theoretical yields. After rinsing the flour and resqueezing, the apparatus is opened and the almost "dry" spent grains are released. This process, too, can be automated, dropping the spent grains onto a conveyor belt and removing them from the system.

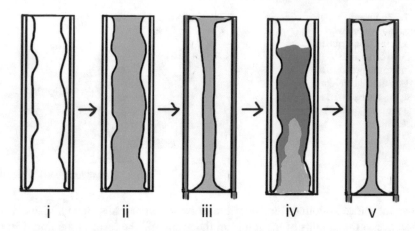

Fig. 7.23 The mash filter. The mash is added to the apparatus (ii), and the bladders are then inflated (iii). The bladders are deflated (iv) and rinse water is added. The mash and sparge water are resqueezed (v) to remove the second runnings. The process is repeated multiple times to remove all fermentable sugars from the mash

While this seems like it might be a rather new technology in mashing, the Meura mash filter, one of the hallmark versions of the mash filter, was actually invented in 1901 by Phillippe Meura. The success of the apparatus has ensured its place in the brewing industry. Slight improvements since then have been made, but the basic apparatus has been used since then.

The apparatus looks like a typical plate-and-frame filter (which we'll discuss later in this text). Basically, a series of bladders are bolted together in a tall rectangular shape. The success of the filter in improving efficiency of extraction has resulted in its use across the world. In fact, this system is responsible for the production of more than 1/3 of all the beer consumed annually.

7.5   When Do We Stop Sparging?

Hot liquor used in sparging should be hot (about 80 °C or 175 °F). The temperature is extremely important in the initial stages of this process. Keeping the temperature hot continues to denature the proteins and destroy the enzymatic activity. In addition, the hot water aids in the extraction of the fermentable sugars from the remnants of the grains. As we noted earlier, some research has been done on the use of sparge water with different temperatures. Those studies have shown that only about 1/3 volume of hot liquor is needed. Any remaining sparge water need not be heated to continue and conclude the sparging process. In spite of this information, it is very common for the brewer to maintain the hot temperatures during the entire process. In most cases, the addition energy required to heat the volume of water to that temperature does not seriously impact the brewer's budget. However, as the volume of sparging grows, this can be a significant cost.

Another feature of the sparge water that must be maintained is the pH. If the pH of the sparge water becomes too alkaline (above pH 6) while still being hot, the extraction of tannins becomes much greater. Thus, care must be taken to make sure that all hot liquor in the brewery is slightly acidic. The risk of tannin extraction may one day cause brewers to consider colder water for the sparge.

Chapter Summary

Section 7.1

Fly sparging and batch sparging are two common methods for removing the sweet wort from the spent grains.

The mash tun can also be used for separating the liqiuids from the spent grains.

Section 7.2

Pascal's law states that the pressure difference between two points in a fluid depend only the height, or elevation, difference. This pressure difference is due to the weight of the fluid contained between the elevation differences, $\Delta P = \rho g \Delta h$.

Section 7.3

Bernoulli's equation, $P_1 + \rho g y_1 + \dfrac{1}{2}\rho v_1^2 = P_2 + \rho g y_2 + \dfrac{1}{2}\rho v_2^2 = \text{constant}$, allows us to understand fluid flow and pumping systems in the brewery.

Darcy's law explains the relationship between bed permeability, flow rate, and pressure across a grain bed during filtration.

Modeling the lauter tun allows us the opportunity to consider modifications to the inlet and outlet and the resulting flow patterns.

Section 7.4

The lauter tun is much wider than the mash tun, allowing the grain bed to be significantly less deep.

The lauter tun employs rakes that can be used to increase bed permeability.

The grant can be used to estimate or calculate the pressure during the sparge.

Questions to Consider

1. High winds, such as from severe thunderstorms and tornados have the potential to remove roofs from houses. Assuming that air has a density of $2.3 \times 10^{-3}\ \dfrac{\text{slugs}}{\text{ft}^3}$ and that the wind speed across the top of a roof is 80mph (117 ft/s), use Bernoulli's principle to find the pressure difference $P_1 - P_2$ on either side of the roof. If the roof as an area of 1300 ft^2, find the total force on the roof.

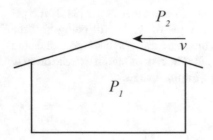

2. Water with density $\rho = 1000 \text{ kg/m}^3$ is passed along a pipe of radius $r_1 = 6$ cm. The pipe tapers to a smaller size, radius $r_2 = 4$ cm. A U-tube, partially filled with an unknown fluid that is immiscible with water of density $\rho = 1390 \text{ kg/m}^3$ is connected to the two sections. If the height difference h is 1 cm, determine (A) the velocity of the fluid in the narrowest part of the device, and (B) the volumetric flow rate.

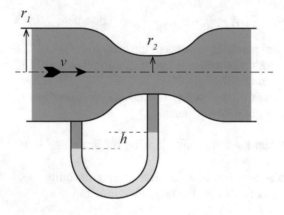

3. A person with advanced arteriosclerosis has an accumulation of plaque on the inner walls. Due to the Bernoulli principle, the increased velocity of blood flow through the restriction and the subsequent reduction in pressure, the artery will collapse and momentarily stop blood flow. When the flow stops, the artery opens again and the process repeats. Such a condition leads to vascular flutter. Assume that diameter of an artery is 6 mm, and that at some point it is restricted to 3 mm. If blood flows through the larger part of the artery at 5 cm/s at a pressure of 100 mmHg (13 kPa), what is the pressure in the restricted portion.

4. The continuity equation also applies to junction of pipes as well as reduction of sizes. Consider a 3″ diameter pipe which "T's" into a 1″ diameter pipe and a 2″ diameter pipe. If the flow rate through the 3″ pipe is 4 gallons per minute, what

is the flow rate in the other two pipes assuming that the velocity of the fluid is the same in all three pipes?

5. Graph the data presented in Table 7.1 and extrapolate head loss values for flow rates of 4 L/s and 8 L/s. Looking at your graph, what can you conclude about the effects of friction as the flow rate increases? If we double the flow rate, do we also double the friction loss?

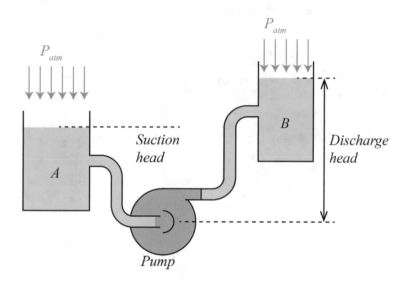

6. Consider the pumping system illustrated above. All piping connections use 2" HDPE pipe. Note that there are two 90° elbows in both the suction piping and discharge piping. The suction head is 3 ft. above the pump inlet and we are moving 160 °F water with a vapor pressure $h_{vp} = 11.2$ ft. What is the maximum volumetric flow rate in ft³/s keeping the NPSH greater than zero?

7. Repeat problem 6, and determine the maximum volumetric flow rate when the water is 20 °C, assuming the vapor pressure is $h_{vp} = 1.2$ ft.

8. In your own words, describe the relationship of flow rate to pressure across a grain bed. What would happen to the pressure and flow rate if the bed was ½ the original depth? …if the bed was 2x as compacted? …if the bed was raked periodically with a mash rake?

9. Outline the pros and cons for the use of a mash filter. Be sure to include at least three pros and three cons.

10. What is the head, in meters, if the pressure is observed to be 4200 kPa? What is the pressure in bar? …in psi?

11. Which has a greater pressure, the bottom of a column of air that is 20 miles deep or the bottom of a column of water that is 2 m deep? ($\rho_{air} = 1.225$ kg/m³; $\rho_{water} = 1000$ kg/m³).

12. Adapt your Darcy's law – sparging spreadsheet to handle situations where the permeability and/or the viscosity is different. This is easiest by referencing another area of the spreadsheet that will contain the k values, such as the example shown. You will model a situation where the viscosity is lower. This means that the k value in Eq. 7.63 is larger. Make your grid 14 × 15 as in the above examples. Assign most of the k values to 1, but make a "channel" from the source to the exit with $k = 4$. How does the flow through the grain bed compare to the situation where $k = 1$ everywhere?

	A	B	C	D	E	F	G	H	I	J	K
1			Potential map						k values		
2	=B2	=(I2*C2	=(J2*D2+H2*B2+2*I3*C3)/(J2+H2+2*I3) +2.3	=(K2*E2	=D2		1	1	4	1	1
3	=B3	=(H2*B2	=(I2*C2+J3*D3+H3*B3+I4*C4)/(H3+I2+J3+I4)	=(J2*D2	=D3		1	4	1	1	1
4	=B4	=(H3*B3	=(I3*C3+J4*D4+H4*B4+I5*C5)/(H4+I3+J4+I5)	=(J3*D3	=D4		1	4	1	1	1
5	=B5	=(H4*B4	=(I4*C4+J5*D5+H5*B5+I6*C6)/(H5+I4+J5+I6)	=(J4*D4	=D5		1	4	1	1	1
6	=B6	=B5	0		=D5	=D6		1	1	4	1	1

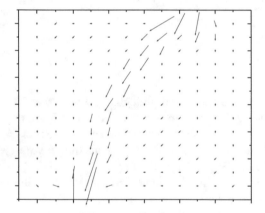

13.   Describe the differences between a mash tun and a lauter tun in terms of their use in filtering the hot wort away from the spent grains.
14. A liquid must be moved from a storage vessel to another. The bottom of the first vessel is 3 m below the bottom of the second vessel. The pump is 3 m below the first vessel. If the pipes attach to the bottom of the vessels, what is the suction head on the pump? What is the required delivery head for the pump?
15. In Fig. 7.9, the inlet pipe to the pump is placed near the top of the supply vessel. What would be the effect on the calculation of the horsepower required for the pump of extending the inlet pipe so that it reached the bottom of the vessel?

16. Repeat question 12, but consider that a dough-ball has been allowed to set up in the middle of the lauter tun. Set the value of k to 0 in the middle-most cell of the table. What is the effect of this dough-ball?

17. What is the effect of having exit points at every location on the bottom of the grain bed during sparging?

18. What would you qualitatively expect to determine for the gravity of the first wort, second wort, and third wort in a batch sparge process? ...how would this compare to a fly sparge that collected the same volumes?

19. Describe the operation of a grant in your own words.

20. Examine the figure below. What is the pressure across the grain bed at the given flow rate? What would happen to the pressure if the viscosity of the wort dropped from 1050 kg/m³ to 1030 kg/m³?

21. Assume a pump has a suction head of 1.5 m and a discharge head of 4.0 m. If the density of the liquid being pumped is 1040 kg/m³, what is the horsepower required by the pump?

22. What is the volumetric flow rate of a sweet wort ($r = 1060$ kg/m³) flowing at 20 L/min? What would be the volumetric flow rate in a pipe that is 2 cm versus one in a pipe that is 5 cm in diameter?

23. In this chapter we noted that Darcy's law is only an approximation of the flow across a grain bed, but that it works fairly well for sand. What differences exist between grain and sand that might result in deviations from Darcy's law? For each, describe the effect that would result.

Laboratory Exercises

Exploring Darcy's Law

In this experiment, we'll explore some of the parameters of Darcy's law by modeling it in the lab. Our "grain" will be pebbles, rocks, and sand, and our "wort" will be water.

Equipment Needed

50 mL funnel with a stopcock – at least 1 cm in diameter
Ring stand and clamp to fit the funnel
Cotton ball
Thin glass rod
Masking tape
100 mL graduated cylinder
100 mL beaker
Clock with sweep-hand or stopwatch
Sand, pea-sized gravel, marble-sized gravel
Water

Experiment

The setup is to clamp the 50 mL funnel in an upright position. Place a 100 mL graduated cylinder underneath the setup so that it collects everything that drips out of the end of the cylinder.

Tear a cotton ball apart to obtain a small piece of cotton. Place this into the funnel and using a glass rod, push it into the small neck at the bottom of the funnel. The point of the cotton is to stop sand and gravel from becoming lodged into the stop-cock when it is opened. Be careful when using the thin glass rod as any lateral force could cause it to snap, resulting in cuts to the hand.

Obtain a portion of sand and add it to the funnel with the stopcock closed so that the funnel is filled ¼ to 1/3 of the way. Use a small piece of tape to mark the top of the sand.

Then, use the 100 mL beaker to fill the entire funnel with water. Make sure that no air bubbles exist in the funnel by holding your hand over the top of the funnel, inverting it, and then reclamping it in place.

Then, at the same time, open the stop cock and start the timer. Collect at least 20 mL of water into the graduated cylinder and stop the timer when that occurs. Note, more accurate readings are obtained when the volume collected is increased.

Repeat the experiment at least three times by closing the stopcock and refilling the water. Record the average of all times for sand. Then, repeat the experiment by washing out the funnel to make sure all of the sand has been removed, and replacing the sand with pea-sized gravel, and then repeat again with marble-sized gravel. Record your results and make a statement about the size of the particles versus flow rate (in these cases the pressure across the bed will be relatively constant).

Repeat the experiment by placing one marble-sized stone in the middle of a batch of sand and determining the time. Then, repeat the process with 3 marble-sized stones. Finally, write conclusions about the application of this experiment to Darcy's law.

Wort Boiling

8

8.1 Why Boil the Wort?

After the wort has been obtained from the spent grains by lautering, sparging, and other processes, the brewer sends the sugar-rich liquid to the boil kettle. The wort contains many of the necessary components needed by the yeast for active fermentation, but potentially lacks the desired flavor profile, and may contain microorganisms and organic and inorganic compounds that are undesirable in the fermentation vessel. Thus, the wort is boiled to adjust the chemistry and biology of the liquid.

While it is often thought that the main reason for boiling the wort is to sterilize it (reduce the number of living microorganisms) or just to allow the hop acids to form the desired bitterness for the product beer, there are a multitude of reasons for heating the wort to these high temperatures. The brewer would not spend the energy, time, and money to perform these tasks if an alternative was available or if a good saleable beer could be made in a different way. As we'll see, the number one reason for boiling the wort is the reduction in the number of bacteria, fungi, and mold that may have remained after the mash.

Sterilization of the Wort It is very possible that microorganisms have been introduced into the wort during the mashing process. Often, the mash tun is not completely sealed, the lauter tun may be open to the atmosphere, the mash paddle (if one was used) may not have been perfectly cleaned, and any other number of introductions of airborne microbes may have occurred. While many of these microbes likely do not survive the higher temperatures of the vorlauf, it is quite possible that some have. Heating the wort to boiling for at least 15–20 min ensures that the number of living organisms in the wort is reduced to almost zero. The high sugar content of the wort also plays against the typical microbe, causing it to be so stressed that application of heat for a short time results in its death.

M. Mosher, K. Trantham, *Brewing Science: A Multidisciplinary Approach*,
https://doi.org/10.1007/978-3-030-73419-0_8

Increased Maillard Reactions Heating the sugary wort solution increases the rate at which Maillard reactions occur. We learned about these reactions in Chap. 5 when the malt was kilned; however, the same process can also occur in boiling wort. The wort is packed with sugars, proteins, and amino acids that can undergo the same reactions that happen in malt. In boiling, these reactions produce organic molecules that have flavors resembling caramel, toast, and other rich deep flavors the brewer may want for the particular style of the beer that they wish to make. In addition, the Maillard reactions also produce molecules that are fairly dark in color. The result is an improvement of the flavor and an overall darkening of the color of the wort. See Chap. 5 for the reactions involved.

Denaturation of the Proteins Any proteins and enzymes that remain after the mash are heated in the presence of relatively acidic (pH 5.0–5.5) water. This results in the reaction of water with the peptide bonds and cleavage of the proteins into smaller pieces. Even those proteins that are only broken a few times are also significantly disrupted such that they cannot form the active sites needed to act as enzymes. The result of this process is the formation of large quantities of amino acids needed for yeast health during fermentation and smaller proteins that can be removed at the end of the boil as trub.

Degassing the Wort While just a side result of the boiling, hot wort can undergo reactions with oxygen in the air quite quickly. Luckily, the brewer knows that oxygen is not very soluble in hot liquids, such that at boiling temperatures, there's almost no oxygen dissolved in the wort to react with any of the sugars or other compounds.

Reduction in Volume After sparging, the wort may not have the correct specific gravity desired by the brewer for the style of beer that they are making. Boiling reduces the volume by evaporating the water in the wort. Typical evaporation rates of 10% per hour for an open boiling kettle also ensure that the right level of Maillard reactions are occurring. In some brewing systems, the same effect can be obtained with a greatly reduced boil off of only 2–4%.

Elimination of DMS Dimethyl sulfide (DMS) results from the decomposition of S-methylmethionine (SMM) under aqueous acid conditions. This small organic molecule has a fairly low flavor threshold and, unfortunately, tastes a little like canned corn. While some styles of beer are acceptable with some DMS, it isn't a very pleasant flavor in all styles. The precursor molecule to DMS is SMM. SMM is a naturally occurring amino acid found in malt, especially undermodified malt. Boiling the wort causes the SMM to release the DMS, and the evaporation of the wort during the boil also carries much of the DMS away. In a typical boiling process, the maximum amount of DMS possible in a beer is often reduced by 50%. Fermentation further reduces the amount of this compound. Some of the DMS produced is also oxidized to dimethylsulfoxide (DMSO), which has a much higher flavor threshold and is difficult to taste at the levels produced. The DMSO can also be reduced in concentration by the evaporation of some of the water in the wort.

S-methylmethionine (SMM)   dimethylsulfide   dimethylsulfoxide
                                     (DMS)   (DMSO)

Formation of Trub   Yes, that greenish brown icky stuff at the bottom of the wort after the boil is a necessity. It results from the complexation of proteins and polyphenols (tannins) that can be extracted during the mash. Boiling the wort increases the rate of formation of these complexes. The precipitate begins to form initially as the wort is heated to boiling. The so-called hot break is the formation of large colonies of the trub at the elevated temperatures. The result is an improved clarity with a reduction in the tea-like flavors from the polyphenols.

Reduction of Wort pH   As the temperature increases, the reaction rate of the calcium ions (Ca^{2+}) from the water and the diphosphate ions ($H_2PO_4^-$) from the malt increases. The result is the formation of calcium phosphate and hydrogen ions. The calcium phosphate precipitates and is collected with the trub; the hydrogen ions slowly lower the pH of the wort 0.1–0.4 units depending upon the initial pH of the wort. The calcium ions can also form insoluble complexes with larger proteins and help remove them from the wort as well:

$$3\,Ca^{2+}\left(aq\right) + 2\,H_2PO_4{-}\left(aq\right) \rightarrow Ca_3\left(PO_4\right)_2\left(s\right) + 2\,H^+\left(aq\right).$$

The utility of calcium ions in the wort means that some brewers will purposefully add calcium chloride or calcium sulfate to the wort prior to the boil in order to help increase the drop in pH to the desired level. We want to make sure not to reduce the pH too much so that it deviates from the optimal 5.2–5.3 range. Too low and the desired hop reactions (see below) will not take place. Too far away from this range on either side will also affect the yeast's ability to ferment, the flavor of the beer, and the other reactions that are needed to produce a particular style. Coincidentally, the reaction of calcium ions and phosphate ions reduces the pH close to that range as long as our initial pH is close.

Addition of Wort-Clarifying Agents Because some of the proteins and protein-polyphenol complexes remain suspended after the boil, addition of a clarifying agent (also known as finings) during the boil is often desired. The higher temperatures of the boil allow the clarifying agent to dissolve completely and help coagulate the proteins, complexes, and other large substances in the wort precipitate. The most common clarifying agent used in this process is carrageenan (when used in raw form it is known as Irish Moss). Carrageenan is a large negatively charged polymer that attracts the positively charged proteins and protein-polyphenol complexes.

Hop Acid Conversion In Chap. 1, we discovered the four main ingredients used in the production of beer. We've explored how water and malt were combined to make the wort and saw the importance of each. During the boil, it is the hops that join the water and malt extract to get us closer to the finished beer. When placed in hot water, the hop oils are extracted from the plant material and interact with the water and other compounds found in the boiling wort. Most importantly, some of the hop oils undergo a dramatic change (as we'll see in Sect. 8.5), to become the bittering and preservative agent that we want. If we simply added the hops to the wort when it was cold (known as "dry hopping"), only a very small fraction of the hop oil undergoes the conversion to the important bittering and protective compounds. In fact, even at boiling temperatures only about 35% of the hop acids undergo conversion after an hour of boiling. As we'll see in Sect. 8.5, this is partially due to the rate of the reaction to form the bittering compounds and partially due to the solubility of the hop acids in acidic water. Interestingly, if the pH is lower than our optimal pH range of 5.2–5.3, the conversion rate is significantly slowed.

CHECKPOINT 8.1

What are three chemical reactions that take place during the boiling of the wort?

8.2   The Equipment of the Boil

Early beer was likely made by heating wort in whatever vessel was available: ceramic, stone, iron or even wooden vessels. However, improvements in heating technology have greatly changed since those initial brews. Brewhouses today typically utilize stainless steel kettles with particular characteristics that enhance the transfer of heat into the wort in the most efficient manner possible.

While there is a cost associated with the four ingredients in a typical beer (water, malt, hops, and yeast), the brewer spends most of the operational overhead on heating water. Any efficiencies that can be had in converting the costly energy source (gas, oil, electricity, etc.) into heat and delivering as much of that energy into the water as possible, will result in a significant reduction in overhead expenses. Ultimately, this ends up as a greater margin of profit for the company.

In this section, we'll explore the characteristics of the boil kettle, their efficiencies, and how it is used in the brewery. Let's start by considering the types of metal that can be employed as heating vessels.

8.2.1   Metals and Heating

When we consider all of the potential options for use as the metal from which to construct our brew kettle, three metals immediately come to mind: aluminum, steel, and copper. Other materials may also be considered based on our experience with cooking in the kitchen, such as ceramic. The characteristics of each of these options for the construction of the brewery must be considered based on the outcome we wish to obtain. In this case, that outcome is to efficiently transfer heat from a source to the wort in the kettle. That heat transfer must be such that our wort can be rapidly warmed to the boiling stage and then maintained there for the duration of the boiling cycle.

Let's examine the characteristics of the four materials we've mentioned and see how they compare. Table 8.1 lists the values for some key parameters that we'll need.

The specific heat capacity, defined in Sect. 8.4, is a measure of how much heat must be absorbed in order to cause the object to get hotter. Thermal conductivity, on the other hand, is a measure of how that heat is transferred across the metal and into the liquid it touches. Thermal conductivity is measured in watts per meter kelvin. The flow of heat is directly proportional to the thermal conductivity but inversely proportional to thickness of the material. So, as the thermal conductivity gets larger (or the material thickness is reduced), the flow of heat gets greater.

From Table 8.1, we note that ceramics have a fairly large specific heat capacity. In other words, the ceramic must absorb a large amount of heat in order to get hot. They also have a fairly low thermal conductivity, and transfer of heat through a ceramic and into the wort is slow. Overall, ceramics tend to be poor choices for boiling kettles. When heat is applied to them over a long period of time, they do get hot but they stay hot for a long period of time. While this might be good if you're making a casserole, it is difficult to adjust the temperature of the ceramic. Moreover, ceramics cannot be shaped easily by hammering (they aren't malleable), and joining two ceramics together is only possible using special glues or mortars. They are

Table 8.1   Key characteristics of materials for heating

Material	Specific heat capacity (J/kg°C)	Thermal conductivity (W/mK)	Corrosion resistance	Malleability	Weldability	Acid resistance
Pyrex ceramic	~800	30	Excellent	No	Poor	Good
Aluminum	902	205	Excellent	Excellent	Acceptable	Poor
Copper	385	380	Acceptable	Excellent	Good	Acceptable
Steel	450	50	Poor	Good	Good	Poor

corrosion and acid resistant, but the drawbacks concerning their brittleness, difficulty in shaping, and heat transfer characteristics make them not a good choice for a traditional boil kettle.

Aluminum is a common metal used in inexpensive cookware. While it too has a fairly high heat capacity, the excellent thermal conductivity value and the cost to the consumer are often the reason we see it in the stores. Unlike ceramic, however, aluminum is fairly strong when it is hammered thin, and so a vessel could be made from aluminum. If it's too thin, however, it may bend or stretch unexpectedly during the normal wear and tear of the brewing process. Yet, aluminum's main issue is that it is not very acid resistant and slowly dissolves into the liquid with time. These added metals in the water can impact yeast health later, and may have a noticeable flavor profile that is not desired. In short, brewers stay away from aluminum pots.

Copper, on the other hand, has very good heat transfer characteristics. It has a relatively low specific heat capacity and a fairly high thermal conductivity. So, it doesn't take much heat to get it hot, and the flow of heat through the metal is very fast. It is malleable and it can be welded together to make a boiling kettle. Copper does have some issues with corrosion (we've all seen the green patina on a copper roof), and it does react with acidic solutions. But, in some cases, the cost of copper and the heat transfer characteristics make this a viable metal to use in the brewhouse.

Steel is not a good option. It does have a low specific heat capacity, but the flow of heat through the metal is very slow. In fact, it's really not much better than ceramic. The overall heat transfer rate (joules per second) can be made better than ceramic since a steel vessel can be made much thinner. But rusting is a serious issue that is difficult to control, and it's susceptible to the acidic conditions of the wort that would increase the level of iron in the wort to unacceptable levels. So, even though it is fairly inexpensive and can be made into brewhouse vessels, it is not a good choice for a metal.

While some breweries are perfectly happy replacing their copper brewing kettles every so often, most brewers work with a different option that isn't in Table 8.1, stainless steel. Stainless steel is an example of an alloy. Let's look more at alloys and see how they stack up to the metals we've just encountered.

Alloys Working with metallurgists, the brewer has found a series of different options based on two of the metals in Table 8.1. The results are mixtures of the metals. These mixtures, called alloys, are simply made by measuring out the correct mix of metals needed, and then melting them in a furnace until they are well mixed together. Alloys have been known for quite some time. In fact, the period from 3300 to 1200 BCE is known as the Bronze Age because of the discovery and use of the very important alloy known as bronze. The good news is that alloys can be formed from an almost infinite set of combinations of metals, and by choosing the metals and their amounts carefully, we can arrive at an alloy that has the properties that we need for our brew kettle.

The metals are mixed together by melting them together in a furnace. The resulting alloy has a much different set of properties than either of the two metals used to make the alloy. For example, when copper is alloyed with tin, we get the alloy known as bronze. Bronze is much harder than copper and as our ancestors found out, it is able to stay sharp when fashioned into a dagger or sword. A copper dagger, on the other hand, would bend too easily and constantly require sharpening. If we mixed zinc with our molten copper, we would end up with the alloy known as brass. Brass is easily molded and can be hammered easily into different shapes. In the right proportions, the brass can even take on the appearance of gold. Both brass and bronze have useful properties; brass is sometimes encountered in fittings used in the brewery. These fitting can be corrosion resistant but sometimes have issues with cracking due to stress.

Steel itself is an alloy of iron and carbon (and other elements in small quantities). However, if the iron is alloyed with at least 11% chromium, it forms a special alloy known as stainless steel. There are a multitude of different stainless steel alloys each given a separate designation known as an SAE steel grade (Table 8.2). The SAE grades were developed by the international Society for Automotive Engineers (SAE International) organization. The most commonly used stainless steel is 304 stainless steel that is an alloy of iron containing 18% chromium and 8% nickel. It has fairly good corrosion resistance and is somewhat resistant to acids. The second most common stainless steel is 316, an iron alloy containing 18% chromium, 8% nickel, and 2–3% molybdenum. The added molybdenum greatly improves the acid resistance of the metal. In fact, where concentrated sulfuric acid would react with 304 stainless at room temperature, 316 stainless is unaffected.

The overall heat capacity and thermal conductivity for stainless steel is not significantly different than plain steel. The relatively low heat capacity means that it will heat up quickly, thus absorbing a relatively small amount of heat energy as it warms. The ability to transfer heat to the liquid it contains is not as good as other metals. However, since stainless steel is relatively strong compared to the other materials mentioned above, the vessel can be made thinner, improving the over energy transfer.

For those particularly corrosive environments, it is possible to also get a stainless steel that has a very low carbon content (<0.03%). Those stainless steels are denoted with an "L" following the code name, e.g., 304 L. These products are very resistant to corrosion compared to the higher carbon content steels, but, for most, if not all, brewery applications, the added expense of this type of stainless steel is not worth the return on the investment.

The stainless steel alloys, in general, are corrosion resistant because the chromium in the metal reacts with oxygen in the air to form an impervious layer of chromium oxide on the surface of the metal that is firmly adhered to the surface of the metal. If the stainless steel is scratched, the layer reforms and the metal is protected again. This is unlike "normal" steel, where the surface of the iron reacts with oxygen to form iron oxides (rust) that flakes off of the metal allowing more of the steel to reaction with the oxygen again.

CHECKPOINT 8.2

Which of the metals would you use in a pot for making soup? Which metal would not be suitable for extended and repeated use, if the food you were making was fairly acidic (such as tomato soup)? Why are aluminum cooking pots so prevalent on the market?

Table 8.2 Selected stainless steel alloys

Group name	Common SAE steel grades	Composition	Pros	Cons
Martensitic	410, 420	12–18% chromium 0.2–1.2% carbon	Hard Corrosion resistant	Poor welding properties
Ferritic	409, 430	12–18% chromium <1% carbon	Corrosion resistant	Thin sheets can be welded
Austenitic	304, 316	18% chromium 8% nickel 2–3% molybdenum <0.08% carbon	Corrosion resistant Excellent weldability Useful temperature range	Oxidizes above 925 °C Chloride ions can etch

8.2.2   Corrosion

Corrosion is the reaction of a metal surface with oxygen from the air to form a metal oxide. If that metal oxide is not firmly attached to the surface of the metal or flakes off of the metal by mechanical or other means, a pit is formed on the metal. Repeated oxide formation in the same location can result in a hole, crack, or general damage to the structural integrity of the metal vessel.

The brewhouse is a wet corrosive environment for metals. And many different types of corrosion can attack those metals. Those types of corrosion include:

- **General corrosion**. In this process, the entire surface of the metal reacts with oxygen and slowly corrodes away. The result is a thinning of the metal across the entire surface of the metal.
- **Galvanic corrosion**. When two different metals are touching because of a weld, close contact, or otherwise fixed to each other, they can transfer electrons from one of the metals to the other. This enhances the corrosion of the metal that gives up the electrons to the other metal. For example, physical and prolonged contact of zinc metal and steel results in corrosion of the zinc.
- **Erosion**. If a liquid is physically pushed against the wall of a vessel or pipe, the constant action of the liquid can erode the interior wall.
- **Cavitation**. The collapse of bubbles against the wall of a vessel or pipe can act similarly to the physical erosion of the metal. The force of the bubble collapse can be significant enough to rapidly cause a weakening of the wall of the vessel or pipe.
- **Weld corrosion**. When stainless steel is welded, the metal at the point of the weld is heated to approximately 1450 °C. The temperature of the metal depends upon how close the region is to the weld. And where the metal is about 850–450 °C, a chemical reaction takes place. At this temperature range, the carbon in the steel reacts quickly with the chromium to form $Cr_{23}C_6$. Above 850 °C, the $Cr_{23}C_6$ breaks down. Because the chromium is tied up with carbon at this distance from the weld, it cannot react with oxygen to form the protective layer of chromium oxide. Therefore, at this distance from the weld, there is an increased risk of corrosion of the stainless steel.

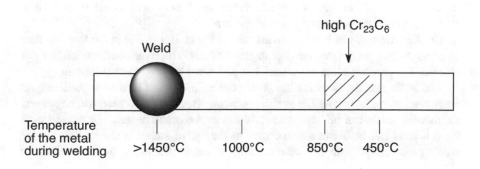

– **Microbe corrosion**. Microbes can form colonies that stick to the surface of the vessels in the brewery. If that colony isn't destroyed, it can form a hard plaque that adheres tightly to the surface of the vessel. The microbes can produce corrosive chemicals and slowly corrode the stainless steel in those locations.

The presence of corrosion on a stainless vessel can be very damaging to the vessel. It can form pits on the surface of the vessel that are unsightly, or it can continue hidden in cracks and crevices on the vessel until a fracture occurs. That fracture could result in catastrophic failure of the vessel. Imagine standing next to a 300-barrel fermenter when corrosion causes failure of the vessel! Therefore it is very important that every brewery has protocols in place to inspect and address corrosion as early as possible.

8.2.3 Methods for Heating

Once the wort is in the boil kettle, it is heated until it reaches its boiling point. How the heat is delivered to the vessel is very important to the brewer. In addition to the cost to produce the heat, the protocols, safety, and design of the vessel are considered. In the typical brewery, two main methods are used to generate the heat: steam and direct-fire.

Steam is produced using a boiler either by heating a container of liquid using gas burners or using electrical heating elements (much like an electric hot water heater). The water boils and is converted to steam under a slight pressure. The steam is then carried through insulated pipes to the boil kettle. This is a more expensive initial investment and requires additional safety protocols and training for employees on the handling of steam; the overall process has some very useful benefits. First, it is less expensive to operate in the long term. Second, and likely the most useful to the growing brewery, additional boiling kettles need only have a piped connection to the steam source to begin heating.

The design of some boil kettles allows them to be heated directly. In these systems, the gas burners are located underneath the boil kettle and the flames of the burners directly contact the bottom of the boil kettle. This direct-fire method for heating costs less to install, but over time can be more expensive to operate than steam. Moreover, if additional vessels are used in the brewhouse, each must have its own burners and its own supply of fuel to burn.

The disadvantages of both steam and direct-fire kettles reside in the fact that they do generate flue gases from the combustion of fuels. The steam boiler is often more efficient and produces less emissions, but both must be properly ventilated.

Two additional methods for the production of heat must also be considered. As mentioned above, it is possible to either generate steam using an electric heater and a direct-fire kettle can be heated directly using an electric heater. The operation of these heaters can be about as expensive as the use of steam or direct-fire, but the safety protocols for the use of electricity in close proximity to water must be followed. Moreover, electricity is not always produced free of flue gases – electrical power plants might use coal, diesel, or other fuels to generate the electrical energy.

The other method for producing heat is currently being explored: microwave heating. While this method does appear to show promise, it is still in its infancy in the brewing world.

Heat Transfer Characteristics It is important that any heat applied to the boil kettle is transferred efficiently into the wort inside. Heating too slowly and the boiling process can take a very long time to achieve. Heating too quickly and the wort may scorch providing off-flavors and increased color to the finished beer.

When the brewer applies heat to the vessel and it is transferred into the wort, the characteristics of the vessel's construction and the rate of heat transfer can result in one of two main types of boiling processes. In the extreme, the surface of the vessel is considerably hotter than the wort. If the metal is not very "wettable" – the liquid is not attracted to the surface of the metal – this extreme heating tends to be favored (see Fig. 8.1). Known as film boiling, the surface of the vessel next to the wort becomes superheated and a layer of steam from the wort forms along the vessel, but doesn't form into bubbles easily. This steam can be super-heated well beyond the boiling point of the wort. In film boiling, the risk of scorching is significant (see Fig. 8.2).

At the other end of the spectrum, the surface of the vessel is barely hotter than the wort and heat is slowly applied. If the metal is wettable – the liquid is attracted to the surface of the metal vessel – and bubbles of wort-steam easily form. In this case, the bubbles tend to be fairly small and release quickly from the vessel surface. Known as nucleate boiling, this provides the most efficient transfer of heat into the wort.

Stainless steel, despite all of its advantages, is not wettable. The use of this metal as the location to provide heat transfer to wort requires that the heating source is not much hotter than the wort so that scorching is reduced. In addition, adequate mixing of the wort must occur to further avoid the Maillard reactions from overtaking the wort and that the area of the vessel being heated is not too great. Copper metal, on

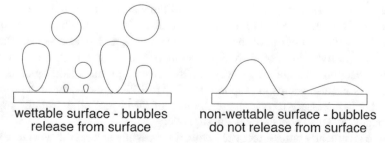

wettable surface - bubbles   non-wettable surface - bubbles
release from surface   do not release from surface

Fig. 8.1 The extremes of heating liquids in metals. The wettable surface allows the bubbles of wort steam to release from the metal. The non-wettable surface holds the wort vapor against the metal causing the vapor to increase in temperature well above the boiling point

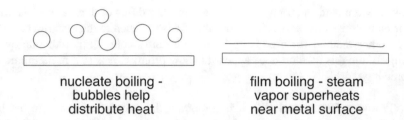

nucleate boiling -
bubbles help
distribute heat

film boiling - steam
vapor superheats
near metal surface

Fig. 8.2 Nucleate versus film boiling. The action of the bubbles in nucleate heating helps cause the liquid to rise in the vessel and evenly distributes the heat throughout the wort. Film boiling does not do this – instead the steam near the surface becomes superheated, causing the liquid to scorch

the other hand, is very wettable. Heat transfer across a copper vessel generates excellent nucleate boiling. While copper's stability toward corrosion and acid are not ideal, the use of copper in a vessel as the area for heat transfer clearly provides a benefit for its use.

8.2.4 Direct Fire Vessels

A direct-fire vessel is heated by the energy released from combustion of a fuel. Historically, these vessels were made of copper and a fire was built underneath them using wood, peat, coal, and other combustible materials. In today's world, the heat is applied by combustion of methane (natural gas), propane, butane, diesel, or heating oils. These copper kettles tend to have a concave bottom or shielding that can trap the heat and keep it in contact with the bottom of the vessel to improve heat absorption by the kettle. Control of the heat is accomplished by regulating the temperature of the firebox underneath the vessel (Fig. 8.3).

While copper boiling kettles still exist, many of the newer direct-fired vessels are stainless steel (typically 304). Heat transfer is not as efficient and the risk of film boiling is very significant in these cases. Ensuring that the temperature of the steam is not too much hotter than the boiling wort is needed to avoid scorching.

The heat transfer into the wort causes the warmed liquid to become less dense and rise to the surface of the vessel. This causes colder wort to move into the space vacated by the warmed liquid. Thus, mixing of the wort occurs as it moves from the exterior walls of the vessel and rises to the middle of the vessel. The area of contact of the wort with the bottom of the vessel can be an issue. In these vessels there is almost always some form of scorching, caramelization of the sugars, or unwanted Maillard reactions that make their way into the finished product. Boiling of the wort in these vessels also results in evaporation of up to 10% of the water in the wort per hour. This is necessary to encourage the loss of DMS and other volatile organic compounds, and also concentrates the sugars.

While technically not a direct fire kettle, it is possible to heat the wort in a kettle of similar design using steam (Fig. 8.4). These steam-fired kettles are surrounded by jackets that circulate steam around the bottom of the vessel and transfer the heat into

Fig. 8.3 Direct fire boil kettle. Heat is applied by combustion of a fossil fuel or organic material in a chamber at the bottom (1), where baffles and an angled bottom help hold the heat and promote circulation currents (8) in the wort. The vessel is insulated (2) to encourage the circulation of the wort during boiling (red arrows). Multiple ports (3 and 4) are placed in the vessel to hold a thermometer, other instrument or an exit/fill port. The kettle usually includes a manway (7), a steam vent (6), and a CIP spray ball (5) for cleaning

Fig. 8.4 The steam-fired boil kettle. Steam is circulated in jackets around the exterior of the stainless steel vessel (1 and 3). The temperature is monitored (2) via a port through the steam jacket (6). In addition to the vent stack (8), manway (7), and CIP spray ball (6), the wort can be drained through a port in the side wall and through the port in the center bottom of the vessel (5)

the wort directly through the sides and bottom of the vessel. Typically, these vessels are constructed of stainless steel and necessitate careful control of the heating process to reduce the degree of film boiling that is likely to occur. This type of vessel is very commonplace in the US steam-fired microbrewery.

> **CHECKPOINT 8.3**
> Provide a list of pros and cons for the use of a natural gas direct-fired kettle. …for a steam heated direct-fired kettle.

8.2.5 Calandria

A technological design that incorporated the advantages of steam heat and the use of copper materials was used extensively until the mid-1900s. Still used in the United States today, the internal calandria (Fig. 8.6) allows efficient nucleate boiling of the wort. In this system, the wort fills a stack of copper pipes that are surrounded by a stainless steel steam jacket (Fig. 8.5). Steam heat then enters the wort through the copper pipes, causing the warmed liquid to rise in the pipes. Colder wort enters the pipes at the bottom of the stack and rises as it is heated. This pulls more wort into the bottom of the calandria, etc. The movement of the rising wort as it

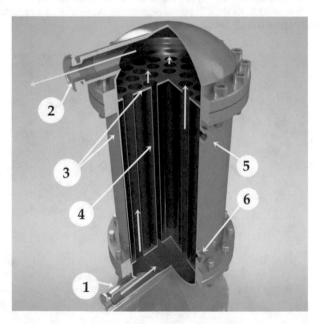

Fig. 8.5 The calandria. The wort enters through the bottom of the calandria (1) where it passes through copper pipes (4) that are heated by a steam jacket (3). The steam enters through a port near the bottom (6) and leaves through a port near the top (5). Eventually, the hot wort exits the calandria (2) either through a port in the case of an external calandria, or directly into the wort vessel in the case of an internal calandria

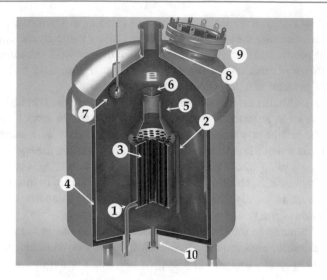

Fig. 8.6 The internal calandria. The copper pipes (3) in the internal calandria must be covered prior to heating the wort. Steam heat enters the calandria through a port (1) and exits through another (2). The gap between the calandria and the wort spreader (5 and 6) allows efficient distribution of the heat by admitting additional wort into the spreader. The vessel contains a manway (9), a steam vent (8), a CIP spray ball (7) all located within a jacketed vessel (4). The wort can be drained through a port in the center bottom of the vessel (10)

reaches the boiling point eventually becomes significant, such that a deflector plate is required to spread the wort out and direct it back into the bulk of the vessel. In this way, wort circulates up into the calandria, reaches its boiling point, and shoots upwards to the deflector plates. The wort then splashes back to the edges of the vessel and travels downwards toward the bottom of the copper pipes. The wort in a boiling kettle follows this flow pattern on its own. This process, where the wort circulates through the calandria on its own, is known as *thermosiphoning*.

Some disadvantages exist with the use of the internal calandria. First, the calandria should be covered with wort. If the steam is applied too early, the wort will boil and not have enough force to eject itself from the copper pipes. Thus, it will overheat and scorch. The time required to fill the vessel with wort can result in a delay in production. Second, the efficiency of heat transfer across copper means that it is very possible that there is some scorching of the wort inside the copper tubes. The resulting layer of caramelized sugars on the copper surface, known as fouling, reduces the efficiency of heat transfer. And third, cleaning of the internal surfaces of the copper pipes inside of the boil kettle is difficult, even when using clean-in-place (CIP) technology.

Workarounds for the disadvantages of the internal calandria have been explored. For example, the wort can be pumped into the base of the copper pipes. In this way, the entire boil kettle does not need to be filled in order to begin the heating process. This system, known as the Stromboli system, works well to decrease the overall processing time, and reduces fouling of the copper pipes because of the constant action and flow rate of the wort caused by the action of the pump. Two disadvantages still exist; cleaning the system inside of the boiling kettle is not addressed with the use of the pump, and the action of the pump on the hot wort can cause problems with adequate trub formation (the force of the pump can disrupt the formation of the complexes of proteins and polyphenols.)

In the mid-1900s, the calandria system moved outside of the boil kettle (Fig. 8.7) to become known as an external calandria. In this system, the wort is pumped into the base of the calandria where it rises into the copper tubes. It is heated and then rushes out of the top of the calandria. It is then piped back into the boil kettle and up into the deflector plate where it splashes into the boil kettle. Actual boiling takes place inside the kettle, but heating occurs exterior to the kettle. In this way, cleaning of the boil kettle is greatly simplified, and because the calandria itself is easily accessed outside of the vessel, it is also much easier to clean.

The use of a pump in the system, and other modifications to the copper piping inside the calandria, also provide useful advantages. For example, the pumping action helps to keep any fouling of the copper tubes to a minimum. This advantage

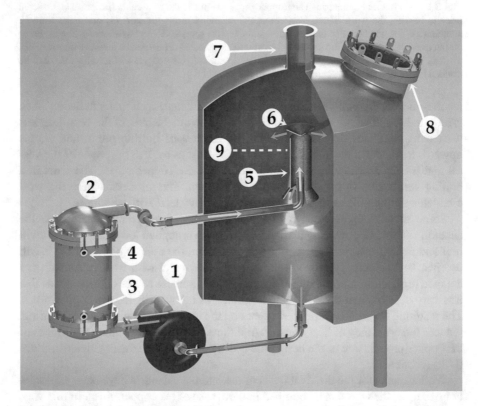

Fig. 8.7 The external calandria (2) is fed by steam through ports (3 and 4). A pump (1) withdraws the wort into the calandria and ejects the wort into the wort spreader (5 and 6). As with other vessels, a manway (8), vent (7), and CIP spray ball (9) are included in the vessel. The advantages of the external calandria are the main reason that this heating system is very widely used in brewhouses

is clear when one considers that about 10 heating cycles (10 batches of beer) can be performed before the calandria must be cleaned. Also, because the wort is pumped into the calandria, the boil kettle needn't be full in order to begin the heating process. To aid in appropriate trub formation, the pumping action can be stopped once the wort is at boiling temperature. The wort, at this temperature will thermosiphon through the external calandria.

By far the biggest advantage to the use of the external calandria is that the evaporation of water from the wort can be reduced to about 5% per hour while still preserving the necessary reductions in volatile organic compounds such as DMS.

8.2.6 Other Heating Systems

New Belgium Brewing Company, in Fort Collins, Colorado, for example, utilizes a different heating system for their boil kettle. The system, known as a Merlin® efficiently boils wort by pumping it from a whirlpool vessel and up through the center of the heating vessel (Fig. 8.8). The wort is then deflected and runs down a steam-heated cone on the inside of the heating vessel. The greatly increased surface area of the wort in contact with the hot cone allows for very efficient evaporation of the wort. Then, after dripping off of the cone, the heated wort is pumped back into the whirlpool vessel to settle.

The wort is injected at an angle on the whirlpool vessel such that the entire contents of the vessel are mixed. The swirling action of the wort causes the trub to move to the middle of the vessel and precipitate. Thus, the trub is removed from the wort while it is being heated.

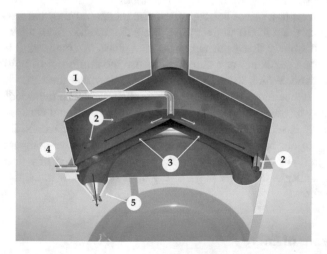

Fig. 8.8 The heating portion of the Merlin® heating system. Wort is pumped from a whirlpool vessel into the system either through the top (1) or through a pipe that ends at the top of the heating cone (3) from the bottom of the vessel. The wort is heated as a thin film as it passes down the cone (3) that is heated by a steam port (4). The wort exits through ports (2) into the collection port (5). The hot wort is dropped into a vessel where it is swirled so that the trub can be separated as it forms

Evaporation rates in the Merlin® system are very low (~4%), but contain the same or better reduction in volatile organic compounds. And thorough heating and mixing of the wort occurs with each portion of the wort passing over the heating cone 5–6 times. Because the system is coupled with the whirlpool, minimal time is needed to remove the trub from the wort at the end of the boil.

> **CHECKPOINT 8.4**
> Compare and contrast the use of an external calandria with an internal calandria. What major advantage does the calandria provide over the direct-fired boil kettle?

8.3   Heat and Temperature

So far we've discovered that heat and temperature are important concepts in the design of a boil kettle. This begs some important questions, what is heat? and what is temperature?

Heat is the transfer of energy from one object to another. Temperature, on the other hand, is a measure of the amount of thermal energy stored in an object. In other words, an object with a greater temperature has more energy in it than one that does not. And when an object with a higher temperature is in contact with one that has a lower temperature, some of that thermal energy can be transferred as heat.

Let's consider an example to illustrate the difference between heat and temperature. Assume we're planning making some noodles for dinner. We start by adding the dried noodles to a pot of boiling water, and then, after they have cooked to our liking, we pour them into a strainer to separate the noodles from the water. During this process it's likely that a mishap could take place. We could spill a drop of the boiling water on our hand, or a slip could cause the entire pot of boiling water to pour on our hand. Which would hurt more? We know that the water in both cases has the same temperature (its nearly boiling when we strain the noodles). But the larger quantity of water would cause a greater burn to our hand because it has more thermal energy based on the quantity of the water. In other words, it can transfer more energy to our hand than can a drop of the water. If we compare the energy transferred in a pot of warm water to the energy transferred in a drop of boiling water, the quantity of the water plays a very large role in how much thermal energy can be transferred.

8.3.1   Types of Energy

There are a variety of forms of energy found in nature. The most fundamental form is energy associated with motion, known as kinetic energy. Anything that is in motion will have associated energy and the amount of energy can be calculated using the equation:

$$KE = \frac{1}{2}mv^2$$

where

m = the mass of the object in kilograms
v = the velocity of that object in meters/second

The result is a value for the kinetic energy in units of $\frac{kg \cdot m^2}{s^2}$. This collection of units is known as the unit for energy, the Joule. In other words, $1\,J = 1\frac{kg \cdot m^2}{s^2}$. As an example of the use of the equation, let's consider a 1200 kg (2646 lb) car traveling at 20 m/s (45 mph). The kinetic energy of the car is then:

$$KE = \frac{1}{2}(1200\,kg)\left(20\frac{m}{s}\right)^2$$

$$KE = 240000\frac{kg \cdot m^2}{s^2} = 240000\,J$$

> **CHECKPOINT 8.5**
> Which has more kinetic energy, a compact car (1045 kg) traveling at 10 m/s or a speeding bullet (0.1 kg) traveling at 620 m/s?

Other forms of energy are categorized as potential energy, which we can think of as stored energy. There are several mechanical examples, such as a compressed spring, or a weight suspended very high. All have the potential to 'do' something or cause another object to move. Electrical energy, for example, can be viewed as charges in motion (kinetic energy), and chemical energy can be viewed as separated charges; e.g. a battery. There is energy associated with sound waves and finally, thermal energy.

The concept of thermal energy can be viewed as the kinetic energy of individual molecules and atoms. When we think of a container of gas, we envision the individual molecules moving around, bouncing off each other and the walls of the container. Each molecule will have an individual speed, and therefore a kinetic energy. A similar concept is held for solids if we view the atoms or molecules connected with each other by small springs. Each atom vibrates, and due to this motion, they will have an associated kinetic energy. Of course, the atoms in a solid are not connected by springs; however the forces involved between neighboring atoms that hold them together into a solid are similar in concept.

It would be very difficult to describe an object's thermal energy by measuring or reporting the individual kinetic energy of every atom in the sample. It is more convenient to measure the 'average' energy if we want to know the thermal energy of that object (Fig. 8.9). The temperature, then, conveys information about the average energy of the object. Temperature is easy to measure macroscopically with a thermometer.

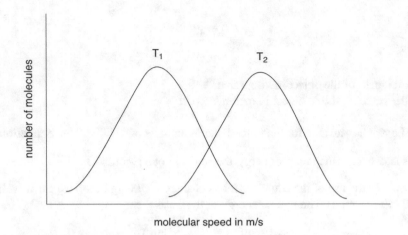

Fig. 8.9 The average molecular speed of the molecules in a hot sample (T_2) is faster than the average molecular speed of the molecules in a cool sample (T_1)

There are three main temperature scales in use across the world. In the US, it is very common to find the Fahrenheit scale where the values are represented in °F. This scale is set such that the melting point of ice is 32 °F and the boiling point of water is 212 °C. The number of divisions between these two values makes it easy to report the temperature; hence, the likely reason that this scale remains in effect in many countries.

The scientific community prefers the use of the Celsius scale where the values for the melting point of ice is 0 °C and the boiling point of water is 100 °C. The majority of the world, in fact, uses this scale. And, as we'll see during our calculations, this is the value that is preferred. As we predict, then, since the mathematical equations use Celsius, and most of the world uses Celsius, this is the scale of temperatures that we will use in this text. We saw this temperature used in the specific heat capacities used to describe metals.

Another scale is also used. The temperature could be reported in Kelvin. This temperature scale has the same divisions as the Celsius scale, but it is offset by 273.15. So, the melting point of ice is 273.15 K and the boiling point of water is 373.15 K. Note that there's no "°" symbol and the numbers are said out loud as "373.15 kelvin." This scale was created to allow scientists to use positive values for the temperature. It's interesting that the coldest a temperature can be (where the object has zero thermal energy) 0 K. This temperature was used in the thermal conductivity used to describe heat transfer in metals.

It's interesting as well to note that when the thermometer was invented in the early 1700s, many brewers were hesitant to use the new technology. They had been brewing beer quite well by boiling water and waiting a certain amount of time for it to cool to a given temperature for mashing. It wasn't until Michael Combrune's manual entitled *Treatise on Brewing* in 1758 recommended its use that many brewers started to use the thermometer. It took many years for all of the breweries to recognize the utility of the instrument.

To take us back to our discussion at the start of this section, when we use the words "hot" or "cold" we are generally referring to an object's temperature. And these words are only relative to some other temperature. But if we consider the atomic model of solids, liquids or gases, these words would then refer to the average kinetic energy or motion of the atoms.

So what would happen if we place a hot object next to a cold object? The thermal motion of the neighboring atoms will start to cause the "cold" atoms to vibrate, transferring energy to the cold object. The word heat describes this process. Not that our definition means that objects do not contain "heat." Rather, an object contains thermal energy.

Note as well that the transport of thermal energy does not depend on the total thermal energy of the two objects. Rather, it depends on the average kinetic energy per molecule (temperature!) of the materials.

8.4   Heat Capacity and Heat Transfer

When thermal energy is transferred to an object, the observed temperature change will depend on how much of the material is present. It will also depend on the type of material. The term "specific heat capacity" addresses how different materials might have different capabilities to store, or give up thermal energy.

As an analogy, let's consider two towels used to absorb a spill. A chamois towel can hold considerably more water than a similarly sized kitchen towel. We might say that the kitchen towel has a smaller "water capacity" than the chamois towel. It's also reasonable to say that as the size of the towel increases the more water it can absorb. For towels, the ability to absorb and store water depends on both the type of material and the size of the towel.

Likewise, when thermal energy is transferred to or from an object, factors that determine the total amount of energy transferred are the mass of the object, the temperature change, and a constant that depends on the type of material that makes up the object. This constant is called the specific heat capacity (c).

The definition of specific heat capacity (or just specific heat) is the amount of thermal energy required to raise the temperature of 1 kg of the object by one degree Celsius. For water, that value is 4.184 J/g °C. Table 8.3 lists some selected specific heats for some common materials. The definition means that we can determine the amount of heat change if we know the mass and the change in the temperature of an object:

$$q = mc\Delta T$$

where

q = heat in Joules
m = mass of the object in kilograms
c = specific heat capacity in J/kg°C

Table 8.3 Specific heat
capacities for selected
materials

Substance c (J/Kg °C)	c (J / kg °C)
Styrofoam™	1300
Aluminum	902
Copper	385
Gold	129
Iron	450
Water	4184
Ice	2030
Malt barley	1674

$\Delta T = T_{\text{final}} - T_{\text{initial}}$, where the temperature is recorded in °C

From Table 8.3, we note that water has a very large value for specific heat. What does this mean? The definition says that it would take the addition of 4184 J of heat to raise the temperature of 1 kg of water by 1 °C. The addition of the same amount of heat to iron would result in a much greater temperature change for the iron. And as we discovered earlier, a larger change in the temperature of the metal can mean that it is able to transfer some thermal energy to the liquid that touches it.

Another important unit a brewer often encounters is the BTU (British Thermal Unit). This, just like the Joule, is a unit used to measure heat. One BTU is equivalent to 1055 J. Burners are often characterized by the *rate* at which they transfer energy. In other words, they are rated by the ability to produce a certain number of BTU per hour. Although it is common to see an advertisement rating a burner at 9000 BTU, this unit is incorrect. The advertiser actually intends to represent the units as 9000 BTU/h, a unit that relates the power of the burner.

8.4.1 Phase Transition – Boiling

We've just discovered the thermal energy changes that are required to increase the temperature of a substance from one temperature to another. However, this only covers the addition (or removal) of that energy while the object stays in its present form. In other words, warming or cooling water is governed by the $q = mc\Delta T$ equation. What happens if the water is converted into steam?

At the boiling point of a liquid, energy that is added into the system is used to convert the liquid into a gaseous vapor. The temperature of the liquid doesn't change during that process, but it still requires energy to continue boiling the liquid. For example, if we heat a pot of water on the stove until it boils, we've added thermal energy into the water to change its temperature to the boiling point of water. We must continue to add thermal energy to the water to keep it at the boiling point; additional energy is used to convert some of the water into steam. A thermometer in the water would continue to read 100 °C (the boiling point of water) until the pot boiled dry (all of the water was converted into steam). If we remove the pot of water from the stove, it immediately stops boiling.

The amount of thermal energy required to change the phase of a substance can be determined using the equation:

$$q = mC$$

where, in the equation…

q = the heat required to change the phase of the entire substance
m = the mass of the substance
C = the latent heat capacity of the substance per kg in kJ/kg

Note that the latent heat capacity (C) is *not* the same as the specific heat (c). The latent heat capacity is not related to a specific change in temperature because the temperature of the object doesn't change as it is converted from one phase or state to another. As we melt ice into water, the temperature remains at 0 °C. As water boils, the temperature of both the water and the steam remain at 100 °C. Table 8.4 lists some useful latent heat capacities and the temperature associated with their phase transition.

For example, let's assume we have 100 kg of water at 100 °C and convert the entire sample to steam. Using the value of the heat capacity for water, we would note that:

$$q = (100\,\text{kg}) \times (2260\,\text{kJ} / \text{kg}) = 226,000\,\text{kJ}$$

In other words, 226,000 kJ of thermal energy would be required to convert all of the water at 100 °C into steam at 100 °C. While a brewer is definitely not interested in changing all of the liquid wort into steam, the concepts are very useful when we consider the operation of a refrigeration system, which we will cover in Chap. 9.

8.4.2 Power

What is power? Power is defined as the amount of energy that can be transferred in a given amount of time. In our example of a 9000 BTU/h burner, the burner is capable of delivering 9000 BTU per hour or 9,495,000 J/h. These units, however, are not the ones typically encountered when measuring or reporting the power of an object. Instead, the SI unit is the watt (W). A watt is equal to 1 J/s. So a 9000 BTU/h burner can also be referred to as a 2638 J/s or 2638 W burner. Mathematically we can represent this as:

Table 8.4 Latent heat and phase transition temperatures

Substance	Latent heat (kJ/kg)	Temperature of transition
Water to steam	2260	100 °C
Ice to water	334	0 °C
R134a (refrigerant)	216	−27 °C
Ammonia (refrigerant)	1369	−33 °C

$$P = \frac{E}{t}$$

where

P = power in watts (W)
E = energy produced or transferred in Joules (J)
t = time in seconds (s)

For example, a 1200 W heater could be used to transfer energy into a bucket of water. If the heater was operated for 1 h, the amount of energy transferred would be:

$$1200\,\mathrm{W} = \frac{E}{1\,\mathrm{h}}$$

$$1200\frac{\mathrm{J}}{\mathrm{s}} = \frac{E}{1\,\mathrm{h} \times \frac{3600\,\mathrm{s}}{1\,\mathrm{h}}}$$

$$E = 4,320,000\,\mathrm{J}$$

Note in the above example that 1200 W is 1.2 kW and that over an hour this is equivalent to 1.2 kW•h. The kW•h is just a unit that relates the energy (in kJ). Electrical companies charge for the use of energy based on how many kW•h are delivered to your home. For our example, at $0.12 / kW•h, we would spend:

$$\mathrm{cost} = \frac{\$0.12}{\mathrm{kW} \bullet \mathrm{h}} \times 1.2\mathrm{kW}$$

$$\mathrm{cost} = \$0.14$$

or 14 cents to deliver 4,320,000 J to our water. This, of course, assumes that no heat was lost to other objects in the transfer, which is highly unlikely. The container, for one, will absorb some of the heat, as will the air around the container.

CHECKPOINT 8.6

If 125 kg of water is heated in a boil kettle and the temperature increased from 60 °C to 95 °C in 1 h, what is the power of the boil kettle in kW?

8.5   Hops in the Boil

As we discovered in Chap. 4, the hop plant is an integral part to our modern interpretation of beer. In fact, almost every batch of beer produced in the United States in today's market contains the compounds found in hops. Knowing about his vital ingredient, the biology of the plant itself and its structure, and how it contributes to the flavor of a beer is very important to understanding the science of brewing.

8.5.1 The Hop Flower Revisited

The female hop plant is the source of the pine-cone-shaped flower that a brewer knows as a hop cone. The male plant is responsible for pollination of these cones, but once it has done so, it greatly reduces the quantity and quality of the oils found inside the flower. For that reason, the hop farmer abhors the male plants and does everything possible to eliminate them from the field.

Hops, *humulus lupulus*, are an example of a bine. They grow on the farm by extending an annual shoot into the air. That shoot has gripping structures that help it climb by wrapping around twine that has been staked from the base of the underground rhizome to a wire about 20–30 ft. in the air. The bine quickly grows up the twine until the plant reaches its maturity and begins producing flowers that look line pinecones.

The structure of the hop flower is interesting. Along the center of the flower is a stick-like structure known as a strig. Extending from the strig are the leaves that encompass the flower and overlap tightly. These leaves, known as bracts and bracteole, protect the oil-bearing portions of the flower. If we peel back the bracts and bracteole, we see these lupulin glands as little yellow dots that at maturity glisten with the oils highly prized by the brewer (see Fig. 8.10).

8.5.2 Hop Oil Constituents

The hop oils contain a very diverse array of compounds. The exact composition of the oils depends not only on the cultivar of hop, but also on the growing conditions for that particular season. The list of compounds contains proteins, water, minerals (such as nickel, zinc, magnesium, etc.), tannins, lipids, sugars, volatile oils, and alpha- and beta-acids.

Volatile Oils The volatile oils include chemical compounds known as terpenes and sesquiterpenes in addition to hundreds of other constituents. These compounds

Fig. 8.10 Hop cone structure. The hop cone has been cut away to reveal the structure of the flower. Note the yellow lupulin glands near the center of the cone

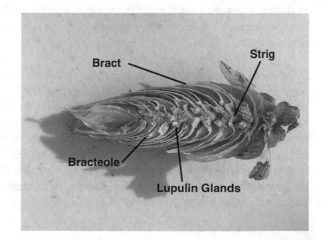

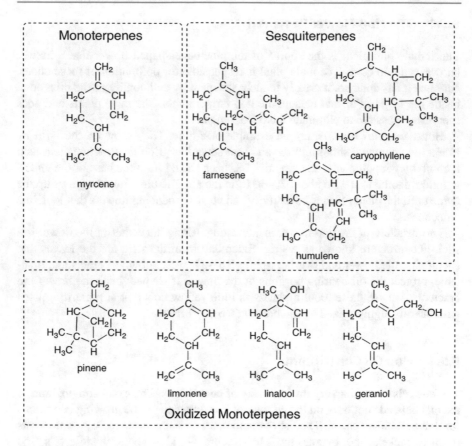

Fig. 8.11 Some of the compounds in the hop oils. These compounds can be broken into two main categories (monoterpenes and sesquiterpenes). Oxidation and maturing of the hop oil often results in a wide variety of partially oxidized compounds. Some common oxidized monoterpenes are shown here. Note the similarities between each structure

impart odors and flavors associated with the hop flower itself. A key terpene found in hops is myrcene (see Fig. 8.11) responsible for a piney aroma. Important sequiterpenes include farnesene and caryophylene, also responsible for a woody, earthy or piney aroma.

Many of the volatile oils have undergone oxidation in the air to form other compounds, such as geraniol (rose-like or flowery aroma), limonene (lemony aroma), and linalool (flowery aroma). In some hop varieties, the oxidation of the volatile oils has resulted in compounds with cucumber, balsamic, or mushroom-like aromas. In overoxidized, or old hops, these compounds continue to change their structures into compounds that have an "old cheese" or "paper" aroma.

Since the volatile oils are volatile, if the brewer adds them to the boiling wort at the start of the boil, most of the oils will be removed with the steam. Thus, the aroma and most of the flavor that the volatile oils contribute will be removed. In other words, if a brewer wants to capture those aromas and flavors as part of the finished beer, they would wait until the last minute to add the hops to the hot wort (or add them after the wort has been cooled back to room temperature.)

Humulones Another class of compounds found in the oils from the hop cone are the humulones, also known as the α-acids (Fig. 8.12). These compounds are prized by the brewer and the hop farmer. In fact, specific cultivars of hops are grown and nurtured to produce extremely high contents of these compounds. They are the compounds that undergo conversion to the bitterness and protective compounds desired in the finished beer.

The general structure of the humulones is illustrated in Fig. 8.12. Three main humulones have been identified that differ in the chain of carbon atoms extending from the central six-membered ring. These compounds are known as humulone, cohumulone, and adhumulone, although the major component of the α-acids is the structure of humulone itself. As the name implies, these compounds are somewhat acidic. And because of their structure, the humulones are not very soluble in the acidic conditions of the wort.

Once added to the hot wort, a small amount of the α-acids dissolve and undergo a reaction. Recent evidence suggests that the reaction requires the presence of the minerals found in the hop cones and in the malt, as the reaction does not proceed if humulone is heated by itself. In any case, the six-membered ring opens during the reaction and the molecule rearranges to a more stable conformation. The ring then closes, but as a five-membered ring system. The result is the formation of

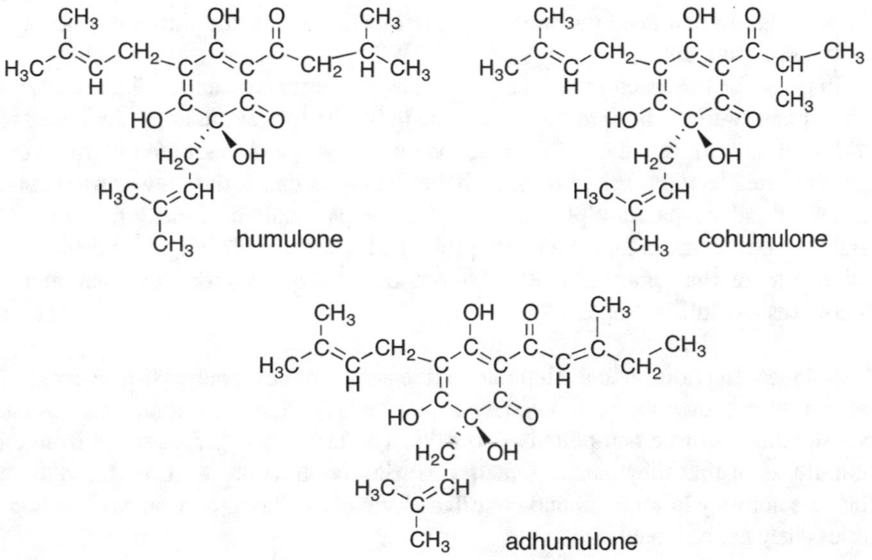

Fig. 8.12 The humulones; humulone, cohumulone, and adhumulone. Can you identify the differences in their structures?

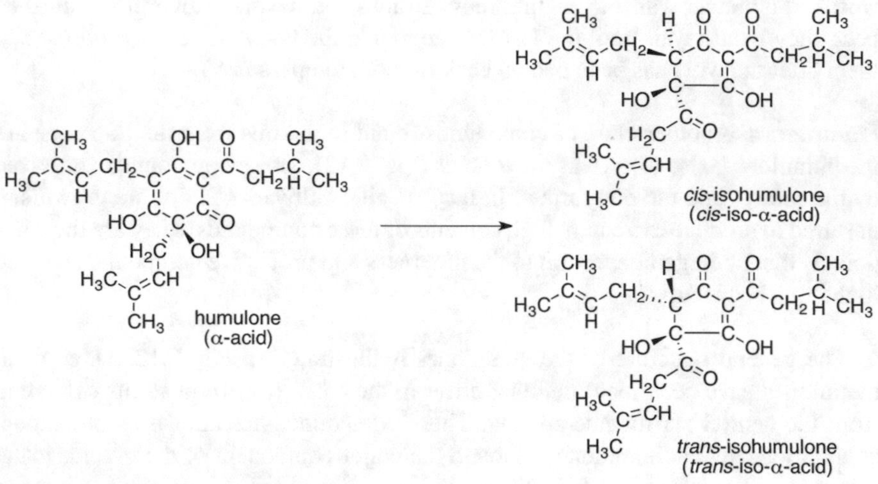

Fig. 8.13 Humulone isomerizes to Isohumulone

cis-isohumulone and trans-isohumulone roughly in a 2:1 ratio depending upon the conditions of the boiling wort (see Fig. 8.13).

The resulting compounds are also known as cis- and trans-iso-α-acids. The iso-alpha-acids are intensely bitter and much more soluble in wort than the alpha-acids. Unfortunately, though, there is a limit to their solubility that depends upon the amount of maltose, sucrose, and other sugars in the wort. About 100–120 ppm iso-alpha-acids appears to be the maximum that can be dissolved. This isn't an issue, because the human taste threshold has a hard time telling the difference between bitterness at this level.

In order for the reaction to take place, heat and time are needed. Therefore, the brewer must add the hops at the start of the boil. The longer the hops are in contact with the hot wort, the more of the reaction takes place, and the more bitterness that gets imparted to the finished beer. It should be noted that if the brewer wants both bitterness and aroma, multiple additions of the hops should be done during the boil. The only drawback is that even with prolonged (>60 min) boiling, it is difficult to achieve more than about 35–40% conversion of the humulone to isohumulone (known as the "utilization rate").

Lupulones In addition to the humulones are a class of compound with a very similar structure known as the lupulones, or β-acids. Just like the humulones, ad- and co- varieties of these compounds also exits. The lupulones differ greatly from the humulones in that they cannot undergo the ring contraction reaction. In addition, they are not very bitter at all and contribute little to the flavor or aroma of the hops, unless they are oxidized.

lupulone

The main use for the lupulones is that they can signal whether the hop is a "noble" hop or a "bittering" hop. Ratios of the humulone and lupulone of about 1:1 indicate a noble hop variety. This can be confirmed by a hop oil containing less than 50% myrcene and a humulene to caryophylene ratio of at least 3:1.

Overall Flavor and Aroma It is the combination of all of the different compounds that give rise to the flavor and aroma of a particular hop oil. With only slight modification of the percentages of the different compounds, the flavor can go from "piney and very harsh" to "grapefruity and pleasant.". It is the job of the brewer to select the hops that are appropriate to the style of beer being constructed and confirm that the hops being purchased match the characteristics that are desired.

8.5.3 Modified Hop Oils

Because the conversion of humulone to isohumulone is time consuming, low yield, and can cause adverse effects in the brew, the brewer may look to a commercial product that provides the same outcome without the use of hops. Research on hop oils and the conversion of alpha acids revealed that the procedure can be completed in the laboratory with excellent efficiency and yield. The result is the modified hop oil.

One modification is to extract the hop oils from the plant material and use the oil directly. This hop extract allows the brewer to reduce the amount of waste products and trub during the boil. The hop oil, when added to the boil, is a little more efficient at the isomerization reaction and can equate to about a 40–45% utilization rate.

Another modification involves the laboratory-based conversion of the humulone to isohumulone in the hop oil extract. The resulting pre-isomerized hop oil is nearly 100% utilized, resulting in a bittering liquid that can be added directly to the wort, or to the finished beer. This product is often used to adjust the bitterness of the beer to move the product into the correct specifications for the style.

Other modifications include treatment of the pre-isomerized hop oil to make either a dihydro, tetrahydro, or hexahydro-isomerized oil (Fig. 8.14). Addition of hydrogen atoms using chemical reagents such as sodium borohydride ($NaBH_4$) or hydrogen gas (H_2) allows the hop producer to control the production of any of these

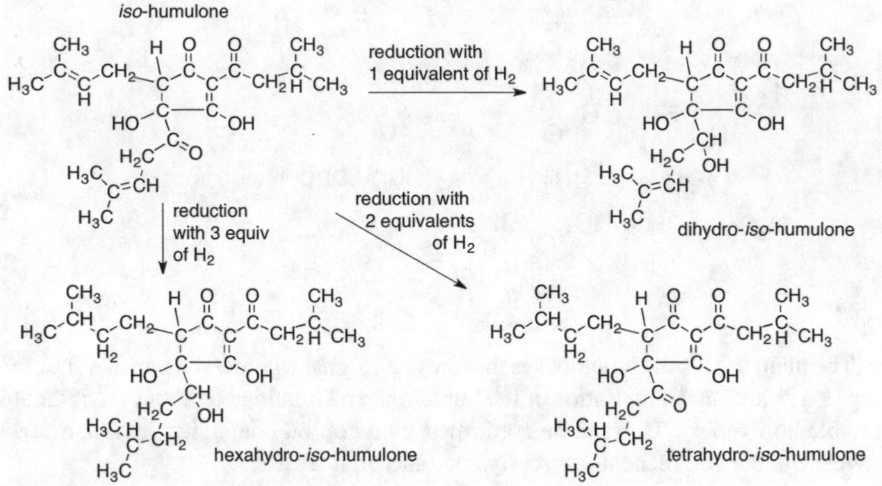

Fig. 8.14 Reduced isohumulones

products. Each of these has a set of very different properties. For example, the
dihydro-iso-humulones (also known as rho-iso-humulones) are not sensitive to the
reaction of light with beer (eliminating the "skunked" flavor associated with a beer
left out in the sun). The tetrahydro-iso-humulones, where the double bonds have
been removed from the molecule, increases the ability of the hop oils to support a
thick foam on top of the beer. The hexahydro-iso-humulones have both the properties of a light-stable hop oil and an enhanced foam on the finished beer.

As we've noted in this section, the structure of a molecule plays a large role in
the properties (such as flavor) of the molecule. And, as we would expect, the bitterness of the reduced isomerized compounds is not exactly the same as before. So
care must be taken by the brewer to add just the right amount of these compounds
to give the desired flavor for the finished beer.

While some may say that additions of modified hop extracts to the beer is cheat-
ing, the results are hard to argue with. Enjoying a cold beer on the beach with a very
stable head of foam is very hard to complain about.

Chapter Summary

Section 8.1

Boiling wort results in many different outcomes. Sterilization, degassing, and
trub formation are accompanied by chemical reactions that improve flavor,
color, and the pH of the end product. Hop oil conversion to the iso-alpha-acids
is also accomplished.

Section 8.2

Stainless steel is typically used as the metal for boil kettle construction. Corrosion
of stainless steel is possible, given the corrosive and acidic environments
found in the brewery.

There are different boil kettle designs in use across the globe, each with advantages and disadvantages. New designs are being explored to improve the efficiency of the boil.

Section 8.3

Heat is the transfer of thermal energy. Temperature is a measure of the thermal energy of an object.

Section 8.4

Thermal energy is a form of kinetic energy.

Heat that is transferred can be calculated using the equation: $q = m\,C\,\Delta T$, where q is a form of energy (E).

Specific heat (c) differs for each type of compound and refers to that object's ability to absorb heat. Latent heat (C) is the heat required to change the phase (from solid $\leftarrow \rightarrow$ liquid or liquid $\leftarrow \rightarrow$ solid) of a substance.

The power of a heating system can be calculated using the equation: $P = E\,/\,t$.

Section 8.5

The hop cones contain lupulin glands that are the source of extractable oils used in brewing.

Hop oil extracts contain volatile oils, resins, and the alpha- and beta-acids.

Isomerization of the alpha-acids results in the bittering and protective agent found in beer.

Pre-isomerized hop oils are commercially available and allow the brewer to know the exact amount of bittering agent added to the finished wort.

Questions to Consider

1. List the outcomes that result from boiling wort.
2. Which of the outcomes that result from boiling wort reduce the amount of volatile compounds in the wort?
3. Describe, in your own words, what happens to the pH as the wort is boiled.
4. Describe, in your own words, how trub forms and explain the compounds that are found in trub.
5. What is a direct-fired boil kettle?
6. What is a calandria? Sketch a drawing of an internal calandria.
7. Why is copper used inside of a calandria?
8. Why is a copper boil kettle not commonly used in a brewery?
9. Describe the pros and cons for the use of SAE 304 stainless steel versus copper.
10. Use the internet to find the exact makeup of 316 L stainless steel and compare the makeup to 316 stainless steel.
11. In a direct fired boil kettle that is made from 304 stainless in two halves, the top half is welded to the bottom half to make the final vessel. The rest of the valves, ports, and other items (such as the legs) are then added. Where would you expect to find rust or stress fractures after extended use of this kettle?
12. What advantages would a Merlin® heating system provide over a Stromboli system?

13. Define "heat" and "temperature."
14. What units are used to record heat and temperature?
15. Explain heat and temperature as they relate to walking barefoot on the sidewalk on a hot summer day.
16. Which metal, assuming the same mass, would you expect to absorb more heat, aluminum or iron?
17. If 4500 J of heat were added to water at 40 °C, what would be the final temperature, in °C? …in °F?
18. How much heat would be required to raise 255 kg of water from 25 °C to boiling?
19. If 2975 J of heat were added to a 1.0 kg block of metal and its temperature increased from 40.0 °C to 47.4 °C, what is the identity of the metal (you may need to use Table 8.3)?
20. Calculate the power, in kW of a burner that would do the work outlined in question 17, if all of the heat was transferred into the water and the burner accomplished the job in 35 min.
21. Recalculate question 20, assuming that only 50% of the heat produced by the burner actually became absorbed by the water.
22. A brewer purchases a burner rated at 22000 BTU. How long would it take the burner to heat a 1 kg aluminum pot containing 20 kg of water from 25 °C to 100 °C?
23. How much heat would be required to change the temperature of 2.2 kg of water from liquid at 75.0 °C to steam at 120 °C?
24. A brewer elects to warm a 145 g sample of water from 34 °C to 55 °C by adding 120 g of water at a higher temperature. What would that higher temperature have to be to accomplish that task?
25. A brewer considers the use of a metal in the construction of a boil kettle. The specific heat capacity of the metal is 240 kg/kJ°C and the thermal conductivity is reported as 405 W/mK. On the basis of these values, would this be a suitable choice? Describe the characteristics of heat transfer for this metal.
26. What is the difference between humulone, cohumulone, and adhumulone?
27. Draw the molecular structure of colupulone.
28. The atoms in the structure of pinene are the same atoms found in the structure of myrcene. Can you identify which atoms in pinene match up with the atoms from myrcene?
29. What flavor would you predict from a hop oil that was rich in pinene? …in limonene?
30. The –OH at the bottom of the six-membered ring in humulone appears to be missing in the five-membered ring in iso-humulone. In the reaction it is changed into something else. Where is that –OH in the iso-humulone structure and what has it become?
31. Why do the lupulones not isomerize? What part of the structure of a humulone, then, would you predict is very important for the isomerization reaction?

32. Identify the similarities between myrcene and limonene, linalool, and geraniol. Can you determine where the "oxidation" of myrcene has taken place to give rise to each?
33. What portion of iso-humulone is likely responsible for causing the light-struck flavor in beer?

Laboratory Exercises

Hop Tea and Identifying Flavors

This experiment is a short introduction to identify some of the key compounds in hops. The hops chosen for the experiment are selected because of the pronounced flavor profile for the particular compound being explored. It is possible, however, to choose different hops based on their availability in a particular area.

Equipment Needed

4 pint mason jars
Potable water
Heater (such as a burner, hotplate, or microwave)
4 samples of hops (recommended: Citra, Cluster, Saaz, Northern Brewer)

Experiment
Boil a suitable portion of water. This can be done for an entire class by the instructor, or each laboratory group can warm their own water. In the absence of a burner, hotplate, or other heating device, it is still possible to extract the hop oils using very hot tap water.

Place a small amount of hops (as pellets) into each of the mason jars (3–5 pellets is enough), and then add enough hot water to fill the mason jar approximately ¼ to 1/3 full. Stir with a spoon or swirl gently. (Caution: hot!)

After 5 min, the aroma of each mason jar should be recorded. Then, after the hop tea has cooled to just above room temperature, a very small sip of the tea is tasted by the students. The aroma and taste should be recorded.

Match the characteristics to of the hops to the compounds that predominate in the hop varieties. Fig. 8.9 will assist with the identification of the compounds in the hops.

Determination of Percent Hop Acids in Hops

This laboratory analysis is based on an ASBC method (HOPS-6) for the determination of hop acid content in hops. The method works by analyzing the absorbance of an organic extract of the hop oil, because the hop acids absorb ultraviolet light.

Equipment Needed

2 × 250 mL Erlenmeyer flasks
2 tight-fitting corks
Pelletized hops
5.0 mL graduated or volumetric pipet and bulb
100 mL volumetric flask
50 mL volumetric flask
Chemicals:
Toluene, methanol, alkaline methanol (0.2 mL of 6.0 M NaOH in 100 mL methanol)

Experiment

Obtain a sample of hops and weigh out 2.500 g of the material (it is important to be as close to 2.500 g as possible). Place the material into a 250 mL E-flask and add 50 mL toluene. Stopper the flask with a cork or other stopper (a rubber stopper isn't recommended) and shake by hand for 30 min. Pass this to your lab partner when your arm gets tired (or use a rotary shaker to do the job for you).

Then, place the flask on the lab bench and work on preparing a second sample of the same hops as you did before. The slurry should stand undisturbed for 30 min. This will allow some of the hop solids to settle to the bottom of the flask and leave some of the clear toluene extract at the top.

Withdraw exactly 5 mL of the clarified toluene extract using a graduated or volumetric pipet and dilute this to 100 mL in a volumetric flask with methanol. Dilute 3 mL of the toluene-methanol solution to 50 mL in a volumetric flask using alkaline methanol (0.2 mL of 6.0 M sodium hydroxide in 100 mL methanol) using a plastic pipet.

Then, measure the absorbance of the solution at 275 nm, 325 nm, and 355 nm using the UV-Vis spectrometer. The instrument should be blanked using a blank made from 5 mL of toluene without the hops added. Take your readings as quickly as possible because the hop acids can degrade quickly in light.

Use your readings to determine the percent of alpha- and beta-acids in the hops. The 1.33 factor in the equation relates the dilution of the liquid to the concentration used in the measurement of the samples:

$$\alpha - \text{acids}, \% = 0.667 \times \left(-51.56 A_{355nm} + 73.79 A_{325nm} - 19.07 A_{275nm} \right)$$
$$\beta - \text{acids}, \% = 0.667 \times \left(55.57 A_{355nm} - 47.59 A_{325nm} + 5.10 A_{275nm} \right)$$

Determination of Wort Viscosity During Boil

This laboratory analysis measures the viscosity of a wort sample as it is being boiled. The effect of evaporation on the sample is observed as are visual comparisons of the wort appearance.

Equipment Needed

250 mL Beaker
100 mL graduated cylinder
1 L tub for 20 °C water bath
Thermometer
Hot plate
Glass rod for stirring
Viscometer (falling ball or Ostwald)
Wort (prepared by mixing dry malt extract with warm water to a gravity of 1.040 g/mL)

Experiment

Obtain a tub and fill it approximately ½ full of tap water. Add either ice or warm water until the temperature of the water in the tub is 20 °C. Maintain that temperature during the course of the experiment.

Obtain 100 mL of wort prepared by your instructor – or prepare it as described above – and place it into the 250 mL Beaker. Place the beaker on a hot plate set no higher than ½ of full-power and stir the solution with a glass rod. While the solution is warming to a boil, measure the viscosity of a water sample using the viscometer. Your instructor will illustrate the method to you. Record the time it takes for the water to pass through the viscometer.

Once the solution on the hot plate has reached a boil, reduce the heat until it just simmers. Occasionally, stir the solution to avoid scorching. Warning, as the solution approaches the boiling point, it may foam significantly. Addition of a drop of vegetable oil, rubbing alcohol, or anti-foam agent will reduce the foaming if it becomes an issue.

After 30 min of boiling, remove the beaker from the hot plate and allow it to cool to room temperature. Then, place it into the 20 °C water bath. Stir or swirl the beaker until the temperature of the wort is the same as that of the water bath.

Then, obtain a fresh sample of wort and determine the time it takes to pass through the viscometer. Repeat that analysis for the boiled wort.

The kinematic viscosity of the wort can be determined by dividing the time for the wort by the time for water and then multiplying the result by 1.0038 (the value of viscosity for water at 20 °C). Use significant figure rules to determine the number of significant digits in the answer.

Compare the viscosities of the pre-boil and post-boil wort. Compare as well the aroma and the color of the two solutions. Is there a difference? Is it what was expected?

Cooling and Fermenting

9

9.1 Setting the Stage

At this point in the brewing process, the brewer has prepared a batch of *bitter wort* (wort that has been flavored with hops), boiled it, and added flavorings such as hops. The wort is hot, sterile, and has just the right color, flavor, and amount of sugars. However, it is too hot to support the growth of yeast. The wort at this stage also has residue floating in it. The residue includes coagulated proteins, polyphenol-protein complexes, hop materials, and likely some grain materials that made it into the boil kettle from the mash tun. This material is known as *trub* (pronounced "troob"). The brewer does not want this in the beer; it simply does not taste good at all.

Getting rid of the un-dissolved material is essential to ensuring that the fermentation in the next step proceeds smoothly. The yeast during fermentation will flocculate (clump and precipitate from the solution) as the process proceeds. The presence of "floaties" in the wort will interfere with this process. In addition, the un-dissolved material that remains in contact with the wort can continue to impart flavors into the wort. This means that the wort could continue to extract tannins and other flavored compounds from the hops. In fact, this could severely damage the finished beer.

So, the brewer works to remove these solids from the wort. The easiest way to do this is to swirl the wort while it is still in the boil kettle. In the smallest vessels, the brewer could simply stir the wort with a large paddle. In vessels of any size over about 1 bbl, pumping the wort from the bottom edge of the kettle and returning it back into the vessel at an angle helps to clarify the wort. Continuous pumping in a tangential manner causes the entire vessel to swirl.

After 10–20 min of whirlpooling the wort while it is still hot, the pumping stops. But the brewer doesn't do anything yet. They tend to wait another 10–20 min to allow the whirlpooling to slow down to the point where the liquid is barely moving. While the wort swirls, the solids move inward to the center of the liquid and then form a pile right in the center. So, how does this work?

© The Author(s), under exclusive license to Springer Nature Switzerland AG 2021
M. Mosher, K. Trantham, *Brewing Science: A Multidisciplinary Approach*,
https://doi.org/10.1007/978-3-030-73419-0_9

As the wort swirls, the water moves very quickly at the outside of the vessel and very slowly at the center. In order to maintain movement in a circle, there must be a force that pushes against the fluid to keep it moving in a circle. In a swirling pool of liquid, that force is provided by the walls of the container – a centripetal force (and not centrifugal). Centripetal means "pointing inward," and an inward force is required to direct the fluid in a circle. These concepts are often confused because it depends on the frame of reference for the observer. If one is moving with the water, one feels a "centrifugal" force. The basic result is that the water becomes deeper on the outside edge than in the center. Try it and see; when you quickly stir a cup of water, the water gets deeper at the edges. This concept is use in centrifuges – a *device* to separate fluid components of different densities. The heavier material tends to settle out at the outside of the spinning circle.

At this point, the suspended solids in the hot wort move slowly toward the center of the whirlpool. If this movement was based only on the "centrifuge effect," we would expect the heavier suspended solids to collect on the outside of the vessel. What's the difference between a centrifuge and the boil kettle? The fluids in the centrifuge cannot mix and those fluids closer to the center of the circle are forced to rotate at the same angular speed as the fluids at the outside. In the boil kettle, due to liquid sheer forces the fluids at center of the circle rotate at a different angular speed. This, in combination with Coriolis forces, will tend to move solid particulates inward. In the end, the particles arrive near the center of the whirlpool and settle out to the bottom of the vessel. For the brewer, the whirlpool results in a cone-shaped pile in the center of the vessel that contains of all of the sediment. The liquid can then be removed from the outer edge of the vessel and pumped to the next step in the brewing process. The sediment remains behind and can be separated by withdrawing it from the valve located at the center of the vessel. Whirlpooling is a very necessary step that doesn't require any additional equipment.

9.2 Wort Chilling

After the trub has been removed from the wort, the next step in the process is to ferment the sugars. However, this occurs at a much lower temperature than the boil. Often the temperatures needed for the start of fermentation are in the range of 18–22 °C (65–72 °F), very much dependent upon the style of beer and yeast used in fermentation. The cool down from boiling (100 °C) must occur aseptically and rapidly.

As the bitter wort cools, it is vulnerable to infection from microbes that would ruin the finished product. In fact, the bitter wort is the ideal growth medium for wild yeasts, bacteria, and other microorganisms. We don't want these microbes anywhere near the wort, because they can begin growing in the nutritious liquid. At the very best, any of these microbes would change the flavor of the beer. At the worst, the microbes could be harmful pathogens (very rare). The exception to the rule to keep everything out of the wort resides in the style of beer we want to make. Farmhouse ales were and continue to be traditionally made by spontaneous fermentation with airborne yeast. In such a case, by all means, we would cool the hot wort in open tanks, invite whatever microbe that happens to be in the area, and hope for the best in the final taste. This style aside, brewers typically like to have full control over the final taste profile.

So, the bitter wort should be cooled as rapidly as possible, to avoid inoculation with some unwanted microbe. This process is achieved through the use of heat exchangers.

9.2.1   Heat Exchangers

A heat exchanger, quite simply, is a device that transfers heat from a hotter fluid to a cooler fluid by heat conduction. If we wanted to cool down a tepid glass of water, we could just drop ice into it. But we don't want to water down our beer with ice! So, the heat exchanger was designed to keep the two fluids separate. For example, the two liquids could be hot wort and cold water as in the case of wort cooling, cold glycol and fermenting beer, or it could be our final beer and steam in the pasteurization process. In all of these cases, we are transferring heat energy from one fluid to another *by conduction* through a material that separates them. So, we'll start by reviewing the physical process of heat conduction. We initially discussed this in Chap. 8, but here we'll delve into the subject in event greater detail.

Thermal heat conduction between two materials, that is to say energy transfer, occurs when the materials are in physical contact. The second law of thermodynamics requires this energy to flow from the object with the greater temperature to the object with the lower temperature. It stands to reason that there will be zero heat transferred if the two objects are at the same temperature.

Let's consider a simple system in which we have a rectangular bar of material in contact with a hot object on one side and a cold object on the other side as illustrated in Fig. 9.1. These hot and cold "objects" will later be our hot wort and cold water,

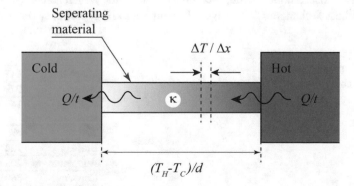

Fig. 9.1 Simple system to illustrate heat conduction physics in a rectangular bar of material separating a hot object and a cold object

for example. The rate of heat transfer, defined as the heat energy Q moved in time t and measured in Watts, is given by *Fourier's heat equation*:

$$Q/t = -\kappa A \frac{\Delta T}{\Delta x} \tag{9.1}$$

where Q/t is the heat energy (Q) moved in time (t) with units of Watts (W)

ΔT is the temperature difference across some slice of the bar
κ is the thermal conductivity
Δx is the thickness of the material

If the material is a simple rectangular bar, these become the hot and cold temperature difference and the length of the material (d in Fig. 9.1), respectively. The cross-sectional area of the bar is given by A. The final term, κ, will be explained later. The minus sign is in the equation to ensure that a positive heat flow goes into the cold object for a temperature difference (ΔT) defined as the hotter temperature minus the colder temperature.

There are a few things we can learn immediately from Fourier's heat equation. If we make the material separating the two objects very thin, i.e., make d very small, then we make the heat transfer process faster and therefore more efficient. Alternatively, we could make the cross-sectional area of the connecting bar very large. A very basic heat exchanger system might have the hot fluid moving through the inside of a tube, and the cold fluid on the outside of the tube. So, we can make the heat exchange faster and more efficient if we used a very long (large area), thin-wall (small d) tube.

Finally, the symbol κ is the thermal conductivity of a given material. This was first introduced in Chap. 8. This is a material-specific value which must be determined empirically. A list of thermal conductivities for some common materials is shown in Table 9.1. The thermal conductivity value tells us how well a certain material will conduct heat, and generally the larger the number the better the material acts as a heat conductor. Looking at Table 9.1, silver is one of the better heat conductors. It is also a noble metal, meaning that it is not very reactive. So why not make our heat exchangers from silver? In addition to its expense, silver is not very

Table 9.1 Thermal conductivities of common materials at 20 °C

Material	κ (W/m·K)
Water	0.6
Stainless steel	14
Tin	67
Brass	122
Aluminum	205
Copper	380
Silver	429

strong. So, the material would need to be thicker to improve its expense. This definitely underscores the impracticality of a silver heat exchanger.

Stainless steel, a common material that you would find in the brewery, has the poorest thermal conductivity. So why make a heat exchanger from stainless steel instead of copper, which is the next best thermal conductor? The reasons mimic those that we explored in the hunt for the perfect metal to make our boil kettle from. Stainless steel is relatively easy to clean, and is not reactive with the somewhat acidic wort and beer that we pump through the system. Moreover, since stainless steel is stronger than copper, we can make the separating material much thinner than if we use copper. So, we can make the net thermal conduction approximately equivalent.

As an example to highlight an important concept later, let's consider a special container in which we can place a thin rectangular plate of dividing material. On either side of the dividing material, we have identical isolated voids to hold water. On one side of the divider we hold 1 kg of water at 100 °C and on the other side we hold 1 kg of water at 20 °C. The question is, due to heat conduction through the dividing material what is the final temperature of both sides and how long does it take for this to happen? We are going to make some assumptions, which will not take away from the final message. First, we'll assume that the water on either side is well mixed so we don't worry about heat conduction through the water. We also assume that there is no heat conduction to the rest of the world. Finally, to answer this question we will need to make some assumptions about the divider. Let's assume it is a piece of thin (0.1 mm) copper with a surface area of 10 cm². Eq. 9.1 tells us the *initial* energy transfer rate since the temperature difference is 80 K:

$$\frac{Q}{t} = \frac{-401\,\text{W}}{\text{m K}} \times 0.001\,\text{m}^2 \times \frac{80\,\text{K}}{0.0001\,\text{m}}$$

$$= -4010 \times 80\,\text{K} \tag{9.2}$$

$$= 320800\,\text{J}\,/\,\text{s}$$

So, in 0.05 s 16040 J of energy is transferred from the hot side to the cold side (i.e., 320,800 J/s × 0.05 s = 16,040 J). What does that do to the temperature of each side? It will raise the temperature of the cold side and lower the temperature of the hot side. We can find by how much. Recall that the specific heat of water $c = 4181$ J/kgK, and the temperature change is related to the quantity of heat energy removed or added:

$$Q = mc\Delta T. \tag{9.3}$$

CHECKPOINT 9.1

If 16,040 J of energy is added to the cold side at 20 °C, what is the new temperature? If 16,040 J of energy is removed from the hot side at 100 °C, what is the new temperature?

So, now what is the heat transfer rate with the new temperatures? Will we still transfer 320,800 J/s? Because the temperature difference is smaller, the heat transfer rate will be smaller. This further means that the change in temperature will be smaller. If we continue this analysis using 0.05 s time steps, it is easy to develop a spreadsheet to calculate the temperatures after every time step. Table 9.2 shows a partial listing of such a calculation and Fig. 9.2 illustrates this table graphically. *Note that we will need to wait a relatively long time to reach thermal equilibrium.* The main point of this exercise is to underscore that the heat transfer rate is greatest when the temperatures of the two different fluids are greatest. But, as we exchange heat, as in a heat exchanger, the temperature difference becomes smaller and smaller which then means that the transfer rate also becomes smaller.

There are a variety of heat exchanger configurations. Again, recall that the idea is to exchange heat energy between two fluids without mixing the two. Let's start with the simplest heat exchanger construction in which the cold fluid flows through the inside of a tube, and the hot fluid is constrained to flow along the outside of this tube. This basically means we have a double-wall tube arrangement. We now encounter our first design consideration. We could have the two fluids run parallel to each other, or in opposite directions. These two different types of heat exchangers are classified as either concurrent-flow or counter-flow (respectively), as illustrated in Fig. 9.3.

In the concurrent-flow heat exchanger, the rate of energy transfer is greatest at the point where both the hot fluid and cold fluid enter the tubes because the temperature difference is greatest. However, as the temperatures become closer together, the rate of heat transfer is reduced. The rate of heat transfer keeps getting smaller as the temperatures get closer and closer. This means that the exit temperature of the hot fluid $T_{H,\text{out}}$ can never be lower than the exit temperature of the cold fluid $T_{C,\text{out}}$. Therefore, the heat transfer is restricted by the cold fluid's outlet temperature.

In the counter-flow heat exchanger, there is always a temperature difference along the length of the exchanger. One benefit of this arrangement arises because the temperature difference is more-or-less the same along the length of the system. Therefore, the thermal (mechanical) stresses in the device are reduced. This net temperature difference also means that it is possible that the exit temperature of the hot fluid can be *less than* the exit temperature of the cold fluid. And this must be considered, because if the temperature of the hot fluid gets too cold, it may freeze into a solid. This would be possible if the cooling fluid's inlet temperature is lower than the freezing point of the hot liquid. It doesn't seem like this would be an issue,

Table 9.2 Estimated temperature as a function of time for two 1 kg portions of water in thermal contact with a thin copper sheet

Time (s)	T_{hot} °C	T_{cold} °C	Time (s)	T_{hot} °C	T_{cold} °C
0	100.000	20.000	3.55	60.031	59.969
0.05	96.164	23.836	3.6	60.028	59.972
0.1	92.695	27.305	3.65	60.025	59.975
0.15	89.559	30.441	3.7	60.023	59.977

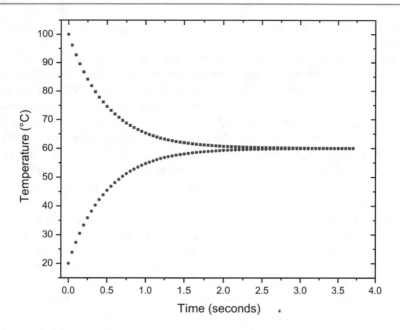

Fig. 9.2 Estimated temperature as a function of time for two 1 kg portions of water in thermal contact with a thin copper sheet. (based on data from Table 9.2)

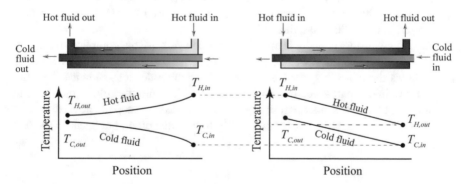

Fig. 9.3 Counter-flow (left) and concurrent-flow (right) heat exchangers. Note that the inlet temperatures are the same for both configurations

but often cold beer needs to be cooled to just below 0 °C (32 °F). It shouldn't freeze at this temperature because it is a mixture of alcohol and water, but if the temperature gets much lower than this, it could.

Generally speaking, a counter-flow heat exchanger will be more efficient at heat transfer for a given flow rate through the tubes. This does not mean that we would never use a concurrent-flow exchanger. The large temperature difference at the inlet means that we can achieve a very rapid change in temperature. If either of the fluids is changing phase, such as steam condensing into water, then the heat exchange occurs isothermally (at the same temperature) and it doesn't matter which exchanger we use.

There are a variety of mechanical implementations of heat exchangers and they are classified by their construction. The two basic types are the tube and shell heat exchanger and the plate heat exchanger. An example tube and shell heat exchanger is shown in Fig. 9.4. It is constructed of a bundle of small tubes that carry one of the fluids. This assembly is contained within a shell, and the second fluid flows around the outside of the tubes. Sometimes these heat exchangers will have extra baffles to direct the second fluid around the tubes. The path of the fluids shown in Fig. 9.4 doesn't quite fall into either a counter-flow or concurrent-flow arrangement. While the second fluid generally flows counter to the first fluid, its path is classified more as a *cross-flow* as it flows around the baffles.

Among the tube and shell heat exchangers, we can have variety. By adding baffles and changing where the fluids enter and exit, we can create, for example, the two-pass tube side heat exchanger in Fig. 9.5.

A plate heat exchanger is constructed from many thin plates, stacked together with a small space between each plate. Gaskets placed around the edges of the plates, as shown in Fig. 9.6, maintain the spacing between each plate. The plates have four openings at each corner, and the gaskets are placed such that the flow of hot and cold fluid alternates between plates. Because the surface area of the plates is quite large, and because the plates can be made from very thin material, plate heat exchangers can be incredibly efficient at transferring energy. Plate type heat exchangers are easier to disassemble and clean compared to a shell and tube heat exchanger. This can be a benefit to a brewery.

Fluid flow inside a plate heat exchanger appears to be complex. It is not. Figure 9.7 illustrates the flow in a plate heat exchanger. We can think of how the liquid flows through the system as if we had a double-decker sandwich. The hot liquid flows in one part – i.e., one half of the sandwich, and the cold liquid flows in the other part – i.e., the other half of the sandwich. Gaskets keep the flow of the two liquids separate so that they never physically touch each other, but because the two

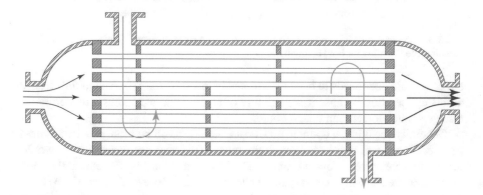

Fig. 9.4 Shell and tube heat exchanger

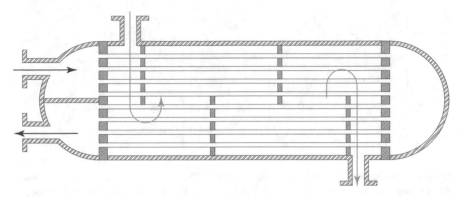

Fig. 9.5 A two-pass shell and tube heat exchanger

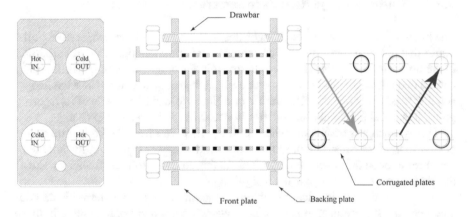

Fig. 9.6 Plate type heat exchanger. The red and blue arrows show the flow of the cold and hot fluids (respectively) in alternating plates

liquids are in contact with the same plate (the middle piece of bread in our double-decker sandwich), heat transfer can occur. The plate heat exchanger can be set up to allow a counter-current flow or a concurrent flow pattern.

Typical plate heat exchangers are arranged so that the hot wort (still above 190 °F after the whirlpool) flows into the exchanger and are cooled to 70 °F without much reduction in the flow rate. Because of the small openings and winding path of the wort as it goes through the plate heat exchanger, a large backpressure develops when the liquid flows (i.e., the pump must develop a large positive delivery head). Care must be taken to ensure that the pumps used are capable of supplying this head without causing cavitation.

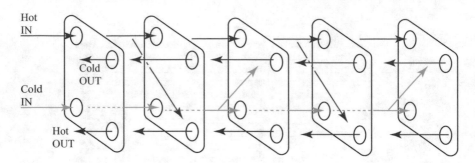

Fig. 9.7 Flow patterns in a plate-type heat exchanger. The red and blue arrows show the flow of the cold and hot fluids (respectively) in alternating plates. Is the overall system shown here representative of a counter-current flow or a concurrent flow?

9.2.2 Multiple-Stage Heat Exchangers

The biggest issue with a heat exchanger lies in the control of the temperature of the liquid that leaves the exchanger. For example, it is possible that a small variability in the flow rates of the cold liquid could result in the temperature of the liquid being cooled to deviate from its final temperature by a degree or two. If the heat exchanger is being used to change the temperature of a liquid to within a degree of its freezing point, it is a real possibility that the liquid could freeze in the heat exchanger.

This is much less of an issue when cooling bitter wort from the boil kettle as it is passed to the fermenter. Often, in that transfer, the brewer is looking for final temperatures to be in the 60–70 °F range. This is far removed from the freezing point of the liquid, so a small change is not a significant issue.

However, this is a significant problem if the liquid being transferred is carbonated beer and the destination is the cold maturation or conditioning tanks. In those cases, the brewer is often trying to get the beer cooled to within a single degree of its freezing point. Any deviation then could result in the beer freezing – and in doing so, the beer would decarbonate, expand its volume, and potentially cause damage to the heat exchanger, piping, or other equipment.

The solution is to use multiple heat exchangers. The first knocks the temperature down to within 5–10 °F of the final temperature. There's little risk of a variability in the flow rate or temperature of the coolant causing the beer to freeze at this point. Then, the beer enters a second heat exchanger where the temperature is reduced to the desired temperature. The second heat exchanger only has to reduce the temperature a couple of degrees, so the final temperature can be more easily controlled.

Often, the second heat exchanger is actually part of the first heat exchanger. As shown in Fig. 9.8, two different coolants are used as the beer runs through the device. The first coolant drops the temperature most of the way to the desired final temperature. Then, the beer enters the region of the heat exchanger that contains the second coolant, where the temperature and flow rate of that coolant are carefully controlled.

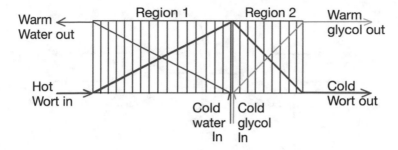

Fig. 9.8 Dual stage heat exchanger. The liquid to be cooled enters the first region that is roughly cooled by the first coolant, in this example water. Then, when it is within a few degrees of the final temperature, it enters the final coolant region of the heat exchanger where it is cooled by cold glycol

9.3   Equipment Used in Fermentation

Once the wort has been cooled to an appropriate temperature to support yeast growth, oxygen gas is added into the liquid. Because gases are much less soluble in hot liquids than cold liquids, the boiling process has removed all of the oxygen from the wort. The addition of oxygen is necessary to support the growth stage of the yeast that will be added. There are many different ways to add oxygen to the wort. These methods include: (a) pumping the cool wort into an open vessel and allowing air to dissolve into the wort, (b) splashing or spraying the wort into a vessel that is full of air, (c) bubbling oxygen gas into the fermenter, and (d) passing the wort through a *Venturi aerator* (Fig. 9.9). Each of these methods attempts to bring the oxygen level in the wort up to about 8–10 ppm.

Some brewers prefer using a Venturi aerator to oxygenate their worts. This device is basically a tube that has a restriction in the flow of the wort by changing the pipe diameter. As the liquid enters the smaller pipe diameter, the velocity of the liquid increases. Because the velocity of the liquid increases, the pressure of the liquid decreases. A small hole in the center of the small pipe is used to inject oxygen into the flowing wort. The low pressure where the oxygen is injected means that bubbles of air (or oxygen) enter the wort at this stage. Immediately after the small pipe diameter, the pipe returns to its normal size. The velocity of the wort slows down and the pressure increases. Because the pressure increases, the tiny air or oxygen bubbles that are dispersed in the wort quickly dissolves.

The best way to inject oxygen into the wort, however, is by bubbling it directly into the fermenter using an "oxygenation stone." The stone is basically a porous tube that forces the oxygen to bubble as very tiny bubbles. These bubbles have a much better chance of being dissolved in the wort.

If air is used to deliver oxygen to the wort, the maximum amount that can be physically dissolved is only about 8 ppm. Other issues with the use of air exist. For example, if air from the room is used, a sterilization process must be put into place so that no bacteria or other flavor-harmful microbes get added into the wort. If pure oxygen is used, the amount of gas soluble in the wort rises to about 9–10 ppm. Keep

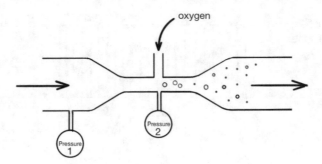

Fig. 9.9 Venturi aerator. The pressure at point 1 is greater than the pressure at point 2 because the liquid flowing in the pipe must increase its velocity in the restriction. This makes the perfect location in which to add oxygen

in mind that oxygen is not terribly soluble in wort, so it is very difficult to get above this concentration of oxygen. And if a brewer tries to do so, the oxygen will simply bubble out of solution in the fermenter. This is likely not a great thing, because a lot of oxygen trapped in the headspace above the fermenting wort may result in oxidation of the wort.

Once oxygenated, the wort is pumped into the fermenter and yeast is added. The style of fermenter plays a very large role in the flavors of the finished product and in how yeast are collected for use in other batches. In the remaining parts of this section, we'll uncover the different types of fermenters.

9.3.1   Refrigeration

Many times in the brewing process, we need temperatures cooler than the ambient surroundings. When cooling our hot, bitter-wort from the boil kettle, it's a straightforward process to use a heat exchanger, and use available room temperature water to cool it. Speaking from a thermodynamics perspective, we are transferring thermal energy (heat) from the hot, bitter-wort to the much cooler water. The second law of thermodynamics allows this process, transferring energy from the hotter object to the cooler object. But, what if we wanted to cool something, like a fermenter or a storage room, to a temperature *less than* the ambient surroundings? The second law of thermodynamics does not permit heat energy to spontaneously and freely move from a cooler area (e.g., the storage room) to a warmer area. As we'll see in a moment, moving heat energy from a colder environment to a warmer environment requires effort (work).

Let's consider an analogy of this process. Assume a small tube and a valve connect two identical containers of water. One container has much more water in it, and thus the water level is much higher in the container. When we open the valve, the water in the container with the highest level begins to flow until the levels are equal. We would never expect water to flow "uphill" in the other direction. This idea is the same at heat transfer. In fact, the equations describing this transfer of water are identical to the example we discussed surrounding Fig. 9.2. If we wanted water to flow from the lower level to the higher level, we need to install a pump to do this work. The same idea is true in *thermodynamics*. We will use a "heat-pump" to move heat energy from a region of lower temperature to that of higher temperature against the natural tendency of heat to flow from the higher temperature object to the lower temperature object.

A "heat-pump" that removes heat energy from an object (thus cooling it) and moving it to somewhere else is commonly called a *refrigerator*. To understand how these devices work, let's explore some general thermodynamic processes first. A useful place to start is on the subject of thermodynamic engines, which convert heat energy into useful work. Further, we will simplify our discussion for the moment by assuming that we are working with an ideal, mono-atomic gas.

9.3.1.1 Introductory Thermodynamics. State Variables and Processes

Consider a cylindrical chamber that has a movable piston, which can change the internal volume V as in Fig. 9.10. We'll assume that the chamber does not leak, and contains a given number of moles of gas atoms, n. Further, we will assume that the gas is at some temperature T, and exerts a net pressure P on the walls of the chamber. Assuming that the gas is ideal, these parameters are related through the ideal gas law:

$$PV = nRT. \tag{9.4}$$

where R is the ideal gas law constant, $R = 8.314 \ \text{m}^3 \ \text{Pa/mol K}$. V is in cubic meters, P is in Pascals, n is in moles, and T is in kelvins.

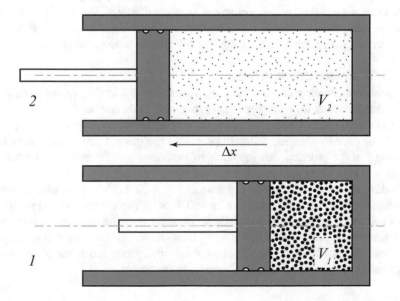

Fig. 9.10 An ideal gas in a cylindrical piston

In thermodynamics, the variables (T, P, V, and n) in the ideal gas law equation (Eq. 9.4) are called *state variables* and the equation itself is an *equation of state*. As the term implies, these variables describe the state of the gas as it is and do not depend on the process that was taken to arrive at that particular set of conditions for the gas. We'll introduce more state variables later, but a complete set of state variables completely describes the gas independent of its past history.

9.3.1.2 Internal Energy and the First Law of Thermodynamics

As long as the gas in Fig. 9.10 is not at absolute zero (0 K), the molecules of gas in the piston will exhibit random motion. The distribution of the "random" speeds of the gas molecules will depend on the temperature and mass of the molecules. This is the origin of thermal energy – the internal motion of the molecules. The gas can also have other avenues of storing and exchanging energy, such as rotational motion for the case of diatomic and larger molecules, or vibrational motion; it's as if the molecules were connected by springs. All of these various internal forms of energy are symbolized by a state function that has the symbol U, *internal energy*. When we *heat* (Q, thermal energy) the gas, we expect the internal energy to change. But we might also use this internal energy to do useful *work* (W). The *first law of thermodynamics* is basically a rule about the conservation of energy:

$$\Delta U = \delta Q - \delta W. \tag{9.5}$$

This equation says, in words, that the *change* in internal energy of a system is equal to the incremental amount of heat put into the system, minus the incremental work done by the system on the environment (and not the other way around). Since the laws of thermodynamics were developed in connection with developing efficient steam engines, it would make sense to think about the work <u>output</u> of a system. Note as well that we can only address changes in internal energy. We usually don't care about the total stored energy in a system, because the changes in energy are what we can observe. Finally, the quantities δQ and δW are not state variables.

9.3.1.3 Thermodynamic Processes

In engines and refrigerators, the working material (a gas in the present discussion) will undergo cycles, following specific processes. Let's consider a variety of process and then see how putting these together will give us an engine or refrigerator. This can be done by applying different processes to a gas in a piston and seeing what comes out of the process. Our selection of processes will be cyclical in nature so that the gas ends up in the same state as we started.

To start, let's first reconsider our piston in Fig. 9.10. Let's assume that the chamber holds 0.1 moles of gas atoms (i.e., $n = 0.1$ mol). The piston starts in position 1 with a volume of 0.01 m^3 (i.e., $V_1 = 0.01$ m^3) and we want the gas to expand, moving the piston to position 2 where $V_2 = 0.1$ m^3. If we do this in a way such that the pressure remains constant, the process is called *isobaric*. Note that as the piston moves from position 1 to position 2, the volume increases.

CHECKPOINT 9.2

In the isobaric process discussed above, what must happen to the temperature of the ideal gas as the volume increases while the pressure stays the same? If we assume that the pressure $P_1 = 20{,}000$ Pa, what are the two temperatures (beginning of the process and end of the process)? (Use the ideal gas law).

Since the gas exerts a constant pressure as it moves, we can find the work done on the external environment. From the definition of work (see Chap. 7),

$$W = F \cdot \Delta x = P \cdot A \cdot \Delta x$$
$$W = P \cdot \Delta V \qquad (9.6)$$

In this equation, A is the cross-sectional area of the piston. In this equation, A is the cross-sectional area of the piston. The important point is that the gas has done useful work on the piston, which could be connected to something else and do work on the outside environment. As we discovered in the checkpoint, there must be an increase in temperature for this to happen, which will require the addition of some thermal energy to the gas. This change in *state* of the system is commonly shown graphically on a pressure-volume (P-V) diagram, such as the isobaric process in Fig. 9.11. Note that, considering Eq. 9.6, the area under the curve from state X to state Y (or position 1 to position 2) is equal to the work done. Also, this is <u>positive</u> work (on the environment) since the force exerted on the environment (the piston) is in the same direction as the displacement.

The amount of heat energy required for this process can be calculated based on the heat capacity ideas presented Chap. 8. For an ideal, mono-atomic gas, the constant pressure specific heat capacity$c_P = \dfrac{5}{2} R$, so the amount of heat energy which

Fig. 9.11 State variables as an ideal gas undergoes an isobaric process from volume V_1 to V_2. The *states* are labeled X and Y, respectively

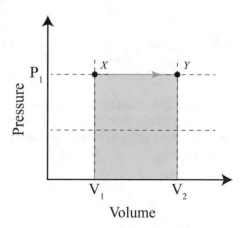

must be added into the system for isobaric processes is given by inserting this equation into a version of the heat capacity equation ($Q = m c \, \Delta T$):

$$\delta Q_P = n\left(\frac{5}{2}R\right)\Delta T, \tag{9.7}$$

where δQ_P is the change in heat energy when the pressure is constant.

An *isochoric* process is one in which the *volume* is held constant. Imagine now that we lower the pressure in the piston chamber, but manage to hold the volume constant. Reviewing the ideal gas law, this means that we will need to lower the temperature of the gas in order to change the pressure of the system. The result would be that we would have to remove thermal energy. We will address that in a moment, but for now consider the work done on the external environment. Since the change in volume ΔV is zero, then the work done is zero. The change in state variables P and V for the gas are shown in Fig. 9.12. It is important to note that the gas is not physically moving around on the PV diagram; the only thing changing is the pressure and temperature of the gas.

For an ideal, mono-atomic gas the constant volume specific heat capacity $c_V = \frac{3}{2}R$, so the amount of heat energy that must be added in order to get an isochoric process to occur is given by

$$\delta Q_V = n\left(\frac{3}{2}R\right)\Delta T. \tag{9.8}$$

Finally, let's consider a process in which both the pressure and the volume change, but do it in such a way that the temperature remains constant. Processes that keep the temperature constant are called *isothermal* processes. Rearranging the ideal gas law,

$$P = nRT\left(\frac{1}{V}\right), \tag{9.9}$$

Fig. 9.12 State variables as an ideal gas undergoes an isochoric process from Pressure P_1 to P_2. The *states* are labeled Y and Z, respectively. Zero work is done on the environment

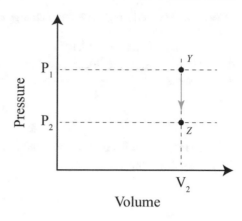

Fig. 9.13 State variables as an ideal gas undergoes an isothermal process from state Z to state X

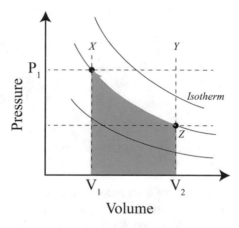

we see that the pressure is proportional to $1/V$. A representative isothermal curve is shown in Fig. 9.13. Actually, we can generate different families of curves that have different temperatures. Just keep in mind that any point on a given curve on the PV diagram has the same temperature. Each of these curves is known as an *isotherm*, where the temperature remains constant.

CHECKPOINT 9.5

Of the three isotherms (lines of constant temperature) shown in Fig. 9.12, which one has the greatest temperature? ...the smallest temperature? What is the temperature for the path indicated from state Z to state X?

The work done in an isothermal process is a bit more difficult to calculate, as it requires integral calculus. The steps are shown here for those that are interested, but the final result is the most important thing to remember. Starting from the definition of work and considering very small changes in volume dV, a small amount of work done is given by

$$\delta W = P \cdot dV. \qquad (9.10)$$

Next, we insert the ideal gas law relationship for pressure. So, the total work done is

$$W = \int_{V_i}^{V_f} nRT \left(\frac{dV}{V} \right). \qquad (9.11)$$

Carrying out the integration from the initial volume (V_i) to the final volume (V_f), we get the important result:

$$W = nRT \cdot \ln \left(\frac{V_f}{V_i} \right). \qquad (9.12)$$

This is the area under the isotherm in Fig. 9.13. In this particular process from state Z to state X note that the initial volume is larger than the final volume.

CHECKPOINT 9.6

Since the initial volume in state Z is 0.1 m^3 and the final volume in our isothermal process, state X, is 0.01 m^3, is the work done positive or negative? What is the work done? Since the internal energy depends only on temperature, what is the change in internal energy (ΔU) for an isothermal process? Given the first law of thermodynamics, what is the change in heat energy for the gas?

If you have been keeping track with the processes to the example gas in our piston, we have returned the gas to the exact same state. At the end of this one cycle, we have absorbed 4500 J of heat energy from a heat source, rejected a total of 3161 J to a cooler object (recall, we must have a temperature difference for heat to flow), and did 1339 J of work in the process. This is summarized in Table 9.3 and graphically in Fig. 9.14. We have just described what is known as a *heat engine*; a system that takes heat energy from a hot object or environment does useful work, and then outputs heat to a colder environment. The heat engine described in the

Table 9.3 Summary of changes made to the example gas

State	State variable P	V	T	Work	Heat
X	20,000 Pa	0.01 m^3	241 K		
Isobaric process to:				1800 J	4500 J
Y	20,000 Pa	0.1 m^3	2406 K		
Isochoric process to:				0 J	−2700 J
Z	2000 Pa	0.1 m^3	241 K		
Isothermal process to:				−461 J	−461 J
X	20,000 Pa	0.01 m^3	241 K		

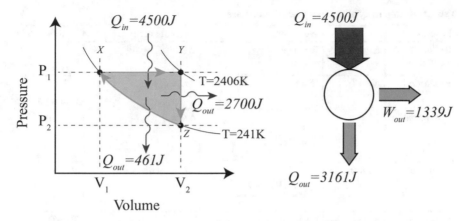

Fig. 9.14 Summary of isobaric, isochoric, and isothermal processes applied to the example mono-atomic gas

example is a thought experiment. No mention was made on how to actually implement this.

The efficiency of most processes can be defined as the ratio of "what you got out" to "what you put in (or paid for)." In the heat engine, the work is what we got out of the device and we had to put in a certain amount of heat energy. So, we can define efficiency (η) as

$$\eta = \frac{W}{Q_{in}},\qquad(9.13)$$

and in the present example the efficiency is $\eta - 0.42$ or as a percent; 42%. All heat engines must reject heat to a cooler environment; this is unavoidable.

Table 9.4 summarizes each of the equations that we've discovered during our exploration of a heat engine. Two additional processes, adiabatic and isentropic, are also listed in the table for reference.

9.3.1.4 Reversible and Irreversible Processes in Thermodynamics

Most laws of physics do not depend on the direction of time. For example, if you watched a video clip of a billiard ball collision (and couldn't see anything else in the video), you might have a difficult time telling if the video was being played forward or backward. In other words, the laws governing the collision do not depend on the direction of time. As another example, the moon orbiting the earth looks the same either forward or reversed. The laws of attraction and motion work for either direction of orbit and are physically indistinguishable.

However, we know that some processes cannot be reversed. For example, a snowflake lands on your hand and melts. This process never happens in the reverse direction *on its own*. As another example, a deck of cards falling off of a table cannot go in reverse on its own (the scattered cards won't restack themselves on a table). Or even more unlikely, dropping a deck of cards and they "randomly"

Table 9.4 Summary of thermodynamic processes

Process		Work done	Internal energy ΔU	Heat δQ
Isobaric	$\Delta P = 0$	$W = P(V_f - V_i)$	$\Delta U = \delta Q - \delta W$	$\delta Q* = nR\dfrac{5}{2}\Delta T$
Isochoric	$\Delta V = 0$	$W = 0$	$\Delta U = \delta Q$	$\delta Q* = nR\dfrac{3}{2}\Delta T$
Isothermal	$\Delta T = 0$	$W = Nk_B T \cdot \ln\left(\dfrac{V_f}{V_i}\right)$	$\Delta U = 0$	$\delta Q = \delta W$
Adiabatic	$\delta Q = 0$	$W = \dfrac{K\left(V_f^{1-\gamma} - V_i^{1-\gamma}\right)}{1-\gamma}$	$\Delta U = -\delta W$	$\delta Q = 0$
Isentropic	$\Delta S = 0$	Reversible adiabatic		$\delta Q = 0$

*Assuming a mono-atomic ideal gas

reshuffle into the perfect poker hand. Consider a car skidding to a stop. Where does the kinetic energy of the car go? The skidding of the tires will heat the tires and pavement. But we know that this process will never spontaneously evolve in reverse. In other words, the reverse process – a car accelerating from rest (but the tires are not rotating) and leaving the road surface colder – *does not naturally happen*.

Natural processes such as these tend to move toward a greater disorder. Anytime there is friction, dissipating other forms of energy into heat energy, or if there is heat transfer from hot to cold, the process is *irreversible*. Dropping and breaking a dish is an irreversible process and illustrates nature's tendency toward increasing disorder. A broken plate on the floor has the same energy as an un-broken plate; same energy but one has greater disorder in the system. As a final example of this process, if we put lukewarm beer in a cooler full of ice, we would never expect the beer to become warmer and the ice to become colder. It is important to underscore that all of the above processes do not violate the first law of thermodynamics, the conservation of energy. So, we need another rule that somehow measures the ways energy can be distributed in a system.

All of the above processes could be reversed, but at great expense. We could build a device that will warm the beer and make ice colder, or we could repair the broken plate but this will require energy and effort from the outside world. All we are saying is that nature does not do these things spontaneously. It turns out that there is another state variable that measures the number of ways energy can be distributed (ordered or not) in a system. It is called *entropy* (S). Just as with the internal energy of a system, the total amount of entropy in a system doesn't matter so much as the changes in entropy in the system. The change in entropy of a system is defined as the amount of heat energy delivered at constant temperature:

$$\Delta S = \frac{\delta Q}{T}. \tag{9.14}$$

Entropy is often tied to the "degree of disorder" of a system or the dispersion of internal energy. The more ways energy can be distributed within internal macrostates of a system, the higher the entropy. Let's consider a basket of tennis balls. If we pour the balls out of the basket, the balls tend to disperse. This analogy is applicable to the diffusion of thermal energy. Consider the warm beer in the ice cooler. As the heat leaves the beer and warms the ice, the heat energy is dispersed among the total contents of the cooler – the entropy of the total system is increasing.

We can use the beer and ice example to show how entropy changes using the above definition. Let's assume that we have 5 kg of ice at 0 °C (273 K, 32 °F) and 5 kg of beer initially at 40 °C (313 K). We put both into a very well insulated cooler. We can then calculate (a) how much ice is left, and (b) the entropy change of the beer, ice, and the system. We start by assuming there is enough ice that the final temperature will be 0 °C, and find the amount of heat required to cool the beer. The amount of heat to be removed from the beer is

$$\delta Q = 5\,\text{kg} \times 4181 \frac{\text{J}}{\text{kg}\cdot\text{K}} \times (-40\,\text{K}) = -836,200\,\text{J}, \tag{9.15}$$

This means that 836,200 J of energy is added to the ice. Considering the latent heat of fusion for water (334 kJ/K), the energy required to melt 5 kg of ice at 0 °C, is

$$Q = 334,000 \frac{\text{J}}{\text{K}} \times 5\,\text{kg} = 1,670,000\,\text{J}, \tag{9.16}$$

so, there is enough ice.

CHECKPOINT 9.7
How much ice of the original 5 kg melted into water and how much remains as ice?

The entropy change of the ice is straightforward to calculate since melting occurs isothermally. The change in entropy is positive since the ice gained thermal energy,

$$\Delta S = \frac{836200\,\text{J}}{273\,\text{K}} = 3063\,\text{J}/\text{K}. \tag{9.17}$$

The entropy change of the beer is a bit more difficult to find since the temperature was not constant. This will take a bit of calculus to show the details, but the final result is useful. Starting with

$$\Delta S = \frac{\delta Q}{T} = \frac{mc \cdot \Delta T}{T}, \tag{9.18}$$

with c the specific heat and m the mass. Approaching the Δ's as small differentials,

$$\Delta S = \int_{Ti}^{Tf} \frac{mc \cdot dT}{T} = mc \ln\left(\frac{T_f}{T_i}\right). \tag{9.19}$$

So, the entropy change for the beer is.

$$\Delta S = 5\text{kg} \cdot 4181 \frac{\text{J}}{\text{kg} \cdot \text{K}} \ln\left(\frac{273}{313}\right) = -2858 \text{J} / \text{K}. \tag{9.20}$$

We note that the entropy change for the ice is greater since it absorbed the same heat energy as the beer lost, but at lower (smaller) temperature. Also note that the *total entropy change for the system* is +205 J/K. As mentioned before, this process is irreversible; we never see beer getting warmer and ice getting colder in a closed system on its own. Although energy is conserved (the net change in energy is zero!), entropy has increased as it always will for isolated, irreversible processes.

We are now in a position to state the second law of thermodynamics in a variety of forms. The Clausius statement (after the German scientist Rudolf Clausius) of the second law is

No process is possible who sole result is the transfer of heat from a colder to a hotter body.

An equivalent form of the second law stated by Lord Kelvin is

No process is possible whose sole result is to transfer a quantity of heat from a body and convert it entirely to work,

which is the same as taking unorganized heat energy and making a more organized energy in the form of work. Notice the words, "sole result." It is possible to move heat energy from a cold object to a warmer object, but this requires an input of work. These two statements are also equivalent to the Planck version of the second law:

Every process occurring in nature proceeds in the sense in which the sum of the entropies of all bodies taking part in the process is increased. In the limit, i.e. for reversible processes, the sum of the entropies remains unchanged.

This essentially means that $\Delta S_{\text{system}} \geq 0$ with the equal sign applying to reversible process. This implies that energy spontaneously disperses from being localized to becoming spread out if it is not hindered from doing so.

9.3.1.5 The Most Efficient Cycle: The Carnot Cycle
We will consider another cycle before tackling the refrigerator. The *Carnot Heat engine* is a four-step cycle where we extract work in exchange for heat. It is another theoretical construct, and our assumptions and the processes in the cycle are such that all steps are reversible. This cycle is important since it determines the absolute theoretical maximum efficiency that is possible within the constraints of the laws of

thermodynamics. We will use the example piston and cylinder again. The Carnot cycle has two isothermal processes which are assumed to occur "slowly" so the gas in the cylinder is always in thermal equilibrium. The cycle also has two *adiabatic* processes. An adiabatic process is one in which the gas absorbs no heat energy. In other words, if we are compressing or expanding the gas we assume that the walls of the cylinder are so well insulated that no heat enters or leaves the system; $\delta Q = 0$. Another way of saying this, an adiabatic process occurs so quickly that the system does not have time to absorb or release heat. This then implies, from the first law of thermodynamics that $\Delta U = -\delta W$. It turns out that the pressure and volume for this process is related by

$$PV^\gamma = \text{constant} = K, \qquad (9.21)$$

where $\gamma = c_P/c_V$. The work done during an adiabatic process is calculated by integrating this over the change in volume; the result of this integration is the equation shown in Table 9.4.

We will assume that the adiabatic process happens in such a way that it is reversible. A *reversible* adiabatic process is also known as an *isentropic* process, where the change in entropy is zero ($\Delta S = 0$).

Using the PV diagram shown in Fig. 9.13, the four steps in a Carnot cycle are:

- **Isothermal expansion**. The working gas is allowed to absorb heat Q_H from a hot object isothermally. During this step, the gas does work on the environment. There is positive work on the environment and the change in entropy for this step is $\Delta S_1 = Q_H/T_H$.
- **Isentropic expansion**. No heat energy enters or leaves the gas during this step. Since this is an expansion (volume increases), the temperature decreases to the temperature of the colder object. Considering that this is an expansion ($V_f > V_i$) and looking at the area under the curve, additional positive work is done on the environment.
- **Isothermal compression**. Starting with this step, the environment does work on the gas, compressing it. It is assumed that the gas is always in thermal equilibrium with the colder object and heat Q_C is transferred. The change in entropy for this step is $\Delta S_2 = Q_C/T_C$ and is the same as ΔS_1.
- **Isentropic compression**. Again no heat energy enters or leaves the gas in this step. Since this is a compression, the temperature of the gas increases to that of the hot reservoir. Just like step 3, there is negative work done by the gas on the environment.

At the end of the cycle, there is a net positive amount of work done on the environment. The work done is the area of the squashed "rectangle" enclosed by the paths drawn in the PV diagram shown in Fig. 9.15.

Carnot was able to show that this cycle has the maximum possible efficiency under the restrictions of the second law of thermodynamics. The efficiency of such a heat engine is given by the equation:

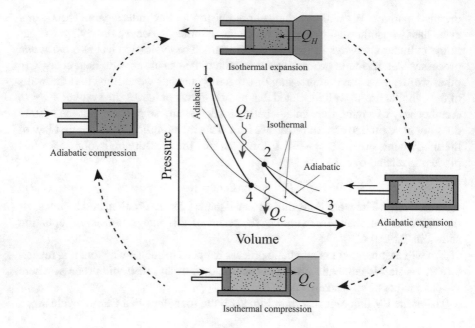

Fig. 9.15 Carnot cycle (heat engine) on a PV diagram

$$\eta = \frac{W}{Q_{in}}$$
$$= 1 - \frac{T_C}{T_H},$$

(9.22)

where T_H is the temperature of the hot object
T_C is the temperature of the cold object

CHECKPOINT 9.8
What is the efficiency of a Carnot heat engine when the temperatures of the hot and cold objects are same? What temperatures would you chose to maximize efficiency?

The Carnot cycle is completely reversible. To illustrate this, let's consider the above steps, but in reverse. Now the environment does work on the system (or the system does negative work on the environment) and we are moving heat energy from the colder object to the hotter object. This is the same process that needs to occur in a refrigerator; heat is moved out of the cold interior of the refrigerator and moved to the outside of the refrigerator where it is hot already. Using the PV diagram shown in Fig. 9.16, the four steps in a Carnot refrigeration cycle are:

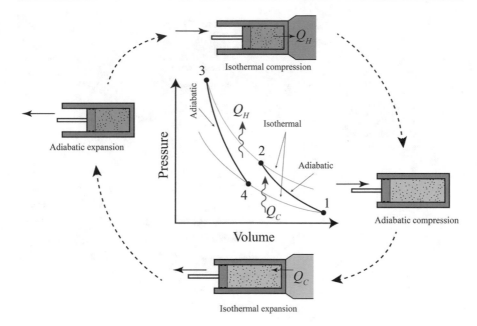

Fig. 9.16 Carnot refrigeration cycle on a PV diagram. Note that the starting point has changed from Fig. 9.15

Starting from the bottom of Fig. 9.16 at point 1, **Isentropic compression**. No heat energy enters or leaves the gas. Since this is a compression, the temperature increases to the temperature of the hotter object.

Isothermal compression. Now the environment does work on the gas, compressing it. It is assumed that the gas is always in thermal equilibrium with the hotter object and heat Q_H is transferred.

Isentropic expansion. Again no heat energy enters or leaves the gas. Since this is an expansion, the temperature of the gas decreases to that of the hot reservoir.

Isothermal expansion. The working gas is allowed to absorb heat Q_C from a cold object isothermally.

Since these processes are either constant temperature or constant entropy, it is more convenient to illustrate the process on a temperature – entropy (T-S) diagram, as in Fig. 9.17, instead of a P-V diagram.

Using the notion of efficiency, we can define a "coefficient of performance" (COP) for a refrigeration cycle by the ratio of "what you get" versus "what you pay for" (see Fig. 9.18). What we get in a refrigeration system is an amount of heat Q_C removed, and we had to add in a certain amount of work. Since, by definition, work is defined as positive when work is done on the environment by the system, the COP is

$$COP = \frac{Q_C}{-W}.$$

(9.23)

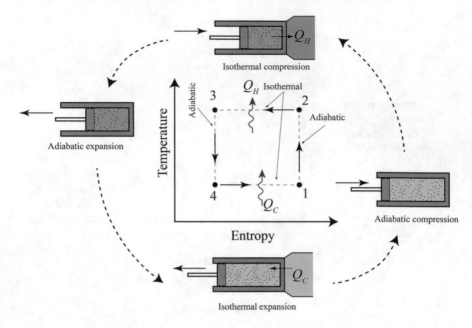

Fig. 9.17 Carnot refrigeration cycle on a TS diagram

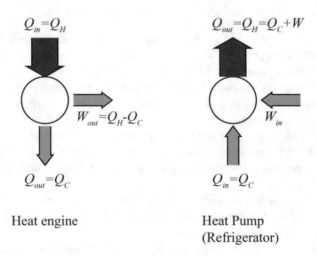

Fig. 9.18 Effective operation of a heat engine or refrigerator

Using the Carnot cycle, it can be shown that the maximum *COP* is given by

$$COP = \frac{T_H}{T_H - T_C},\tag{9.24}$$

which can be greater than one.

What steps do we need to do to implement the Carnot refrigeration cycle and use it to cool our beer? First, we need to find a process that absorbs or releases heat at

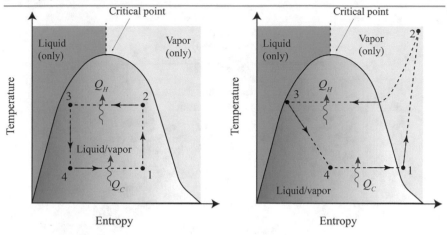

Fig. 9.19 Left: Ideal Carnot refrigeration cycle. Right: practical implementation

constant temperature. Fortunately, when a substance changes from a liquid to a vapor, the temperature remains constant. (You can verify this by measuring the temperature of ice water as the ice is melting. It stays constant the entire time until all of the ice is gone.) The energy added to cause this change does not affect the temperature. Instead, it separates the molecules of the substance so that the liquid becomes vapor. The latent heat of vaporization is the value used to determine the amount of energy required to convert the liquid into vapor per unit of substance. So, for the isothermal processes, we just need to "boil" (evaporate) the fluid during the isothermal expansion (4→1), and "condense" the fluid in the isothermal compression (2→3). Physical implementation of isothermal heat transfer is easy with heat exchanger.

Figure 9.19 shows a typical phase diagram for most substances, with the Carnot cycle superimposed. To the left of the bell-shaped curve, the fluid is a liquid and to the right it is a pure vapor. But at intermediate entropies inside the bell-shaped area, it is a mixture of liquid and vapor. We've placed the Carnot cycle in this area since we want the isothermal processes to just change the phase of the fluid. However, there are issues with actually implementing this for the compression and expansion phases. For instance, during the adiabatic compression (1→2 in Fig. 9.19) we would require a device that could handle both liquid and vapor at the same time. A compressor that can handle both phases would be relatively expensive. The same is true for the adiabatic expansion (3→4). We would require a device to extract work from the fluid, such as a turbine. Again, an extra device that could handle both phases would be impractical. So, we'll make some compromises. First, we'll use a free-expansion for the adiabatic expansion portion. This is easy, just let the working fluid flow through a restriction and expand into a lower pressure area. Unfortunately, this process is not reversible and therefore not isentropic, but it is mechanically simple and cheap. This modifies the cycle diagram slightly as shown in Fig. 9.19. Secondly, we design the system so that the isothermal expansion ("the boiling") completely converts the working fluid into a vapor. Then we just need a relatively simple and

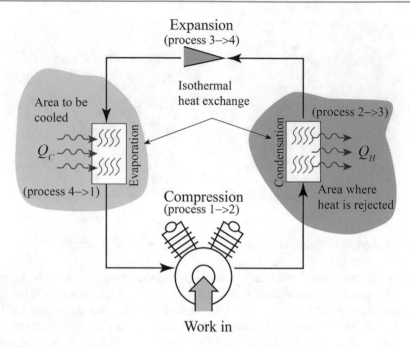

Fig. 9.20   Physical implementation of the approximate Carnot refrigeration cycle

inexpensive compressor that will only need to handle the gas phase of the working fluid. Again, this will modify the final cycle, but the solution is practical and easy to implement (Fig. 9.20).

When working with refrigeration cycles, it is convenient to work with another state variable known as, enthalpy. *Enthalpy*, *H*, is thermodynamically equivalent to the heat content of the fluid. Rearranging the first law of thermodynamics, we get

$$H = U + P\Delta V. \tag{9.25}$$

Enthalpy can be viewed as the energy needed to create our system in a certain state (the internal energy) plus push away the environment to make room for it (the work, $P\Delta V$ term). Specific enthalpy is another useful term. It is simply the enthalpy divided by the mass of the liquid (in kg). So specific enthalpy has units of J/kg:

$$h = \frac{H}{m}. \tag{9.26}$$

The nice thing about working with enthalpies is that these values are already known for many different refrigerants. In this context, the state diagram for the working fluid is plotted on a pressure-enthalpy (P-H) diagram as in the example shown in Fig. 9.21.

A few words are in order for the P-H diagram. It is similar to the T-S diagram in that the various states of the working fluid are separated: liquid, liquid and vapor, and just vapor. Looking at Fig. 9.21, the "tongue-shaped" curve represents the

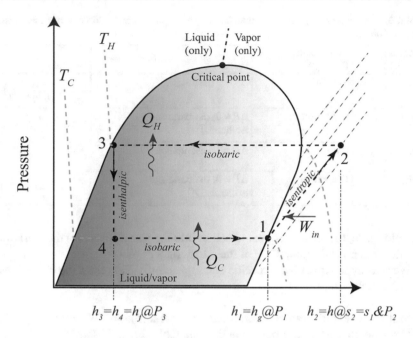

Fig. 9.21 Modified Carnot refrigeration cycle – the vapor compression cycle

boundary for a saturated liquid on the left side, and for a saturated vapor on the right side. The modified refrigeration cycle we discovered in Fig. 9.19 is also shown in the figure. We still consider the process 1→2 as an isentropic ($\Delta S = 0$) compression. However, we know that the expansion 3→4 is not isentropic because it is not reversible. But it is *isenthalpic* ($\Delta H = 0$) since there is no heat exchange with the environment. Also shown in the figure are representative lines of constant temperature (isotherms), and lines of constant entropy.

The calculation of the rate of heat removal, or work done by the system, is fairly easy to calculate once each of the enthalpies is known. The *rate* that energy is exchanged with the system is commonly expressed in terms of the *rate* of mass flow:

$$\frac{dE}{dt} = \frac{dm}{dt}\left(h_A - h_B\right), \tag{9.27}$$

using the change in specific enthalpy at two states A and B. Energy could be the exchange of heat or work done depending on the process in question (Table 9.5).

Let's consider an example and calculate some values for our refrigerator. Consider a refrigeration cycle that uses the refrigerant 1,1,1,2-tetrafluoroethane (known as R-134a). It operates on the vapor-compression cycle like that described above in Fig. 9.21. The high-pressure side of the cycle (process 2→3) is 0.9 MPa and the refrigerant leaves the evaporator (at point 1) at −20 °C. If the mass flow rate of the refrigerant around the loop is 3 *kg/s* what is (a) the rate of heat removal from

Table. 9.5 Summary of first and second laws applied to an ideal, closed vapor compression refrigeration cycle

Process in Fig. 9.18	Component in Fig. 9.19	Process	Formulae
1→2	Compressor	$\Delta S = 0$ (isentropic)	$\dfrac{dW}{dt} = \dfrac{dm}{dt}(h_2 - h_1)$
2→3	Condenser	$\Delta P = 0$; (isobaric and isothermal)	$\dfrac{dQ_H}{dt} = \dfrac{dm}{dt}(h_2 - h_3)$
3→4	Expansion	$\Delta H \cong 0$; (isenthalpic)	$h_3 = h_4$
4→1	Evaporator	$\Delta P = 0$; (isobaric and isothermal)	$\dfrac{dQ_C}{dt} = \dfrac{dm}{dt}(h_1 - h_4)$

the cold space, (b) the rate of heat rejection into the warm space, (c) the work input into the system, and (d) the COP of the refrigerator?

We start by looking up the specific enthalpies for R-134a from the data tables to describe each state.

- **State 1**: The refrigerant is a *saturated vapor* at this point. We are given that the temperature is $-20\,°C$. Looking at the data tables for R-134a, we find:
  - $P_1 = 132.8$ kPa.
  - $h_1 = h_g$ at this pressure and temperature, so $h_1 = 238.43$ kJ/kg.
  - $s_1 = s_g$ at this pressure and temperature, so $s_1 = 0.9457$ kJ/(kg K). We will need this for state 2.
- **State 2**: The refrigerant is a *superheated vapor* at this point. But, since we assume that the process 1→2 is isentropic, we conclude:
  - $s_2 = s_1 = 0.9457$ kJ/(kg K).
  - Now we look up the enthalpy based on the pressure at this point and specific entropy. Since $P_2 = 900$ kPa is given in the statement of the problem and we now know s_2, we extrapolate from the data on the Internet to get $h_2 \cong 276.3$ kJ/kg and $T_2 \cong 42.5\,°C$.
- **State 3**. The working fluid is now a *saturated liquid*. At the pressure given, $P_3 = P_2 = 900$ kPa we can get from the data tables:
  - $T_3 = 35.5\,°C$.
  - $h_3 = h_f$ at this pressure and temperature so, $h_3 = 101.62$ kJ/kg.
- **State 4**. After the expansion, the refrigerant is a *mixture* of liquid and vapor at the lower temperature ($-20\,°C$). Since the process is isenthalpic:
  - $h_4 = h_3 = 101.62$ kJ/kg.

We are now in a position to calculate the requested information:

(a)   …the rate of heat removal from the cold space

$$\frac{dQ_C}{dt} = \frac{dm}{dt}(h_1 - h_4)$$

$$\frac{dQ_C}{dt} = 3\frac{\text{kg}}{\text{s}}\left(238.43\frac{\text{kJ}}{\text{kg}} - 101.62\frac{\text{kJ}}{\text{kg}}\right) = 410.43\frac{\text{kJ}}{\text{s}} = 410.43\text{kW}$$

(b)   ...the rate of heat rejection into the warm space

$$\frac{dQ_H}{dt} = \frac{dm}{dt}(h_2 - h_3)$$

$$\frac{dQ_H}{dt} = 3\frac{kg}{s}\left(276.3\frac{kJ}{kg} - 101.62\frac{kJ}{kg}\right) = 524.04 kW$$

(c)   ...the work input into the system

$$\frac{dW}{dt} = \frac{dm}{dt}(h_2 - h_1)$$

$$\frac{dW}{dt} = 3\frac{kg}{s}\left(276.3\frac{kJ}{kg} - 238.43\frac{kJ}{kg}\right) = 113.61 kW$$

Note that we could have obtained this from $\dfrac{dW}{dt} = \dfrac{dQ_H}{dt} - \dfrac{dQ_C}{dt}$

(d)   ...the *COP* of the refrigerator?

$$COP = \frac{Q_C / dt}{-W / dt}$$

$$COP = \frac{410.43 kW}{113.61 kw} = 3.61$$

9.3.1.6 Type of Refrigerants

CHECKPOINT 9.9
Using the temperatures in the example, calculate the theoretical maximum *COP* assuming a Carnot cycle.

The working fluid in a refrigerator needs to have a high latent heat of vaporization, and a boiling temperature near the lowest temperature expected. Early refrigerants used ammonia, sulfur dioxide, methyl chloride, or propane. These substances are either extremely toxic or flammable and no longer widely used. The first "safe" refrigerant was invented in 1928 and marketed under the trade name Freon (R-12), a chlorofluoro-carbon (CFC). It was discovered that when CFCs were released into the atmosphere, they generated free chlorine atoms (chlorine radicals). This has a serious detrimental effect to the ozone layer protecting the Earth. Newer versions of refrigerants are hydro-genated CFCs (HCFC), such as R-22 and R-123, or hydrofluorocarbon (HFC) such as R-134a. These compounds tend not to release chlorine atoms as readily as the CFC's, but it has been recently recognized that these gasses are significant greenhouse gases. Greenhouse gases in essence reflect the heat generated by the Earth back to the surface of the Earth causing it to warm. Given these problems, US manufacture of CFC-based refrigerants has been banned since 1996. Further, HCFCs are being phased out with an anticipated complete phase out by 2020. The search is on for better refrigerant.

9.3.1.7 Mechanical Implementation of Refrigeration. Glycol Circulation

Closed loop, vapor evaporation refrigeration systems tend to use refrigerants that are either toxic, harmful to the environment, or both if the working fluid is vented to the atmosphere. So, how do we cool our fermenters and other equipment? We need to be able to configure, and reconfigure how the devices are being cooled in the brewery. Making and breaking connections with refrigerant in them is not possible (unless one is certified and using special equipment, it's against the law!). Also, who wants CFCs or ammonia accidentally getting into their beer?

The solution is to separate the working fluid of the refrigeration cycle by means of a glycol circulation system. Not that glycol is particularly healthy in beer if it accidentally goes in, but it doesn't pose the same environmental risks if it is accidentally spilled or released. Figure 9.22 shows a simplified glycol circulation system. A pump circulates glycol through a heat exchanger with the refrigeration cycle before it is pumped to the device(s) to be cooled.

This type of system is common in breweries, because the primary refrigerant can be contained in a particular area and the secondary coolant (glycol in this case) can be used throughout the brewery by simply pumping it in insulated pipes. The system only requires the addition of one refrigerator to the brewery instead of one for every fermenter.

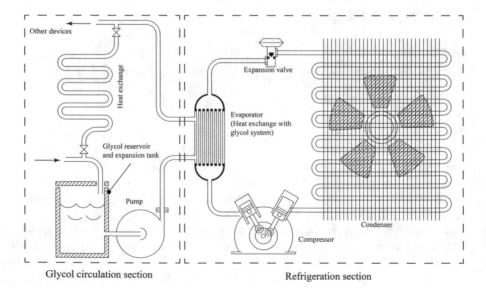

Glycol circulation section   Refrigeration section

Fig. 9.22 Simplified glycol circulation-refrigeration system

9.3.2 Fermenters, CCV, and Round Squares

Early brewers used vats for fermentation. These vessels were open to the air and environment. But this was done because brewers did not understand that yeast were needed to do the fermentation. Any attempt to ferment the wort in a closed vessel didn't result in a good beer. Yeast were added to these open vessels by either spontaneous fermentation (by the natural inoculation of yeast that happened to be floating in the air), addition of *krausen* to the wort (krausen is the *barm* from another wort that was already actively fermenting), or by the innoculation of the wort using an *ale stick* or fermenting vessel that had previously been used to make beer. In the case of the ale stick, yeast from an earlier batch that had become imbedded in the stick were able to start the fermentation. Ale sticks, in fact, were highly guarded family heirlooms that were passed from father to son. Loss of an ale stick was disastrous to the family that needed their beer.

As brewers' techniques improved, vessels were created to not only industrialize and standardize the brewing process, but to increase the efficiency and reduce the risk of contamination from unwanted microbes floating in the air. Many different styles of fermenter were explored. The more successful ones are discussed here.

Open/Square Fermenter One of the first fermenters used by brewers were open fermenters. Open fermenters are exactly what their name says they are. They do not have a top or cover. Although a lid can sometimes be added to them, it doesn't make a tight seal. Instead, if a lid is added, it is meant to keep things out, rather than keep things in. For example, the lid may be closed when a neighboring fermenter is being cleaned. The brewer adds the cooled wort to the open fermenter by splashing it into the vessel. This helps to aerate the wort.

Then, the yeast is pitched by simply pouring it into the vessel. If the vessel is rather large, it can be added as the wort is being added, so that the two mix together uniformly. Fermentation occurs and a thick *barm* forms on top of the fermenting liquid. Barm is the name given to foam containing active yeast that rises above the liquid. If the brewer wishes to harvest the yeast off of the open fermenter, they can scrape off the layer of barm as it forms. The first scrapings are typically thrown away as they may be contaminated. The second scrapings go into a yeast *brink* to save for pitching the next batch.

The yeast brink is a small vessel that can hold suspended yeast. It typically has wheels or can be physically moved. The brink is stored in a refrigerator to keep the yeast cold. A manual stirrer is usually placed into the brink to keep the yeast suspended. In some cases, the brewer may wish to wash the yeast to remove unwanted wort or reduce the number of bacteria. Washing can be done with high-quality liquor or with acidic liquor. The yeast can survive fairly low pH liquor for a short time, where the bacteria cannot.

Fermentation in an open container results in an uncarbonated green beer. Conditioning must still be performed. In addition, since the open fermenter doesn't have a built-in cooling or warming system, the fermentation typically takes place at the temperature of the room. Modern open fermenters are usually square or

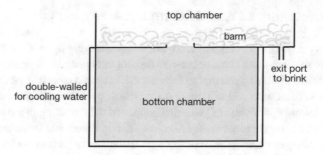

Fig. 9.23 Traditional Yorkshire square design

rectangular in shape, are jacketed so that they can be cooled as needed, and are only 1–2 m deep. This allows regulation of the temperature without much input from the glycol. In addition, the shallow depth to the fermenter increases the amount of esters and other flavors produced by the yeast (compared to deep fermenters that are greater than 2 m deep.)

Yorkshire Square The Yorkshire square was likely invented by Timothy Bentley (1768–1830) around 1795 when he opened the Lockwood Brewery near the present-day town of Huddersfield, West Yorkshire, England. This vessel was originally constructed from sandstone, but slate eventually became the norm. The stone was cut into panels that formed a double-decked square system (Fig. 9.23). The bottom vessel was double-walled with a space between the walls to allow water to pass. The water was circulated through the gap so that the temperature of the square could be maintained.

Yeast is pitched into the wort in the bottom vessel of the square. As it ferments, the yeast barm rises through the manhole in the center of the top deck. Here the barm settles and the yeast can be harvested. Periodically, the yeast is *roused* by pumping it up to the top vessel and spraying it through a fan-shaped sprayer to mix with the yeast. This rouses the yeast and increases the rate of fermentation. Then the manhole is opened and the slurry splashes back into the bottom vessel. This process can be repeated a couple of times until the wort is nearly free of sugars. At this point, the beer can be transferred into casks or kegs to finish the fermentation and carbonate the beer. The yeast can be harvested from the top vessel after the wort has been returned to the bottom half of the vessel.

This system of fermentation experienced widespread use in parts of England. However, development of new methods and processes has resulted in the replacement of this fermenter with other technology. A few breweries still use the Yorkshire square, but they tend to have only a cult following. Due to problems that include cleaning the lower chamber, modern squares tend to be made from stainless steel with a jacketed lower chamber. These modern systems (known as "round-squares") also tend to be round in shape to aid with cleaning.

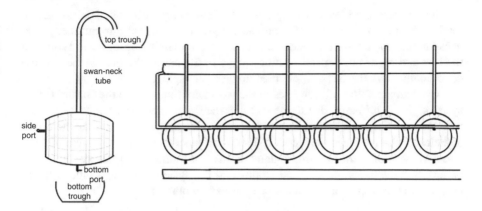

Fig. 9.24 Burton Union system. Barm rises through the gooseneck (or swan-neck) tube from the cask into the top trough. The liquid settles and is returned to the casks via a side port. Yeast remains in the top trough. Once complete, the beer is drained from the casks into the bottom trough where it runs to the conditioning tank for treatment and packaging

Double-Drop   The issues with the Yorkshire square design led to the development of the double-drop system for fermentation in the late 1800s. In this system, the cooled wort is pumped (or "dropped") into a fermenter. The action of pumping the wort splashes it with air and helps to add oxygen into the wort. About 14 h after pitching the yeast, the actively fermenting wort is dropped by gravity into another fermenter and allowed to finish fermentation. The process removes the actively fermenting wort away from remaining trub and dead yeast, resulting in a cleaner fermentation and clearer beer.

This system isn't extremely common but has found use in many breweries. The second drop re-aerates the wort and encourages further growth, much like the rousing that is done in the Yorkshire square system. And, just like the Yorkshire square system, the double-drop causes the production of esters and other compounds that can contribute to the flavor of the beer.

Burton Union   The Burton Union system was created in the 1830s as a way to perform the same process as the Yorkshire square, but in more readily available wooden casks. This system involves pitching yeast into an open fermenter containing fresh wort. After 12–24 h, the actively fermenting wort is then fed by gravity into a series of wooden casks (Fig. 9.24). A gooseneck-shaped tube is then attached to the cask that ends in a trough above the casks. As the fermentation continues, the barm rises into the gooseneck and drops into the trough. Here it is separated, wort drips back down into the casks, while much of the yeast remain behind in the trough.

This process clarifies the beer quickly. Remaining trub and dead yeast are retained in the original fermenter, and as the yeast are pushed through the goose-necks in the wood cask fermenters, the amount of remaining un-dissolved solids in the beer are removed. Typically, after 6 days in the casks, the beer is moved to a conditioning tank for carbonation, blending, or packaging.

Less than a handful of breweries in the world continue to use the Burton Union system. This is primarily due to the fact that the system requires constant care and maintenance. Cleaning between runs must be done with only hot water as chemicals and cleaners can damage the wood. In fact, Marston's Brewery in Wolverhampton, just northwest of Birmingham, England, holds fast to the use of this system. They believe that their mixture of yeast strains and the system provide a unique flavor profile for their beers that could not be obtained by other methods.

Cylindroconical Vessels (CCV) By far the most common of the fermenters are the closed fermentation systems known as cylindroconical vessels (CCVs) or uni-tanks. These were well on their way toward replacing other fermentation systems, and by the mid-1900s were well entrenched in the fermentation process. These vessels allow the fermentation process to be conducted at a slightly higher pressure (which allows the fermentation to proceed faster). In addition, the shape of the base of the vessel provides a way to cleaning remove any trub, dead yeast, or even harvest clean yeast. Because of this, the CCV can be used as the primary fermenter, secondary fermenter, and even the conditioning tank (hence the term uni-tank).

The vessel, shown in Fig. 9.25, is jacketed to allow the temperature of the wort inside to be controlled. Zones along the sides and bottom of the vessel also provide additional control to the fermentation. During active fermentation, the yeast rise through the center of the vessel and then cascade down along the side of the vessel where the yeast are then deposited on the bottom. The slope of the bottom forces the yeast to collect in the cone. Then, by simply opening a valve, the brewer can collect it as needed.

The CCV also has a CIP ball in place to clean the interior of the vessel with limited labor. In fact, since computer servomotors can control all of the valves, the use of the CCV helps automate the entire brewery. They are inexpensive, easy to maintain, and occupy significantly less floor space than any of the open fermenters. With appropriate insulation, they can even be placed outside of the brewery. The larger of these fermenters can hold 6000 hL or more. The pressure on the yeast at the bottom of the fermenter limits the maximum size of the fermenter.

CHECKPOINT 9.10

Provide a drawing of a typical yeast brink. Plan how you would add the yeast from an open fermenter into the brink, and consider how the yeast could be pitched from the brink into another fermentation.

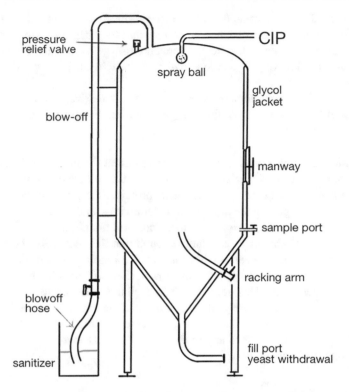

Fig. 9.25 The cylindroconical vessel (CCV). The blow-off tube is connected to a hose that is placed in a bucket of sanitizer. Bubbles of CO2 can be seen exiting through the bucket. The racking arm is a tube that can be rotated to withdraw clear beer from the fermenter without transferring yeast. The bottom port serves as the fill port and the yeast withdrawal port. Zones within the glycol jacket allow different regions of the fermenter to be cooled

9.4 Yeast

Yeast are single-celled microorganisms, classified as fungi. As we've already discovered, there are two major species of yeast that are useful in brewing (and many others that aren't). Those two species are *Saccharomyces cerevisiae* (aka ale yeast) and *Saccharomyces pastoranus* (aka lager yeast). Each of these species has a large number of variants or cultivars. Each of these cultivars has been developed based upon the desired final characteristics of the brewed beer. The characteristics actually define the cultivar and determine if the particular yeast is suitable for a given style of beer. Often these are very proprietary. Breweries that develop a specific cultivar for one of their own brews tend to be very protective.

Yeast are facultative anaerobes. This means that they are able to grow in the presence or absence of oxygen. The brewer uses this to their advantage. When they begin the fermentation process, they add oxygen to the wort. Under these aerobic conditions the yeast begin to grow, increasing their numbers until the oxygen has been consumed. Some carbon dioxide is also produced as they uptake fermentable sugars. At this point, their metabolism adjusts and they begin making ethanol and carbon dioxide. Eventually, the alcohol concentration gets too high or the amount of sugars still in solution gets too low, and the yeast begin to come together and flocculate.

Brewers and microbiologists develop new yeast cultivars to enhance certain properties of the yeast as it goes through fermentation. Focus tends to be on the following key features.

Flocculation   As yeast enter dormancy during and at the end of the fermentation process, they flocculate. Proteins that extend from the surface of the yeast cell point outward when calcium binds to them. These proteins interact with those on neighboring yeast cells, again in the presence of calcium ions, and "stick" together. As the yeast enter dormancy, they gather together and form a large mass, with each yeast cell stuck together. This mass gets large enough that the yeast cells can't stay suspended in the fermenting beer. So they fall out of solution and collect at the bottom of the vessel. If the yeast flocculates too slowly, they will remain suspended in the beer after fermentation. This could result in a cloudy beer at the tap. If the yeast flocculate too quickly, they may not hang around in the beer long enough to convert all of the fermentable sugars into alcohol, or worse, that the yeast don't remain suspended long enough to absorb any of the bad off-flavors that may have been produced during the fermentation.

Attenuation   This is a term that is used to describe how much of the fermentable sugars are converted into alcohol. As yeast conduct their reactions, not all of the sugars are used to make alcohol; some of the sugar is used for other things (as we'll see later in this chapter). In addition, attenuation can also refer to the ability of the yeast to continue to ferment sugars until they are all gone. Some cultivars of yeast turn dormant when the concentration of sugar in the wort gets low.

Operation Temperature   The optimum temperature for the yeast to thrive in the sugar solution is another of the features that can be selected for when new cultivars are prepared. The temperatures can be as low as 40 °F or as high as 78 °F (or even more separated than this). The temperature range for a cultivar determines if that yeast will be used in creating a lager (low temperatures) or an ale (high temperatures).

Alcohol Tolerance   Many yeast cultivars are fairly resistant to the presence of alcohol in the fermenting wort. Others are no so resistant. Alcohol is toxic to yeast. Just look at the ingredients on the hand sanitizer that you use. Some cultivars stay active and survive in high concentrations of alcohols before they flocculate. Others flocculate quickly. The range of alcohol concentrations can go from as low as about 5% all the way to about 14% (or even higher).

Yeast are incredible creatures. They are small enough that the unaided eye can't see them, but cause such significant changes to a batch of sugar water in very short order. They execute their changes by growing and reproducing into the millions per milliliter. It is the sheer number of these creatures that do the work. Let's look into the yeast cell a little closer to learn more about them.

9.4.1 Yeast Morphology

The typical brewer's yeast is roughly 5–10 μm (a μm is a micrometer and often listed as "microns") in diameter. Under the microscope these tiny cells are somewhat visible. They have no color and because they're so small brewers often add a colorizing agent in order to see them better. The typical colorizing agent is methylene blue. This compound stains the cells, and leaves the background wort in which the yeast live slightly blue in color. The methylene blue stain also has another use (in addition to making the yeast visible). We'll uncover that later in this section.

Yeast have a cell wall connected to a membrane immediately inside the cell wall. The cell wall, which is about 25 nm (1 nanometer is 10–9 m), is 0.25–0.50% of the entire distance across the cell. Even though it's relatively small, it makes up about 25% of the dry weight of the cell. The cell wall contains regions that are primarily made up of different glucose polymers. The inner-most region is almost entirely β-glucan. β-Glucan, in fact, is the major component of this and the middle regions of the cell wall. However, at the inner-most region, it is the sole component. As we uncovered earlier in this text, β-glucan is a structural molecule. It has linkages that are not easily broken and the molecule adopts a very linear structure.

The middle of the cell wall is a region that contains some mannan interspersed in the β-glucan matrix. Mannan is a polymer of mannose with α-[1 → 6] linkages along the main chain of the polymer, and branches with α-[1 → 2] and α-[1 → 3] linkages (see Fig. 9.26). In addition, this region of the cell wall contains some proteins.

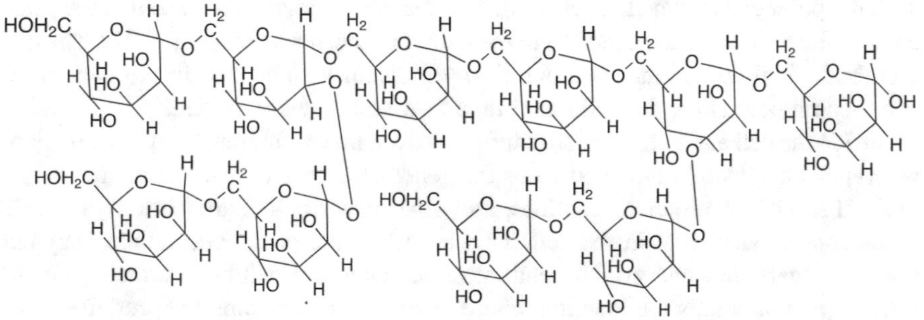

Fig. 9.26 The structure of mannan, a polymer of mannose. Note the similarities and differences to starch (Fig. 6.11)

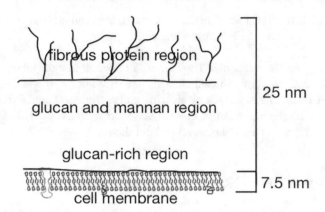

Fig. 9.27 Cell membrane

The outer-most region of the cell wall contains mostly proteins and enzymes. Many of these proteins have small chains of mannan or mannose bound to them. The enzymes in this region include invertase (an enzyme that hydrolyzes sucrose into glucose and fructose), β-glucanase and mannosidase (enzymes that can hydrolyze β-glucan and mannan into smaller pieces; likely, these enzyme help break down the cell wall during the growth stage of yeast), and lipase (which hydrolyzes fats and lipids). The proteins include Flocculation protein 11 (aka FLO11), a calcium-binding protein that helps the yeast adhere to other yeast cells during flocculation.

The cell membrane is about 7.5 nm thick. It is made up of phospholipids, sphingolipids, and sterols. These molecules are relatively linear in their shape. At one end of the molecule are oxygen, nitrogen, and phosphorus atoms. The rest of the molecule tends to be carbon based. The oxygen, nitrogen, and phosphorus atoms tend to form hydrogen bonds with molecules such as water or carbohydrates. These are polar molecules that interact with the polar end of the cell wall molecules. The long carbon chain region of the cell membrane molecules tend to be non-polar in nature and don't interact with water or the cell wall very well.

The cell membrane sets up as two layers of these molecules (known as a lipid bilayer). The first layer is oriented so that the cell membrane molecules are oriented with the polar ends facing the cell wall. The second layer is oriented so that the polar ends of the molecules are oriented to the center of the yeast cell (Fig. 9.27). The cell membrane is fluid and moves around. Any molecules embedded in the membrane can then migrate along the cell membrane to where they are needed.

Inside the cell are a series of structures, like organs in a human body. The nucleus is wrapped in a membrane. It contains the genetic material (DNA, RNA) that define the makeup of the proteins, enzymes, and other material that are made by the cell. Mitochondria also exist in the cell. This is the engine of the cell where enzymes convert sugars into energy. The endoplasmic reticulum, which is attached to the nucleus membrane, is the location where enzymes and proteins are produced. The vacuole can be thought of as a large sack. It serves as a storage vessel for sugars and other compounds that the cell needs as nutrients. Finally, lipid granules can also be found in the cell. These are essentially fat storage for long-term energy needs by the cell.

As the cell grows, it softens an area of the cell wall. Then, it makes more cell wall material and pushes out until a small bud is formed. The vacuole separates into a large number of smaller sacks. Some of these migrate into the bud in order to become the vacuole for the new yeast cell. The DNA duplicates itself and the nucleus splits into two; one stays behind while the other ends up in the bud. Finally, when all of the organs have been made, the cell wall pinches shut and the *daughter cell* is formed. The cell wall, however, is scarred during the process. Repair of the cell wall leaves a ring of chitin (a polymer of aminoglucose) in the cell wall. Similarly, the new daughter cell is marked with a *birth scar*. This hard polymer cannot be broken down. Thus, when the yeast has budded multiple times so that the entire surface of the cell is marked with *bud scars*, it cannot produce another bud. The entire process, in sufficient oxygen, takes 60–120 min.

9.4.2 Yeast Metabolism

Yeast are pitched into the fermenter and become suspended in the sugary wort. We'll recall that the wort at this stage is the perfect temperature for yeast to grow and is laden with fermentable sugars, amino acids, and other nutrients that they yeast need to grow. In addition, the wort is fully oxygenated with approximately 8 ppm to 10 ppm oxygen. The yeast at this point begin to grow, and while the oxygen is present, they enter something we call aerobic metabolism. Aerobic conditions indicate the presence of oxygen.

The yeast absorb oxygen in the solution as they grow. This depletes the concentration of oxygen in the solution. So, after some rapid growth, the yeast have eliminated all of the oxygen in the solution. This triggers the yeast to enter anaerobic metabolism. Under anaerobic conditions (conditions where oxygen does not exist), the yeast continue to consume fermentable sugars but convert them into ethanol and carbon dioxide.

9.4.2.1 Aerobic Conditions

Aerobic conditions refer to the presence of oxygen in the medium. In these conditions, the yeast take up the fermentable sugars in the solution. Those sugars include glucose, maltose, and maltotriose. In some cases, the brewer may have added sugar to the wort. In the first steps, the yeast utilize the sucrose through the action of the invertase that is found in the outside of the cell wall. Invertase hydrolyzes sucrose into glucose and fructose. It has been found that glucose is the first carbohydrate to be taken up by the cell at a rate twice as fast as fructose. Then fructose, maltose, and maltotriose in that order, are consumed. There is some overlap to their rates of uptake.

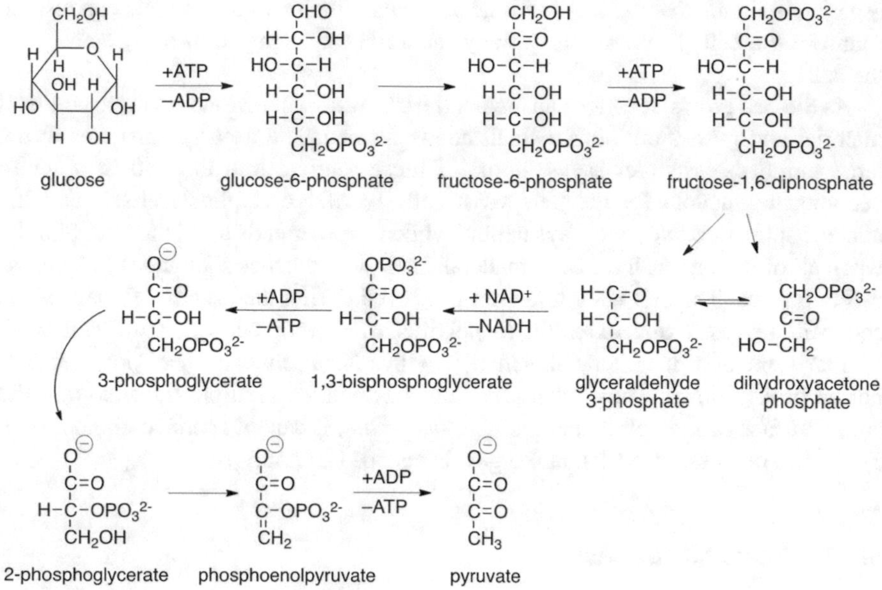

Fig. 9.28 Glycolysis. The input is glucose and the outcome is two molecules of pyruvate

Once inside the cell, a special hydrolase (enzyme that adds water across a bond) converts maltose into two molecules of glucose. Similarly, maltotriose is broken down into three molecules of glucose.

Glucose is metabolized via glycolysis. This is a pathway that converts glucose into small building blocks that can be used to make other molecules that the yeast need to grow. Fig. 9.28 outlines each of the steps in glycolysis. The first step is to add a phosphate group to the molecule to improve solubility. The second step converts the glucose molecule into a fructose molecule. Another phosphate group is added to make fructose-1,6-diphosphate. This molecule is cleaved into two three-carbon pieces (glyceraldehyde 3-phosphate and dihydroxyacetone phosphate). These two pieces can be converted into one another as needed. The glyceraldehyde 3-phosphate is oxidized over the next two steps to make 3-phosphoglycerate. The phosphate is then moved to the 2 position, and then removed with oxidation to end up with pyruvate. If we look through the steps, we can see that one glucose molecule ends up making two pyruvates.

Along the way, high-energy molecules are used to drive the reaction forward. The high-energy molecules are known as adenosine triphosphate (ATP). When they release their energy, they form a molecule of adenosine diphosphate (ADP). Later in the glycolysis, the phosphates in the small molecules are removed and ADP is returned to ATP. One step indicates that a molecule of NAD+ is converted into NADH. In this step, electrons from NAD+ are used to oxidize the molecule of glyceraldehyde 3-phosphate. This step requires that adequate NAD+ be available, and if it's not, glycolysis shuts down entirely. Oxygen can create more NAD+ in order to

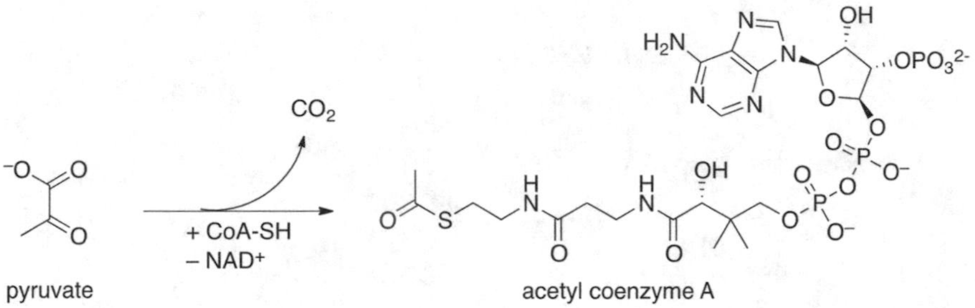

Fig. 9.29 Oxidation of pyruvate to provide Acetyl CoA. The two carbon atoms from pyruvate end up on the end of the acetyl CoA

keep this pathway going, however, when oxygen is scarce (as we'll see below), the NAD+ has to be made from other sources.

CHECKPOINT 9.11

What is the net change of ATP in the glycolysis of one molecule of glucose? How many NADH molecules are formed?

At this stage, the pyruvate can continue on to provide more energy and other starting materials for the synthesis of the molecules that the cell needs to survive. In order for it to move forward, however, it must be oxidized. This is a single step that converts pyruvate into AcetylCoA. AcetylCoA is the shorthand for acetyl coenzyme A, a fairly complex looking molecule. However, the end of the molecule (the two atoms in blue in Fig. 9.29) contains the atoms from pyruvate.

Now, the molecule is ready for the next step, the citric acid cycle. This cycle moves molecules around generating energy and carbon dioxide along the way. Any of the molecules can be removed from the cycle to make amino acids or other compounds useful for the cell. Note that another molecule of NAD+ is reduced to NADH in this step.

The citric acid cycle (aka the Kreb's cycle; shown in Fig. 9.30) begins where the acetyl CoA joins the process. If we color to two atoms from Acetyl CoA blue and follow them around the cycle, we can see that they aren't eliminated in the first cycle. But in the second cycle they can be. Note that succinate is a symmetrical molecule. That means the molecule can be flipped as it moves along to make fumarate. Thus, at this step, the carbons from the original acetyl CoA get distributed among all of the atoms of the fumarate molecule.

This cycle consumes three molecules of NAD$^+$ and one molecule of FAD$^+$ (a compound that also transfers electrons like NAD$^+$). In addition, this cycle forms one molecule of GTP (guanosine triphosphate) which is similar in structure and function to ATP. In other words, a lot of energy is used up in this cycle. So, what's the benefit to doing the cycle? That's the next step in the process.

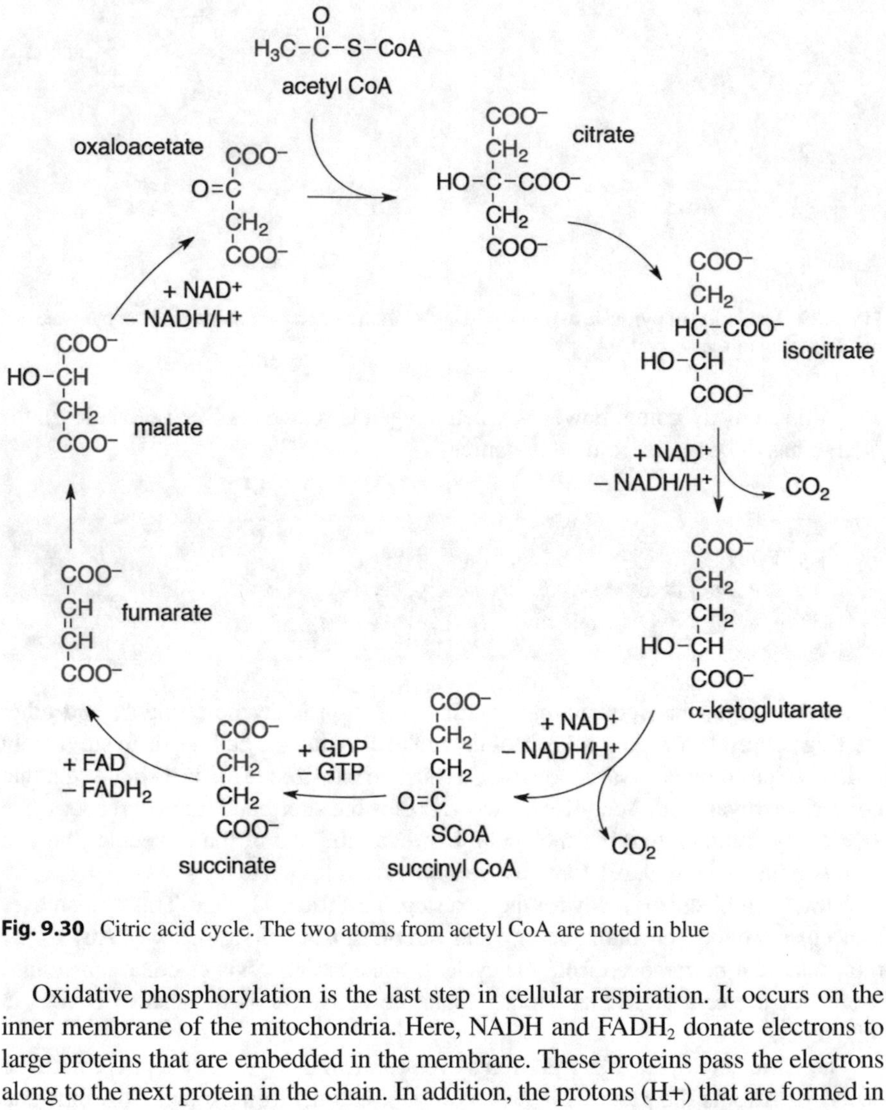

Fig. 9.30 Citric acid cycle. The two atoms from acetyl CoA are noted in blue

Oxidative phosphorylation is the last step in cellular respiration. It occurs on the inner membrane of the mitochondria. Here, NADH and FADH$_2$ donate electrons to large proteins that are embedded in the membrane. These proteins pass the electrons along to the next protein in the chain. In addition, the protons (H+) that are formed in the redox reaction are pushed across the inner membrane of the mitochondria and into a space between the membrane and the outer membrane. The electrons eventually are delivered to molecular oxygen (O$_2$) to make water. The NADH and FADH$_2$ become NAD+ and FAD+ in the process. The large concentration of protons just across the inner membrane represents a very high energy state because anything in high concentration tends to move to areas of low concentration (this is an entropy driven process). As the protons move back across the membrane they pass through an enzyme known as ATP synthase that converts ADP into ATP. The end result is the production of 30–32 ATP molecules for every molecule of glucose that the cell consumes. This represents a tremendous amount of stored energy that can be obtained from one molecule.

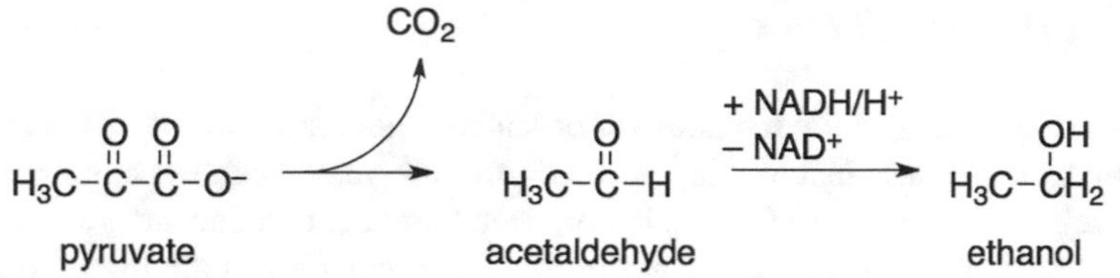

$$H_3C-\overset{\overset{\displaystyle O}{\|}}{C}-\overset{\overset{\displaystyle O}{\|}}{C}-O^- \xrightarrow{CO_2} H_3C-\overset{\overset{\displaystyle O}{\|}}{C}-H \xrightarrow[\;-\;NAD^+\;]{\;+\;NADH/H^+\;} H_3C-\overset{\overset{\displaystyle OH}{|}}{C}H_2$$

pyruvate   acetaldehyde   ethanol

Fig. 9.31 Fermentation pathway in yeast

9.4.2.2 Anaerobic Conditions

However, the yeast cells don't have enough oxygen to continue to use the glycolysis-citric acid cycle-oxidative phosphorylation pathway forever. In fact, while glycolysis and the citric acid cycle can still be used (they don't require oxygen), only glycolysis and a portion of the citric acid cycle are truly available. When oxygen is low, the cell enters anaerobic metabolism and obtains energy from a different pathway.

That pathway is known as fermentation. In fermentation, some of the pyruvate made from glycolysis is used to make acetaldehyde by decarboxylation (Fig. 9.31). A molecule of NADH then reduces the acetaldehyde to ethanol and generates NAD^+ in the process. This also helps eliminate some of the acid (protons, H^+) generated in the cell during the consumption of sugars. In addition, the NAD^+ can be used by the cell to create additional ATP during the glycolysis steps. It is not a lot of energy compared to oxidative phosphorylation, but it is enough to keep the cell living.

Lactobacillus and pediococcus can also ferment sugars. The pyruvate that they make after glycolysis, however, is reduced directly rather than becoming decarboxylated into acetaldehyde before being reduced. The result is the formation of lactic acid ($CH_3CHOHCOOH$).

9.4.2.3 Effects on Metabolism

Two actions that brewers often perform have a measureable effect on the action of yeast. The first of these is known as the *Crabtree Effect*, named after Herbert Crabtree who explained it in the 1920s. In the presence of high concentrations of glucose, even under aerobic conditions, yeast ferment and produce ethanol and carbon dioxide. This effect arises because the large concentration of glucose forces the production of a tremendous number of ATP molecules. Proceeding through the Citric Acid Cycle and oxidative phosphorylation is not necessary. The result is that the yeast growth is slowed or stunted because the glucose is moved exclusively toward ethanol production.

The other effect is known as the *Pasteur Effect*, reported by Louis Pasteur in 1857. In the presence of oxygen, yeast growth is highly favored and fermentation is slowed or stopped. This effect arises because the presence of oxygen allows the yeast to use oxidative phosphorylation to obtain tremendous amounts of energy. So, if a brewer aerates their actively fermenting wort, the yeast stop fermentation and multiply. This is exactly what happens when yeast are roused or dropped in many of the fermentation systems. This effect can be beneficial if the number of yeast in the beer is limited, but definitely not beneficial in the opposite case.

9.4.3   Products of Yeast

With all of the steps in the metabolism of sugars, there are a lot of different molecules that yeast make. Just looking at the figures above, we can see common molecules such as citrate, acetyl CoA, ethanol, acetaldehyde, oxaloacetate, etc…. Some of these are final products that are not further modified. Others are highly modified to form other compounds.

Esters are one class of compounds that are prized by brewers and produced by yeast during their metabolism of sugars. Research is still out on exactly how these form during yeast metabolism, but conjecture suggests that acetyl CoA plays a large role in making esters. An enzyme known as alcohol acetyl transferase (AAT) can catalyze the addition of an alcohol to an acetyl group from acetyl CoA. This type of ester is the most abundant in yeast fermentation. With restricted growth (low oxygen levels) the acetyl CoA is in abundance and can add to ethanol, butanol, and other alcohols found inside the cell.

Fatty acids are molecules containing the carboxylic acid group that have very long carbon chains attached. These molecules are used to make cell membranes, but sometimes can be esterified instead.

Fusel alcohols, alcohols other than ethanol that are present inside the yeast cell, are made from α-ketoacids. The α-ketoacids are byproducts and intermediates in the synthesis of amino acids. In fact, it has been shown that a yeast strain deficient in its ability to make valine and isoleucine (two amino acids) was unable to prepare 2-butanol, 3-methyl-2-butanol, and 2-pentanol. Thus, these alcohols come about as the yeast cell is undergoing rapid growth and producing large quantities of amino acids for the new yeast cells.

Diacetyl and other VDKs are produced during rapid growth as well. Diacetyl itself comes from valine production, specifically from α-acetolactate, a precursor to the formation of the amino acid (Fig. 9.32). The biosynthesis of valine involves four separate steps from pyruvate. Once the a-acetolactate is excreted from the cell, it undergoes a spontaneous oxidative decarboxylation to produce diacetyl. This step requires, as we've discovered previously in this text, the presence of metal cations or oxygen in order to work.

Yeast then slowly uptake diacetyl and other VDKs when the concentration of fermentable sugars is low. Once back inside the cell, diacetyl is reduced by NADH forming NAD^+ and acetoin ($CH_3CHOHCOCH_3$). The NAD^+ is used by the cell (see the figures above). Acetoin is further reduced to 2,3-butanediol by another NADH, resulting in the total production of two molecules of NAD^+ for every molecule of diacetyl.

Each of these molecules would be undetectable in beer if they stayed inside of the cell walls. However, large quantities of them produced inside the cells can either leak through the cell walls or be transported outside of the cell as waste products. Carbon dioxide, for example, is a small molecule that simply diffuses across the cell wall. Acid (i.e., protons, H+) is transported using specialized cellular pores. In the end, if it's inside the yeast cell, it likely will leak out and flavor the beer.

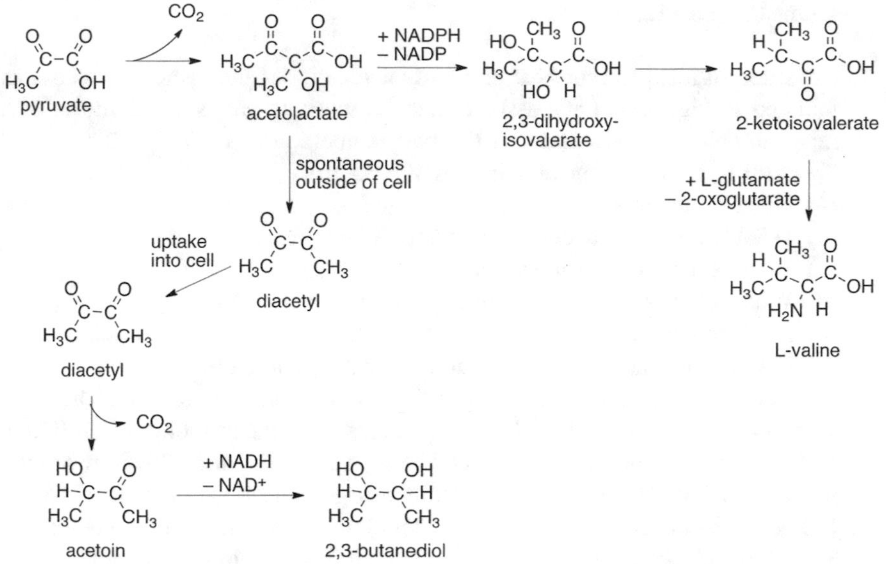

Fig. 9.32 Production and uptake of diacetyl

Chapter Summary

Section 9.2

Fourier's heat equation can be used to determine the rate of heat transfer.

Wort can be cooled using the shell and tube and the plate heat exchangers.

Dual stage heat exchangers allow greater control of the final temperature.

Section 9.3

The ideal gas equation relates the pressure, temperature, amount, and volume of a gas.

The amount of work done by or on a system can be calculated from equations based on the type of process.

Pressure-enthalpy diagrams can provide information about a refrigeration process.

Glycol refrigeration systems are used to cool equipment and vessels in a brewery.

Fermentation vessels include open and closed systems. The CCV is the most common of the fermenters.

Section 9.4

The structures inside a yeast cell have very specific functions.

Metabolism of glucose drives the production of energy in the yeast cell.

The production of off-flavors is due to the overproduction or leaking of compounds from the yeast cell.

Questions to Consider

1. Consider an ideal Carnot heat engine that undergoes the cycle of process as indicted in Fig. 9.14. Let's assume that the working gas is $n = 2$ moles of a mono-atomic ideal gas and that the two temperatures are $T_H = 1200$ K and $T_C = 200$ K. If the volume at point 3 is $V = 0.1$ m^3,
   (a) What is the pressure at point 3?
   (b) What is the pressure and volume at point 4?
   (c) What is the pressure and volume at point 1?
   (d) What is the pressure and volume at point 2?
   (e) Find the work done for all four processes.
   (f) Find the heat energy absorbed and released, Q_H and Q_C.
2. Consider a closed loop vapor refrigeration system that uses R-134a. The pressure in the condenser is 900 kPa and the pressure in the evaporator is 100 kPa. If the work input into the system is 600 kW, what is the mass flow rate around the loop? What is the actual COP for this refrigerator? What is the ideal COP?
3. Provide a list of pros and cons for the Double-drop system of fermentation.
4. What benefit does the yeast cell obtain from the production of ethanol?
5. Why do yeast go "dormant"? Under what conditions does this occur?
6. What problems would exist with a yeast cell that was deficient in the production of valine? How might a brewer overcome this issue?
7. Consider that you have 1.0 moles of an ideal gas with a pressure of 760 mmHg and temperature of 150 °F. What is the volume of this gas?
8. If the gas in question 7 was heated to 100 °C, what would be the new volume? Assume the pressure remains constant.
9. How many liters does 1.0 mole of an ideal gas occupy at standard temperature and pressure (273 K and 1 bar)?
10. What likely happens to any fructose that is consumed by a yeast cell?
11. Why are acetate esters the most common of the esters produced by yeast?
12. Which steps in the metabolism of glucose produce carbon dioxide? Which is the main reason that beer is carbonated naturally?
13. What would happen to a sample of fermenting wort if a large quantity of sucrose were added?
14. A brewer accidentally aerates a fermenting wort when they transfer it to a secondary fermenter. Describe what would happen?
15. What is the ultimate fate of the aldehydic carbon in glucose (the anomeric center)? Describe how this happens.
16. What is the benefit of transferring heat while the substance is a mixture of vapor and liquid?
17. Can a system have a COP of 0.80? Explain your answer.
18. Explain the first and second laws of thermodynamics.
19. What is the change in heat energy in an adiabatic process?
20. In your own words, describe how a Venturi aerator works.

21. What is the rate of heat transfer when a 100 °C wort sample is separated from 20 °C water by a 0.2 mm sheet of stainless steel? ... by a 0.2 mm sheet of aluminum? Which is better at transferring heat?
22. Why does the rate of heat transfer slow down as the hot liquid cools?
23. What is the change in internal energy if a system does 400 J of work and receives 200 J of heat?
24. What is the efficiency of the system in question 23? If the temperature of the cold fluid was 20 °C, what would be the temperature of the hot fluid?
25. Describe the differences between mannan and starch.
26. If splashing wort into an open fermenter results in about 8–10 ppm oxygen, why would anyone want to use a Venturi aerator?
27. What is the change in work to a system if the pressure stays constant at 1 bar while the volume of the system changes from 1.0 L to 1.5 L?
28. What is the definition of a state variable? Is altitude a state variable? Is distance a state variable?
29. In Fig. 9.30, what are the chemical formula for pyruvate and acetaldehyde? Assuming that CO_2 is a product of the transformation, what other atom is missing from the transformation? What is the likely source of this extra atom in the reaction?

Laboratory Exercises

The Effect of Sugars on Fermentation

This experiment is designed to illustrate the differences in fermentation rates using different sugars. The rates can be easily measured by collecting the byproduct (carbon dioxide).

Equipment Needed

Erlenmeyer flasks, 125 mL
Single-holed stoppers to fit flasks
Packet of dry yeast, or uniformly mixed liquid yeast
Stir bars (one per flask) and stirring plates
Glass tube and hose (one set per flask)
Graduated cylinder or gas-collection tube (one per flask)
Large dish to use as cold-water bath
Sugars from the following list: (at least three should be used)

Sucrose	Arabinose	Glucose
Glucose	Invert sugar	Fructose
Fructose	Maltotriose	Lactose
Maltose	Starch	Glucitol

Experiment

Each group should obtain one setup per member of the group. Add 50 mL to each of the Erlenmeyer flasks using a graduated cylinder. Then, measure out 5 g of each of the monosaccharides and 2.5 g of each of the disaccharides. Dissolve each sugar in the water in its own Erlenmeyer flask. Add a stir bar and 0.5 g of yeast to each flask. Push the glass tube into the stopper and attach the hose to the end (CAREFULLY to avoid breaking the glass). Insert the stopper in the flask and the end of the hose into a graduated tube filled with water.

The best way to fill the graduated cylinder with water is to lay it flat in a dish of water, then insert the tube. While keeping the open end under the surface of the water, raise the bottom of the graduated cylinder until it is vertical. Then use a clamp and ring stand to hold it in place. The cylinder should be free of any bubbles and only filled with water.

Turn on the stir plate to vigorously mix the yeast in the sugar water and then measure the time it takes each flask to generate 5 mL of gas. This can also be done by recording the time for every 5 mL of gas that are produced and then creating a plot of the time (x-axis) versus the volume (y-axis).

Comment on the production of gas based on the specific sugar. An alternative to this process is to perform the fermentations at different temperatures or with different yeast cultivars.

Maturation and Carbonation

10

10.1 The Purpose of Maturation

After fermentation, the beer produced is known as *green beer*. This product, while drinkable, is not finished, often is flat, and has a relatively unstable flavor profile. The *maturation* process (sometimes referred to as *conditioning*) is required to convert the flavors into those desired by the brewer and to allow the beer to mature and settle into a more stable flavor profile. This process is akin to the preparation of most food in the kitchen.

You may have heard the phrase *it tastes better the second day*. This is true for the author's chili which mellows and melds its flavors when it rests overnight in the fridge after being cooked. It's also definitely true for the author's pulled pork that requires a little time to allow the spices and smoky flavor to really bring out the flavor of the meat. Let's consider the process to make spaghetti. First, we grab a few ripe tomatoes and cut them open. Then, we scoop the seeds out and throw them into the compost. The meat of the tomatoes is chopped up (or squished between the fingers) until the entire mass has the consistency of a puree. The tomatoes are then added to a pot containing sautéed onions and garlic and the entire mixture brought to a simmer. Some liquid, chicken stock, wine, or water, is added and spices such as oregano and thyme are added. Once hot and mixed together, the spaghetti sauce is done. Just like green beer, it's edible and can adorn a pile of your favorite linguine. However, until it is simmered for a period of time (the author requires at last 2 h) or stored overnight in the fridge, the flavors haven't developed into the recognizable spaghetti sauce that makes your mouth water when you taste it.

It's possible that the chef adds some sugar or baking soda to the sauce once it has started simmering. Just as in the maturation or conditioning of green beer, the brewer may do similar steps to adjust the acidity or flavor of the beer. In other words, maturation is the time and place where the brewer "adds salt to taste."

Before we continue, it might be worthwhile to provide a few words about "stable" flavor in beer. While the beer that has undergone conditioning is ready to be

© The Author(s), under exclusive license to Springer Nature Switzerland AG 2021
M. Mosher, K. Trantham, *Brewing Science: A Multidisciplinary Approach*,
https://doi.org/10.1007/978-3-030-73419-0_10

consumed with a stable flavor profile, the flavors in a beer are never really 100% stable. The brewer knows this, too. Beer does have a fairly limited shelf life. That shelf life can be extended by keeping the product cold and free of oxygen. But even under these conditions, the flavor of the beer is not entirely stable. The flavor will change. The brewer knows this and might even indicate a best by date on the packaging. That date may be weeks, months, or even years from the date it was brewed, but nonetheless, there is a time when the beer will no longer taste as good as the day it was brewed and matured.

Why is a maturation step performed? There are many reasons why the brewer conducts this step; each of those reasons is outlined below. Maturation and conditioning can:

- Induce secondary fermentation to carbonate the beer
- Mature the flavors and odors of the beer
- Reduce or eliminate the potential of the beer to form haze
- Adjust the flavor, color, or aroma in the beer
- Adjust the amount of compounds to improve or reduce foam
- Eliminate or reduce bacterial growth
- Clarify the beer prior to filtration

10.1.1 Secondary Fermentation

Secondary fermentation is often a highly desirable feature of the brewing process and can be considered as a maturation step in the brewery. In this process, the *green beer* is either (a) recirculated within the fermentation tank, (b) transferred by pump or gravity to a new fermentation tank, or (c) transferred to a storage tank fitted with cooling jackets and pressure regulators. The yeast concentration that remains in solution after primary fermentation has finished is often still above a million yeast cells per milliliter of beer. The process of recirculating or transferring the beer within the tank *rouses* the yeast and the fermentation can essentially restart.

The brewer can also allow oxygen to enter the process as the beer is either recirculated or transferred. The addition of oxygen, known as the Pasteur Effect, results in a dramatic increase in yeast growth within the solution. These additional active yeast cells can continue to consume the remaining sugars and remove unwanted flavor compounds in the beer.

The addition of oxygen to increase the yeast concentration is not always a good idea. This could be due to the risk (both perceived and real) of oxidation of the beer that would lead to a poor flavor profile. Thus, when the beer needs an increase in the concentration of yeast to encourage the removal of undesirable flavors, the brewer can perform one of two options.

In the first case, a separate strain of yeast can be pumped into the after primary fermentation. This injection of yeast can boost the level of fermentation and result in the consumption of any remaining sugars and undesirable flavor compounds. For example, if the brewer were interested in making a very high alcohol beer and the first fermentation was accomplished with a standard yeast, the addition of a yeast strain that can continue to work at high alcohol concentrations might be needed to finish off the rest of the fermentable sugars. If the beer will be cold matured (see below) with the additional yeast, the secondary yeast strain is often one that can handle the lower temperatures during that maturation process.

The second option involves the injection of a portion of wort and yeast-containing foam from an actively fermenting batch of wort. Literally, the yeast-laden foam from an actively fermenting wort (known as *kräusen*) is scraped off of a batch and dumped into the fermenter containing the older fermentation. This process is called *kräusening*. Most commonly used at the start of a cold maturation process, the addition of kräusen can help mature the flavor of the cold green beer quickly because a very large injection of fresh yeast is added to the beer.

10.1.2 Warm Maturation

Typically, at the end of primary fermentation, the brewer speeds up the action of the yeast on the beer to help remove unwanted or undesirable compounds that impart a flavor to the finished beer. In these cases, the brewer warms the beer from the standard fermentation temperature by 1 °C to 5 °C (2 °F to 9 °F) and holds the fermentation at that temperature for a period of time. As the temperature increases, the reactions inside the yeast proceed faster. Thus, the yeast cells work harder to uptake the remaining fermentable sugars. And when those sugars are depleted, the yeast uptake other compounds that can provide energy.

One of the more important class of compounds that are consumed by the yeast when the fermentable sugars are depleted are the VDKs (vicinal diketones). VDKs have a similar chemical structure that contains two carbonyl groups (C=O) adjacent to each other. While there are many different VDKs, the main VDKs that form during fermentation are 2,3-butanedione (diacetyl) and 2,3-pentanedione, compounds that have a flavor threshold that is quite low. For this reason, warm maturation is also known as the "diacetyl rest." As we'll see later in this discussion, reducing the concentration of VDKs to as low as possible is very desirable.

As a side note, there are two pronunciations of the word "diacetyl" that pervade the brewing world. The word is actually a chemical term that is derived from the chemical structure of acetic acid (pronounced "a-SEAT-ic"). Thus, chemists know the word diacetyl is pronounced "di-a-SEAT-il." However, in the brewing industry, especially within the United States, the word is often pronounced "di-ASS-it-il." Whether the brewing pronunciation is correct or not is less important than understanding that the two pronunciations refer to the same chemical compound.

The speed of the process is the main reason why a brewer often chooses warm maturation. The same result can be obtained at lower temperatures (i.e., the reduction of the levels of diacetyl, 2,3-pentanedione, and other off-flavor compounds), but the process occurs much slower. Typically, if a beer is matured at cold temperatures, it may require multiple days, weeks, or even months in order to accomplish

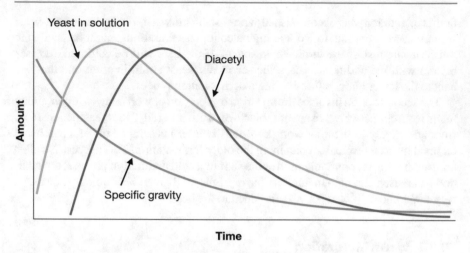

Fig. 10.1 Diacetyl concentrations during fermentation at normal fermentation temperature

the same thing that happens in 12–24 h at a temperature just above the normal fermentation temperature.

Diacetyl and 2,3-pentanedione concentrations begin to rise at the end of the rapid growth phase of the yeast during fermentation (see Fig. 10.1). Diacetyl arises from the production of α-acetolactate during the biosynthesis of amino acids by the yeast. Levels of a-acetolactate can be as high as 200 ppm. As would be expected, the amount of α-acetolactate produced in the fermentation is highly dependent upon the composition of the wort, the strain of yeast, and the temperature of the system. For example, if the wort has a high FAN content (i.e., a high concentration of amino acids), the production of α-acetolactate is limited. If the levels of valine and isoleucine amino acids in the wort are large, their presence suppresses the formation of the precursors to make them (i.e., α-acetolactate). Higher fermentation temperatures also have a positive impact on diacetyl production.

Warm maturation takes place when the temperature of the fermenter is adjusted after the rapid growth phase of the yeast (Fig. 10.2). Note that the increase in the temperature causes a rapid change in the concentration of diacetyl in solution (and in the precipitous drop in the gravity of the wort).

Diacetyl formed during fermentation (assuming there are no bacteria in the mix) arises from extracellular α-acetolactate as shown in Fig. 10.3. After excretion from the yeast, α-acetolactate undergoes a non-enzymatic oxidative decarboxylation to provide diacetyl. The rate of this process is highly temperature and pH dependent. The oxidative step in the process requires the presence of an oxidizer. Dissolved oxygen may serve as that oxidizer; however, the presence of metal cations in the solution may also act as sources of the oxidizer. In particular, Cu^{2+} and Fe^{3+} have been suggested as possible oxidizers.

As fermentable sugars in the wort decline in concentration, yeast begin to uptake extracellular materials to use as energy sources. VDKs (vicinal diketones) are one such source. If the yeast are still in suspension and haven't flocculated too quickly,

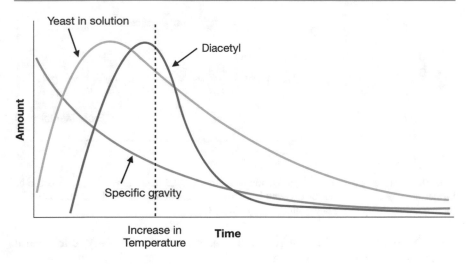

Fig. 10.2 Diacetyl concentrations during a warm conditioning phase

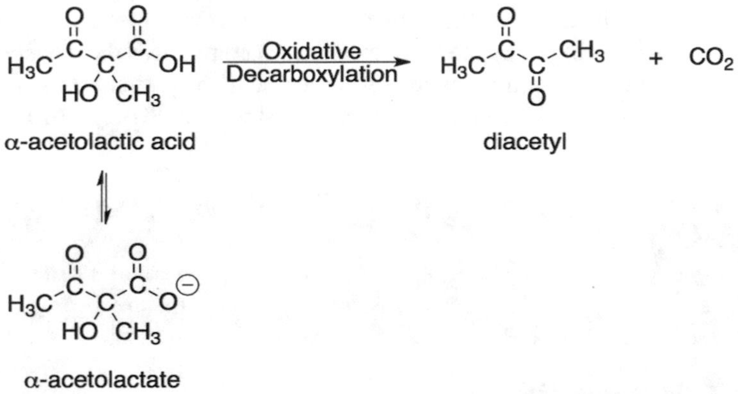

Fig. 10.3 Diacetyl production is an extracellular process. The rate of this reaction is dependent upon the temperature, pH, and presence of metal cations

they will uptake these compounds quickly as shown in Fig. 10.2. If the yeast have already flocculated, had poor health due to stressed growth, or have already entered a cold maturation stage, the rate of diacetyl and 2,3-pentanedione uptake is slow.

Diacetyl that has been taken up by the yeast cell is reduced to acetoin and then to 2,3-butanediol (Fig. 10.4). Both acetoin (pronounced "a-SEAT-o-in") and 2,3-butanediol can be excreted from the cell.

The process of converting diacetyl into acetoin and 2,3-butanediol is extremely important because diacetyl imparts a perceptible buttery or butterscotch flavor to the finished beer even at levels as low as 0.02 ppm. It is true that not every person can perceive the flavor of diacetyl even at elevated concentrations; however, some people are very sensitive to this compound. And, while it may sound like a buttery

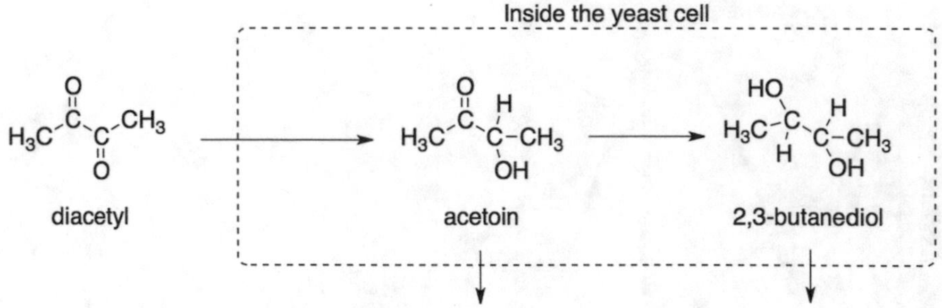

Fig. 10.4 Uptake of diacetyl and production of acetoin and 2,3-butanediol. The products can be further used by the yeast cell or excreted back into solution. However, the flavor thresholds for these compounds are much higher than that of diacetyl

flavor in your beer would be a good thing, after a few sips it's easy to tell that it really doesn't belong in every style.

Alternatively, the brewer can add enzymes directly to the fermenter to help reduce or eliminate undesirable flavors in the beer. In particular, α-acetolactate decarboxylase can be added to the fermenter. This enzyme converts α-acetolactate directly to acetoin. It bypasses the yeast machinery and removes the diacetyl flavor from the beer. This is sometimes quite useful because of the particular conditions or strain of yeast used (e.g., an overly flocculant yeast strain would only slowly remove diacetyl).

CHECKPOINT 10.1
What type of reaction is happening when diacetyl is converted into acetoin? In your own words, explain why diacetyl is formed and then reabsorbed.

10.1.3 Cold Maturation

After warm maturation, it is often advantageous to mature the beer at lower temperatures. There are many reasons for performing a cold maturation on a particular beer brand. Those reasons include the following.

Removal of Suspended Solids Chilling the beer after warm maturation allows any solids that remain to precipitate to the bottom of the vessel. This allows the beer to be clarified without the use of special additives or filtration. And, if the beer is stored for a period of a few days or more at temperatures just below 0 °C, the majority of any solids that would form and make haze in the beer are also removed. The end result is a product with colloidal stability that has a longer shelf life.

Removal of Compounds   Extended time at very low temperatures reduces the concentration of compounds that have an undesirable flavor. The brewer can monitor the beer during cold maturation to observe the loss of those compounds and determine exactly when the flavor of the beer has reached the specifications for that brand.

Stock and Storage   In some cases, the equipment used to package the beer may not be ready to accept more product. Or, the beer may have a small fault (such as a color, flavor, or alcohol content) that is out of specification. In both cases, the beer can be stored at the low temperatures. Then, when needed, the beer can be moved to the packaging or filtration departments in the brewery or blended with other batches of beer to reduce any faults and put the beer back within specifications.

Addition of Carbon Dioxide   The cold maturation vessel can be constructed to allow it to hold pressure. If so, the brewer can add gaseous carbon dioxide and pressurize the vessel to help the carbon dioxide dissolve. Because the beer is cold, the gas dissolves much easier than at room temperature. In fact, the beer can be carbonated to the exact pressure needed in the finished product.

Addition of Compounds   The brewer could take this cold maturation stage and adjust other specifications of the beer. For example, if the beer's color is too light, colored compounds (such as caramel coloring) could be added. If the hop bitterness isn't within specifications, it can be adjusted by adding pre-bittered hop oils. If the beer brand calls for a dry-hopped beer, that can be added at this stage as well.

Often, the cold maturation step is known as *lagering*. When the beer is lagered, it is placed in tanks and cooled to about 10 °C. Then, over a period of days or weeks, the temperature is lowered to just below the freezing point of water (0 °C). While water will freeze at this temperature and potentially damage the equipment, beer does not because it contains a fairly significant amount of alcohol (often 4% alcohol or more).

The slow decrease in cold temperatures causes the fermentation process, which starts out fairly slowly at 10 °C, to slow even more and more as the tanks get cooler. In addition to removing the undesirable flavor compounds, as the tanks get cooler, solid materials that are insoluble at these lower temperatures fall out of solution. The precipitation of these materials helps to clarify the beer. And once the temperature hits the lowest end of the range, even the yeast cells fall out of solution. Thus, the lagering process, once complete, results in a very clear beer.

In fact, while there are many reasons to perform the cold maturation process, the most important reasons are to clarify and improve the flavor of the beer.

10.1.4 Other Adjustments

While removal of diacetyl and other VDKs is one of the most important goals of conditioning the beer, many other things can be done to stabilize the beer and adjust the beer to match the style parameters. Let's look at each of the different things that can be done and see how to accomplish these tasks.

Haze Reduction Polyphenols (tannins) that are present in the beer after the initial stages of fermentation can combine with proteins (typically in the 10,000–60,000 molecular weight range). Any excess oxygen in the beer can react with the tannins and make them even larger. This would increase the rate at which they form complexes with suspended proteins. The protein-polyphenol complexes are held together with hydrogen bonds where the polar groups on both the polyphenols and the proteins interact (Fig. 10.5). Other hazes are also possible. These include the large β-glucan molecules and calcium oxalate crystals. Calcium oxalate crystals tend not to be an issue as they only form when there are significantly high levels of calcium in the beer after the fermentation is complete.

Reducing the haze can be done in the conditioning tanks by cold maturation at temperatures below 0 °C. However, this can take a long time and doesn't always result in complete removal of any haze.

A better way to ensure that the beer's haze has been reduced as much as possible is through the addition of *finings*. Finings are compounds that can selectively bind to the proteins in the beer and form a precipitate. If the proteins in the beer are reduced, then they are unavailable to bind to polyphenols and form a haze. The category of finings includes isinglass (collagen from the swim bladders of fish), gelatin (collagen from animals), and Irish moss (carrageenan). Many finings work by forming a gel that entraps yeast, protein-polyphenol coagulates, and other larger compounds that would normally precipitate if given enough time. These finings,

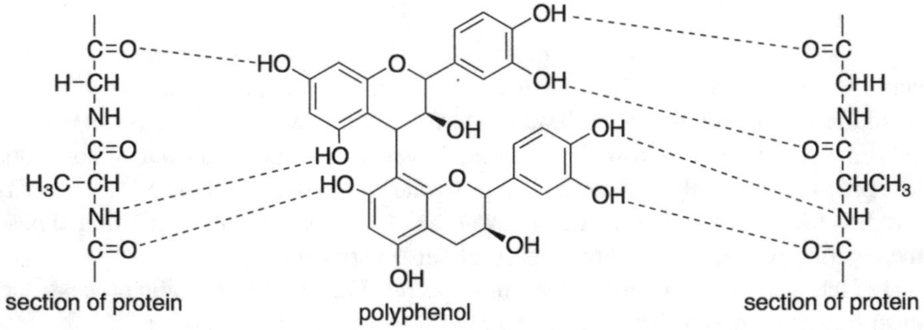

Fig. 10.5 Haze production from polyphenols and proteins. The polyphenols interact by sharing hydrogen atoms with multiple proteins. The result is a large protein that becomes insoluble in the beer. As more interactions occur, the haze becomes more and more stable

however, rapidly speed up the process. Typically, they can clarify the beer to be within very acceptable levels in 4–5 days.

Instead of adding finings, an enzyme can be added to the beer that can break the proteins into smaller pieces that, when bound to the polyphenols, do not make large complexes. In other words, the process of forming the complexes still occurs, but it doesn't make haze. These proteases include papain – a protease isolated from papaya. Papain is particularly good at cleaving protein bonds and breaking down these larger molecules into smaller ones.

Alternatively, the brewer can add *auxiliary finings* to the cold maturation process. These are compounds that result in clarifying the beer but are not collagen or carrageenan. For example, the proteins can be removed by the addition of silica gel (an amorphous form of silicon dioxide) to the beer. The highly polar silica gel associates very strongly with the proteins in the beer. The resulting mass precipitates and falls to the bottom of the vessel where it can be removed from the clarified beer. Advantages to the use of silica include the significant reduction in proteins and β-glucans. Unfortunately, the disadvantages can outweigh the advantages. Removal of too much of the proteins reduces the amount and stability of the head on the finished beer. In addition, many of the flavor compounds have polar groups in them, and if too much silica gel is accidentally added, the result could be a reduction in the flavor of the finished beer.

A common auxiliary fining that reduces the haziness of beer is PVPP (polyvinyl polypyrrolidone). This chemical polymer is somewhat similar in structure to a protein. Thus, it binds to the polyphenols in the beer and removes them by forming a precipitate. Of the auxiliary finings, PVPP removes the astringent polyphenols while leaving much of the proteins behind to maintain a stable head on the finished product. PVPP is fairly expensive, but it can be reused by filtering the beer through a pad containing the polymer. Then, it can be washed for use on another beer.

The downside to the use of finings or auxiliary finings is that the amount required to clarify a beer is dependent upon that specific batch of beer. In other words, it is possible to accidentally add too much or not enough. In addition, it is best to not use finings unless the yeast have accomplished everything for which they are needed, because once added, the yeast will be nearly completely removed from the product.

Carbonation   As we noted above, carbon dioxide (CO_2) can be added to adjust the amount of carbonation in the beer. Typically, CO_2 is added until the beer contains 1.5–2.8 volumes of CO_2. We explore carbonation in much greater detail later in this chapter.

Flavorings   The taste of the beer in the conditioning tank is one of the last places where the beer can be adjusted to give the flavor that is required by the brewer. For example, if the beer is lacking a particular concentration of ester (such as isoamyl acetate), that compound can be added. Adjustment of the beer flavor can also occur through the addition of fruit, spice, or other flavors. For example, a brewer may wish to add cherry flavoring to the beer. Adding artificial or natural cherry flavors while the beer is in the conditioning tank could do this.

Coloring Agents   If the SRM of the beer is incorrect for the brewer's requirements, it can be adjusted in the conditioning tank. Typically, this is done by adding caramel color. Dosing this into the beer provides the appropriate color of the beer without adding additional flavor.

Hop Additions   While hop additions typically take place in the secondary in what is called dry hopping, hop oils and reduced hop oils can be added in the conditioning tank. The addition of hop oils can adjust the flavor components to provide a hoppier flavor to the beer.

Isomerized hop oils can be added to increase the bitterness of the beer to match the brewer's style requirements. Dosing with these hop products will provide the bitterness without requiring the product to be boiled. Alternatively, the use of reduced hop oils can be advantageous. These compounds, which we discovered earlier in this text, can provide additional bitterness while at the same time increasing head retention and/or providing light stability.

Sugars   The sweetness of the beer can also be adjusted at this point. This can be done through the addition of non-fermentable sugars such as lactose, or through the use of fermentable sugars. Though, if fermentable sugars are used, it is imperative that the yeast be removed from the beer or it will ferment again. This can be useful if the beer is to be "naturally" carbonated. But, it would be particularly disadvantageous if the beer were to be packaged with significant quantities of added maltose and still containing yeast.

This problem can be overcome if the beer is to be pasteurized or sterile filtered. Because these methods significantly reduce or eliminate yeast from the finished beer, additional sweetness can be added. Alternative methods to remove the yeast from the beer include adding bacteriostatic compounds to the beer. Typically, this involves the use of SO_2. Doing so, however, requires that the concentration of this compound be clearly noted on the label because some people are unable to consume this compound without significant health problems. In most countries, the limit of SO_2 is 10 ppm.

Yeast   Yes, yeast can be added in the conditioning stage. This would typically be done to add a strain of yeast that is very good at cleaning up or adjusting some other parameter in the beer. For example, it may be added to clean up any remaining VDKs or to provide a yeast that can naturally carbonate the beer with the remaining fermentable sugars. The added yeast can later be removed through the use of finings or by sterile filtration, centrifugation, or pasteurization methods.

10.2   Equipment Used in Maturation

Beer in the maturation tank requires that it be cold and pressurized. The beer should
be able to be adjusted through additions of other materials and should be able to be
removed when finished with maturation. Finished beer from this step will move on
to the packaging line. And it is with the requirements at this step that the equipment
and vessels are designed.

10.2.1 The Cold Maturation Tank

The cold maturation tank, see Fig. 10.6, is typically a vessel that is jacketed in order
to adjust the temperature. It has ports that allow not only the addition of the beer via

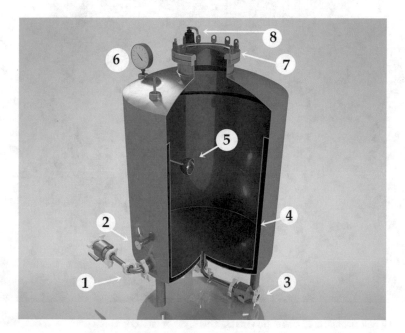

Fig. 10.6 The conditioning tank is a jacketed (4) tank that can be cooled. The green beer enters
through the bottom port (3) and is withdrawn through a side port (1). Other ports (2) are included
to monitor temperature or add additional carbon dioxide. A pressure gauge (6), pressure release
valve (8), a manway (7), and a CIP spray valve (5) complete the vessel

the bottom, but also along the side to allow the withdrawal of the finished beer. One of the ports usually allows the brewer to add carbon dioxide via a carbonation stone. Gauges and monitoring devices can also be found on the vessel. The body of the vessel is also jacketed so that the contents can be warmed or cooled with glycol or some other coolant. If space is an issue, the tank can also be designed in a horizontal fashion (see Fig. 10.7). In this situation, groups of horizontal tanks can be stacked on top of each other. The process and layouts of the horizontal tanks are essentially identical to the vertical tanks.

One important thing to note about the vessel is that it typically has a much flatter bottom than the fermentation vessel. It is still concave to allow the beer to be removed entirely from the vessel, but because there is typically not that much material that precipitates compared to the CCV, the bottom can be flatter. A side port near the bottom is typically the location to remove the *bright beer* after conditioning. This is because any precipitates that form due to additives (such as PVPP used to reduce haze) will collect on the bottom near the central port.

Alternative forms of the conditioning tank do exist. In fact, the CCV can be used as a conditioning flask as long as it can be pressurized safely into the 10–15 psi

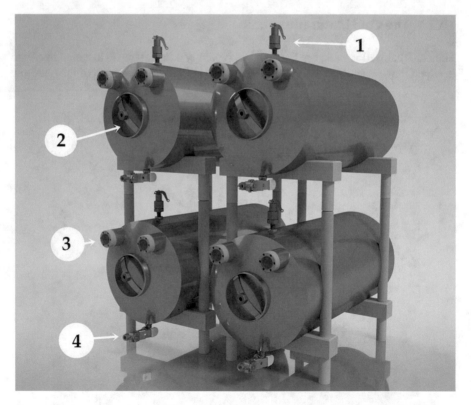

Fig. 10.7 Horizontal cold maturation tanks. This style of tank contains a pressure relief valve (1), manway (2) and fill/exit port (4) in addition to ports (3) for CIP spray balls, temperature and pressure gauges, and sampling

range. In addition, the CCV would work best if it was also jacketed and temperature controlled. This is not the best vessel to choose as the brewer would likely need to add a carbonation stone to the vessel after fermentation were complete – which could result in the loss of a significant amount of beer during the exchange. If the carbonation stone were in place prior to the start of fermentation, it would be possible that flocculating yeast and residual trub could clog the pores of the stone and render it less effective in carbonation. For this reason, should a CCV be employed as a conditioning tank, the brewer would likely naturally carbonate the beer by setting a pressure regulator on the blow-off arm to be equal to the final volumes of CO_2 desired in the beer.

10.2.2 Cask Conditioning

Another alternative to the stainless steel conditioning tank is the cask. Cask conditioned ales are quite popular in some parts of the world (notably in Europe). While many patrons in the United States consider cask ales as a novelty, their popularity is beginning to catch on. With over 180,000 members, the Campaign for Real Ale (CAMRA) in the United Kingdom has worked since 1971 to advocate for cask ales, pubs, and consumer rights. This organization provides a listing of over 4500 pubs in the United Kingdom that support and offer cask ales. They also run an annual festival that highlights more than 900 cask conditioned beers and ciders.

Cask ales are often fermented in a different vessel and then moved to the cask for conditioning. The cask itself has a very useful shape that results from how the cask is used as shown in Fig. 10.8. The specific parts of a cask have very specific names as well. Additions to the cask take place through a *bunghole* that lies on the central band of the cask. The bunghole is simply a hole in the keg that can be sealed by placing a bung (a stopper) into the hole. Once the additions are complete, a *shive* (a bung with a small hole in the center) is pounded into the bunghole. A *stile*, a peg that can be inserted into the shive, is used to control the conditioning process. That control could be to allow an active fermentation to expel barm from the cask, or it could be to pound the stile into the shive to allow the beer to carbonate. Another bunghole

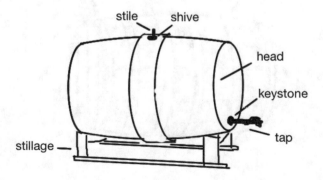

Fig. 10.8 The modern cask and its key features

exists in the head of the cask near the edge. This hole is sealed with a bung known as a *keystone*. When the beer is to be served, a *tap* is hammered through the keystone. This pushes the keystone into the keg and replaces it with the firmly seated tap. Casks are stored on *stillage* while they mature. These racks can be designed to hold multiple casks and even stacked one on top of another to conserve space in the cellar. When the beer is to be served, it is moved to an angled stillage that tilts the cask forward to allow all of the beer to exit through the tap.

As you might imagine, tapping a cask involves a steady hand to hold the tap against the keystone and strike it firmly with a rubber or wooden mallet. The process pushes the keystone into the cask. And, because the cask is under pressure from natural carbonation, the process typically results in a spray of beer and foam. Poorly struck taps can spray a large amount of beer all over the person holding the tap in place.

As we noted much earlier in this text, early casks were wooden (Fig. 10.9). Some brewers continue to use the wooden cask because of the advantages of conditioning in contact with wood. The parts to the wooden cask are each named. The cask itself was made up of a series of *staves* that locked into the *head* boards (the edge pieces were known as cants). The bunghole located near the center was entirely placed in the middle of one of the staves. Another was placed entirely within one of the head boards. This ensured that the holes did not fall on a joint.

Either wooden or metal bands were tightened around the staves to hold them into place. Typically, there were three *hoops* on each half of the cask, the *bilge* hoop, the *quarter* hoop, and the *head* hoop. Construction of the cask was, and still, is an art. Until the casks were soaked in water, they tended to leak between the joints of adjacent staves and the head boards. However, once the wood soaks up enough water, the staves and head swell making a very tight fit.

Modern casks can be made of stainless steel. They have the same basic shape, but because the cask is a single piece of metal, no need for hoops exists. And because

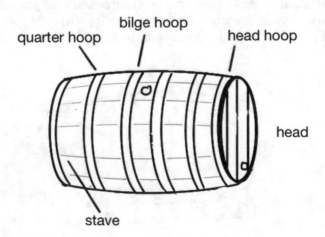

Fig. 10.9 The parts of the wooden cask

there are no joints in the stainless steel casks, they do not leak. They still have the bunghole on the end and at the middle.

Typically, *racking* to a cask takes place while sufficient fermentable sugars remain in the beer to absorb any accidentally added oxygen and lightly carbonate the finished beer. If sufficient sugars do not remain, the brewer can add *priming sugar* to the fermenter immediately before racking to the casks. The priming sugar used for this process tends to be corn sugar, dextrose, or even cane sugar depending upon the specific recipe. In some cases, caramelized sugars can be used to add additional flavor while also carbonating the beer.

If the yeast count is relatively low when the beer is to be racked to a cask, additional yeast can be added. Approximately one million cells/mL is needed to ensure adequate carbonation and conditioning in the cask. This additional yeast is often added to the fermenter immediately prior to casking. Doing so helps to ensure that the yeast are adequately distributed throughout the beer before it is racked. Conditioning in a cask does not limit the addition of other components to reduce haze, adjust flavor, etc., except that these additions tend to be performed in the fermenter immediately before racking to the cask. For example, the addition of isinglass (a fining agent) is typical when the beer is to be casked. This helps to clarify the product as it is conditioning resulting in a clear product.

If the casks are moved prior to serving, the sediment that has formed tends to be stirred up into the beer. Thus, if a clear beer is the desired product, the cask should be set on their *stillage* (the wooden racks that hold the cask firmly in place during serving) for at least 24–48 h prior to tapping. The sediment then can reform and drop out of the beer.

10.3 Carbonation

Carbonating a beer can be accomplished by two main processes. The first, natural carbonation, requires simply closing off the fermenter when the amount of fermentable sugars that remain are just enough to result in the correct amount of carbonation. This requires the use of a vessel that is able to withstand the pressure, and the ability to accurately determine the remaining amount of fermentable sugars that can give the necessary level of carbonation.

The second method is much more commonly used. In this case, carbon dioxide gas obtained from another source is pumped into the beer until the appropriate amount has been added. Known as *forced carbonation*, this process requires that the beer is contained in a vessel capable of holding the specific pressure.

10.3.1 The Principles of Carbonation

Beer is typically presented with carbonation. This is done because the consumer expects that their beer is carbonated. So, why is the carbonation expected in a beer? Carbonation adds tartness to the beer that is quite desirable. The bubbles add a tingling sensation on the tongue. Together, the tart tingling characteristic of the beer is very important to creating the appropriate mouthfeel.

As the beer attemperators, it releases the carbon dioxide that was dissolved. The carbon dioxide forms small bubbles that rise upward in the beer. The action, at first, is considerably vigorous. It is so vigorous that it helps form a foam on top of the beer made from proteins, hop oils, and other materials. And, as the beer sits, it continues to release the bubbles that support the foam on the beer. In fact, without carbonation, it is very difficult to form a foam on the beer.

Finally, when the beer is placed in the appropriate glass, the bubbles add to the attractiveness. To enhance the attractiveness of bubbles in the beer, breweries specify exactly what their glass should look like. In addition, the brewery might also specify that the bottom of the glass is etched. The scratches encourage beer bubbles to form.

The amount of carbon dioxide that ends up in the final beer is governed by four different factors.

Temperature   Carbon dioxide is not very soluble in water or beer. As the temperature decreases, the solubility increases dramatically. In fact, the gas is about twice as soluble at 0 °C than it is at room temperature. Because of this, the brewer will make sure that the beer is as cold as possible before trying to add carbon dioxide. The table below lists the maximum solubility of carbon dioxide in water as a function of temperature (Table 10.1).

pH   Carbon dioxide, when dissolved in beer, forms carbonic acid. In other words, the concentration of carbonic acid is directly related to the amount of dissolved carbon dioxide. At higher pH's, the carbonic acid dissociates into ions that are very soluble in water. As the pH decreases, the amount of carbonic acid increases.

In other words, at high pH, carbon dioxide becomes more soluble as it forms carbonic acid and then dissociates into ions. Conversely, as the pH drops, the carbon dioxide becomes less soluble because the carbonic acid does not dissociate as much.

Density   The specific gravity of the beer does play a role in the solubility of carbon dioxide. As more compounds dissolve in beer, the carbon dioxide becomes less soluble.

Table 10.1 Solubility of carbon dioxide in water

Temperature (°C)	Solubility (g CO_2 / kg water)
0	3.3
5	2.9
10	2.5
15	2.2
20	1.7

Pressure This is the most important of the effects. The pressure of the system is directly related to the quantity of carbon dioxide that can dissolve. Let's explore this effect in much greater detail.

The typical measurement for the amount of carbon dioxide in beer is *volumes of* CO_2. One volume of CO_2 can be thought of as 1.0 l of CO_2 dissolved in 1.0 l of beer. Assuming that CO_2 behaves ideally (an assumption that is relatively acceptable at the pressures and temperatures that we will work with), we can use a chemical law known as the *Ideal Gas Law* to determine how many grams of CO_2 are in a given volume of CO_2. The ideal gas law is:

$$PV = nRT \tag{10.1}$$

where P = pressure in atmospheres

V = volume in liters
n = number of moles
R = universal gas constant (0.082 Latm/molK)
T = temperature in kelvin

Since the number of moles of any substance is equal to its mass in grams (m) divided by its molecular weight in grams/mole (mw), we can substitute this into the ideal gas equation:

$$PV = \frac{m}{mw} RT \tag{10.2}$$

Rearranging gives us a way to calculate the mass if we know the volume of the gas:

$$\frac{PV\,mw}{RT} = m \tag{10.3}$$

Let's assume that we have 1.0 volumes of carbon dioxide in our beer. This means we have 1.0 L of carbon dioxide in 1.0 L of beer. If we assume that we measure 1.0 volume of carbon dioxide at 1 atmosphere pressure and room temperature (25 °C, 298 K), the mass of CO_2 dissolved in the beer can be calculated (note that carbon dioxide has a molecular mass of 44 g/mol):

$$\frac{(1\,\text{atm})(1.0\,\text{L})\left(44\,\frac{\text{g}}{\text{mol}}\right)}{\left(0.08206\,\frac{\text{Latm}}{\text{molK}}\right)(298\,\text{K})} = 1.799\,\text{g}\,CO_2$$

In 1.0 L of beer, 1.799 g is equivalent to 1799 mg. So, this would equate to 1799 mg/L or 1799 ppm CO_2 in the beer. Similarly, 2.0 volumes of CO_2 would equate to 3.599 g/L or 3599 mg/L or 3599 ppm CO_2. This, of course, assumes that the temperature of the beer is 298 K (25 °C) and that the external pressure of the system is 1 atm (760 mmHg). The amounts would be slightly different if the pressure and temperature deviate from these values.

10.3.2 Equipment Used to Carbonate

Forced carbonation can be accomplished using a number of different methods. Commonly, the process takes place in the cold maturation tank or inline as the beer is transferred. There may even be a second or third location where carbon dioxide is added, such as immediately before packaging. These secondary locations are often used to adjust the carbonation of the beer as it finishes the brewing process.

It should be noted that once the beer is carbonated, the brewer must maintain the pressure and temperature of the beer so that it doesn't decarbonate. For example, if the beer is warmed by even a few degrees, it could decarbonate and release some of the dissovled carbon dioxide. This would result in the formation of bubbles in the beer that could damage equipment or even cause pumps to fail. The brewer knows this and takes significant precautions when transferring carbonated beer by ensuring the pressure is sufficient and that the beer has been fully chilled.

There are several types of carbonation devices that are used in the brewery. These include:

- Static systems
- Carbonation stones
- Nozzles
- Venturi tubes

The static systems are exactly like they sound. The beer is placed in a vessel and carbon dioxide pumped into the vessel until a set pressure is attained. Then, over time, the carbon dioxide slowly dissolves in the beer. As it does, the pressure of the carbon dioxide is adjusted. Eventually, the correct volumes of CO_2 is added to the beer.

Static systems are wonderful in that there is no specialized equipment required. Moreover, all that is required is a tank of carbon dioxide and time. Often, the brewer will wait 1–3 days for the carbon dioxide levels in the beer to be adjusted. Because there is no specialized equipment required, this type of carbonation can even be performed on beer that has been packaged in kegs.

The issues with this form of carbonation are only limited to the amount of time that it takes to complete the carbonation. However, in large vessels, the carbon dioxide tends to stratify in the beer. The carbon dioxide concentration at the top of the vessel is much greater than the concentration at the bottom of the vessel. Thus, the brewer may turn on pumps to circulate the beer within the vessel and help speed the dissolution of CO_2. In kegs, the beer can be "stirred" by shaking the keg as the CO_2 is added.

Carbonation stones can also be employed in either a large vessel or in-line within a hard pipe. These are metal tubes that have been sintered at the end revealing a network of very small holes. They are placed at the bottom of a tank and the carbon dioxide is pushed through the carbonation stone. The result is a stream of very small bubbles that rise up through the beer. As the pressure increases, the small bubbles can easily dissolve. And, the action of the bubbles rising in one part of the tank causes the beer to circulate and mix.

Carbonation stones that are placed inline have the same feature. The CO_2 is added as a stream of very tiny bubbles into the beer as it is transferred. This causes the carbon dioxide to dissolve very quickly. The biggest issue with the carbonation stones lies in the fact that they have quite a lot of small holes that can be difficult to clean.

Nozzles can also be installed either in a vessel or in-line in a pipe. Very much like an injector nozzle used in the automotive industry, the carbon dioxide is pushed through the very small opening in the device. This results in fairly small bubbles of carbon dioxide that dissolve quickly due to their size. Often a vessel has multiple nozzles in order to improve the speed at which a beer gets carbonated.

The issues that occur with nozzles are similar to those with the carbonation stones. First, the small hole within each nozzle must be cleaned regularly. In addition, the nozzle doesn't generate the super small bubbles that occur with carbonation stones, so the beer doesn't circulate within a vessel as much as it does with a carbonation stone.

The most efficient way to force carbonate a beer involves the use of a Venturi tube. This device requires that it is placed in-line and that the beer to be carbonated flows through the device. The tube contains a constriction that increases the velocity of the beer as it flows. This also causes the pressure to drop at that location. Carbon dioxide gas is injected into the flow. Shortly after the constriction, the flowing beer returns to its original velocity and original pressure. The gas that's been added to the beer then dissolves as the beer flows.

By adjusting the flow rate through the Venturi and the pressure of the carbon dioxide, the carbonation level can be highly controlled. In addition, the ability to use the Venturi as the beer flows into a vessel means that it can be used at almost any location within the brewery. The disadvantages of this device focus primarily on cleaning the device after use. Because it is in-line, a thorough cleaning of both the gas port and the tube itself is important.

Carbon dioxide can also be bubbled through the beer as a way to scrub other compounds from the liquid. Volatile compounds, such as DMS, H_2S, and O_2, can be removed in this fashion. This is best accomplished by leaving the vessel open to the atmosphere to allow these volatile compounds to escape. Thought of another way, the undesirable compounds can be scrubbed from the beer using carbon dioxide.

If the beer was naturally carbonated in the primary or secondary fermenter, the pressure of CO_2 could be too great for the particular style. So, the brewer could also adjust the volumes of CO_2 by decreasing the pressure on the tank. Care must be taken in this case to make sure that the pressure is slowly reduced in order to reduce the amount of foam that is generated during the process.

10.3.3 Issues with Carbonation

The best place in the brewery to force carbonate the beer is during the transfer to the cold maturation tank, or within the cold maturation tank itself. This is the best place to conduct the carbonation. At this stage, the beer is chilled to be at its coldest. This increases the ease at which the carbonation can occur.

Unfortunately, this also creates the requirement that the brewer maintain the temperature and pressure of the beer until it is packaged. So, at each of the points of the process until packaging, the pressure and temperature of the beer is monitored. In addition, the volume of CO_2 in the beer is also evaluated.

One place within the product stream that causes issues occurs when there is a restriction or constriction within the piping. These constrictions cause the velocity of the beer to increase and the pressure of the flowing beer to decrease. The result may be the decarbonation of the beer.

Additionally, the brewer takes special care with the rate of flow of the beer once it is carbonated. If the beer flows too fast, it becomes a turbulent flow. This causes eddies and currents to result in decarbonation of the beer. Thus, the velocity of the flowing beer is maintained below 1.5 m per second.

For these reasons, the brewer typically has installed secondary locations along the production line where the carbonation level of the product can be adjusted. One common location to adjust the carbonation level is immediately before packaging. This helps ensure that the beer is packaged with the correct level of CO_2.

Chapter Summary

Section 10.1
  Conditioning is performed on beer to ensure flavor stability, reduce haze, and carbonate the beer.
  Lagering is a fermentation process where the temperature is kept low, causing yeast metabolism to slow but dramatically reducing haze.
  Warm conditioning, known as the diacetyl rest, is performed in order to speed the uptake of diacetyl during fermentation.
Section 10.2
  The conditioning tank is a specialized vessel used to finish the beer before sending it to the packaging line.
  Cask conditioning naturally carbonates the beer and allows flavor stabilization in a more traditional method.
Section 10.3
  Carbonating beer requires increasing pressure and decreasing temperature.
  In addition to static systems, beer can be carbonated using nozzles, carbonation stones, or Venturi tubes.
  The velocity of carbonating beer is maintained below 1.5 m/s to avoid turbulent flow.

Questions to Consider

1. Use Fig. 10.1 and add a line that indicates the concentration of ethanol as the fermentation progresses.
2. Use Fig. 10.1 and add a line that indicates the concentration of α-acetolactate.
3. Consider Fig. 10.3. Does this figure explain why the reaction is faster when the pH is lower? Why or why not?
4. Use the Internet and look up the freezing point (i.e., melting point) of ethanol–water solutions. At what concentration of ethanol would a beer need to be in order to be lagered at $-3\ ^\circ C$?
5. Why is the carbonation stone placed at the bottom of the conditioning tank?
6. A brewer at 5000 ft. altitude is carbonating beer. Is the concentration of CO_2 in a bottle of beer less than a beer that is made at sea level?
7. If a beer has 2.2 volumes of CO_2, what is the concentration in ppm? Assume that the measurement is done at 0.80 atm and 25 $^\circ C$. How does this compare to a measurement performed at 1.0 atm and 25 $^\circ C$?
8. A beer is found to contain 1200 ppm CO_2. How many volumes of beer is this?
9. In our calculations of the relationship between volumes of CO_2 and its concentration in ppm, we assume that the gas behaves ideally. CO_2 is not an ideal gas. For one reason, CO_2 interacts with itself (an ideal gas does not). If we were to perform the volumes to ppm calculation assuming CO_2 was a real gas, how would this change the result?
10. Calculate the ppm CO_2 in beer if it contains 2.6 volumes of CO2. Assume the measurement is performed at 25 $^\circ C$ and 1.0 atm. Would this be different if the gas was N_2?
11. A brewer adds silica gel to the beer in the conditioning tank. Why would this be done and what disadvantages are there to doing so?
12. A brewer adds caramel sugar instead of caramel color to the beer in a conditioning tank. What would the effect of this be?
13. If a beer is found to only contain 250,000 yeast cells per milliliter, does the brewer have to add yeast when the beer is transferred to the conditioning tank? Why or why not?
14. A beer is transferred from the primary fermenter to the conditioning tank. It is found to contain ten million cells per pint. Is this enough yeast to perform a natural carbonation?
15. Why is oxygen not added when additional yeast is added to the conditioning tank?
16. Given the answer to question 15, what step in the process of transferring beer into a conditioning tank must be followed?
17. What would a brewer do to ensure the beer maintained carbonation as it passed through a valve that caused a slight constriction in the flow?
18. A brewer wishes to prepare a strawberry flavored ale. This can be done using fresh strawberry puree, strawberry preserves, or artificial strawberry flavoring. Describe the pros and cons of each method.

19. Look up the structure of PVPP on the Internet. Then, indicate how this compound could mimic the structure of a protein.
20. Which do you think would be more volatile and provide an aroma to the beer; diacetyl, acetoin, or 2,3-butanediol? Why did you choose your answer?
21. In the introduction to this chapter, we noted that conditioning can reduce or eliminate bacteria in the beer. Describe how this might occur?
22. Calcium oxalate, also known as beer stone, can be an issue in beer. If the brewer thinks it may be an issue, how would it be removed in the conditioning tank?
23. A brewer forgets to perform a warm conditioning step. Can this step be added later in the primary fermentation? What effect would doing so have on the beer?
24. Why does lagering tend to clarify beer?
25. Why is the cask larger in the center than on the ends?
26. A beer is pumped into a 10 m tall conditioning tank. The pump is 0.5 m below the filling port of the conditioning tank. What is the delivery head required to begin the transfer? What is the delivery head of the pump when the tank is full?
27. Would the delivery head for a pump be less if the beer from question 26 is transferred into the conditioning tank through the CIP ball?
28. What would happen to a carbonated beer in the conditioning tank, if the glycol cooling system breaks and the temperature rises from 40 °F to room temperature (72 °F)?
29. Using the information we've uncovered in this text, what would you do if the bitterness (by IBU measurement) did not conform to the brewer's expectations? Be sure to indicate what to do if it is too low or too high.
30. How many grams of carbon dioxide are dissolved in beer that is reported as containing 2.5 volumes of CO_2?

Laboratory Exercises

Diacetyl Determination in Beer

The measurement of diacetyl concentrations in fermenting wort are required to determine the ending point of the fermentation. In this experiment, diacetyl will be measured in commercial beer, and if possible, in a fermenting sample of wort.

Equipment Needed

2 or 3 12 oz. samples of beer
Distillation setup (250 mL distilling flask, heating mantle, still head, condenser)
Graduated cylinder, 50 mL
Graduated cylinder, 5 mL
Volumetric flask, 10 mL
Pipettes, dropping with a bulb
Diacetyl (aka, 2,3-butanedione) 0.500 g dissolved in 1.0 L water. Store in the dark in a cool location
Water, distilled

α-naphthol solution (4 g α-naphthol in 100 mL isopropanol. Add decolorizing carbon, shake or stir for an hour, then filter into an amber bottle.)

Creatine solution (0.3 g creatine in 80 mL 40% aqueous KOH, store cold.)

Optional

Fermenting wort (prepare a 1.040 wort from dry malt extract and begin fermentation at room temperature 24–36 h prior to lab.

Experiment

Obtain 100 mL of beer (or fermenting wort). Place this into the distillation setup (see Fig. 10.10) and distill at least 15 mL into a 50 mL graduated cylinder con taining about 5 mL water. Once the sample has been collected, turn off the distillation apparatus and dilute the distillate to 25 mL with water. Then, clean out the distillation apparatus and proceed to the next beer sample. WARNING: the samples may foam excessively during the distillation. Monitor the application of heat so that the distillation doesn't boil over.

Prepare solutions of diacetyl as follows. From the stock solution (500 ppm diacetyl), take 1 mL of the stock solution and add 99 mL water. This solution is 5 ppm diacetyl.

Prepare 5 standards from 0.5, 1.0, 2.0, 3.0, and 4.0 mL of the 5 ppm diacetyl solution by adding the indicated amount of the diacetyl solution to a 10 mL volumetric flask. Then add 1.0 mL of α-naphthol solution and swirl. Then add 0.5 mL of creatine solution and swirl again. Finally, add water to dilute the sample to the mark. Invert a few times to mix and then pour the solution into a test tube until ready to measure. Repeat the formation of the other standards by repeating the procedure with the indicated amounts of the 5 ppm diacetyl solution.

Exactly 5 min after pouring into the test tube, read the absorbance of the solution at 530 nm. Once all solutions have been measured, make a plot of the concentration

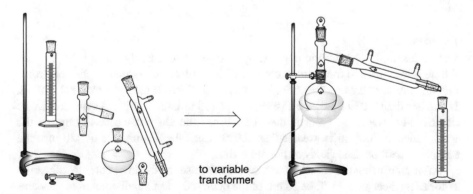

to variable transformer

Fig. 10.10 Distillation Setup. The parts on the left are assembled into the apparatus. Note the addition of the variable transformer to control the heating mantle. The condenser is cooled by attaching a hose to the condenser from the sink and running a hose to the drain

of diacetyl versus the absorbance of the solutions. This is the standard curve (known as the Beer-Lambert plot, or Beer's Law Plot) for the measurement.

Then, each distilled sample is treated and measured. To do this, take 5 mL of the sample and add it to a 10 mL volumetric flask. Then add 1.0 mL α-naphthol solution and swirl. Add 0.5 mL creatine solution and swirl again. Finally, dilute to the mark and invert the volumetric flask multiple times to mix. Pour the solution into a test tube. Exactly 5 min after pouring into the test tube, the sample should be placed into the spectrometer and its absorbance measured at 530 nm.

Determine the concentration of each of the distilled samples from the standard curve by referencing the absorbance of the distilled sample and determining the concentration of diacetyl in ppm.

Optional Experiment

A fermenting sample of wort can be sampled periodically (i.e, every 3 h) over a multiple day period. If this is done, each 100 mL sample collected is placed into a plastic bottle and cooled to 2 °C until measured in the laboratory. If this is done, a plot of the concentration of diacetyl can be made for the particular fermentation.

Adjusting the Color

This experiment is designed to illustrate how the SRM color of a beer sample can be adjusted in the conditioning tank.

Equipment Needed

Beer sample, 12 oz. (best for this experiment if it is a very clear light colored beer)
Caramel color solution (or a degassed sample of a dark and clear beer)
Spectrometer capable of measuring at 430 nm and 700 nm

Experiment

Obtain a sample of beer and degas it by shaking it repeatedly for at least 10 min. Allow it to settle and the foam to collapse. Then, place the sample in the spectrometer and measure the absorbance at 430 nm and 700 nm. If the absorbance at 700 nm is smaller than 0.039 times the absorbance at 430 nm, the sample is considered free of turbidity. The SRM color is then 12.7 times the absorbance at 430 nm. If the absorbance at 700 nm is greater than 0.039 times the absorbance at 430 nm, the sample cannot be used (it is considered turbid).

After measuring the color of the beer sample, use the caramel color to adjust the color of the beer sample. Take 10 mL of the beer and add a small amount of the caramel color (record exactly how much was added). Then, determine the SRM color of the sample. Repeat this at least 3 additional times.

Create a plot of the SRM color of the beer versus the amount of caramel color added. Is the plot linear? Why or why not?

Verify the plot by creating a beer sample with an SRM color of 18 and confirming the color using the spectrometer.

Clarification and Filtration
11

11.1 Introduction

Even in brands that are not perfectly clear or bright, the brewer does take steps to remove most of the solids that result during the brewing process. Those solids can eventually result in a sediment at the bottom of the bottle, can, or keg. And no consumer thinks that sediment is a wonderful part of their experience. Despite the fact that the sediment typically has no impact on the flavor of beer, consumers reject beers with sediment completely.

Thus, brewers work with their product to ensure that the solids are removed. As we'll see in this chapter, those solids can be removed naturally by allowing them time to precipitate to the bottom of the vessel. In some cases, the beer needs to pass through a filtration device in order to remove the tiny solids that remain suspended.

Why does the brewer care about removing these solids? Three very important reasons drive that decision:

- Consumers expect that the beer they drink has the correct appearance.
- Brewers expect the brand to have a particular quantity of suspended solids.
- Beer brands with fewer solids have a longer shelf life.

In some cases, the consumer expectations and the brand specifications are the major reasons to ensure that the beer has the correct amount of suspended solids. But in most cases, extending the shelf life of the beer to the longest that it can be is the main reason that the beer is clarified or filtered. Whatever the reason, this chapter will explore the clarification and filtration processes in detail.

© The Author(s), under exclusive license to Springer Nature Switzerland AG 2021
M. Mosher, K. Trantham, *Brewing Science: A Multidisciplinary Approach*,
https://doi.org/10.1007/978-3-030-73419-0_11

11.2   Colloids and Colloidal Stability

A colloid is a term used by the brewer to describe the physical phenomenon that happens when a beer gets cloudy. In this section, we'll explore the process of haze formation and how it translates to the formation of solids in the finished product. These can be very detrimental to the brewer's and consumer's perception of the beer.

Colloids occur in many different solutions that we are exposed to every day. Milk, orange juice, smoke, and muddy rivers are all examples of colloids. What do these four things have in common? They remain opaque or cloudy for long periods of time. Let's examine them in greater detail especially as they apply to beer.

11.2.1 What Is a Colloid?

All chemical compounds that are found in beer must be soluble in order to result in a clear or bright beer. This includes small molecules such as ethanol and isohumulone, but also those much larger molecules such as yeast, proteins, and polyphenols that arise during the brewing process. The solubility of these compounds is highly dependent upon the number of hydrogen bonds and other interactions that they can form with water, the base liquid in beer. If the number of hydrogen bonds compared to the size of the compound is just right, the compound will dissolve. If the size of the particles is too large or the number of interactions with water is not enough, the compound will not dissolve. Instead it will remain as a solid. For example, ethanol is a small molecule that has strong interactions (hydrogen bonds) with water. Those interactions are so strong that the ethanol dissolves completely in water. On the other hand, a single yeast cell does have some interactions with water, but those interactions don't account for its very large size. Instead, it tends to exist as a suspended solid particle.

When a suspended solid particle is less than about one micrometer in size, it can remain suspended and form a *colloid*. Colloids are basically liquids that have small-sized particles suspended within them. Strong interactions between the particles and the liquid keeps them suspended and they won't settle or precipitate. If the interactions with the liquid are limited or non-existent, the solid will precipitate or settle to the bottom of the vessel. In those cases, a colloid does not form. Instead, over time, a sediment forms at the bottom as the particles fall out of solution.

Solids tend to reflect and deflect light, rather than allow the light to pass through them. Because a colloid is made up of solids suspended in the solution, the solutions appear cloudy or hazy when you look through them. And, because of the buoyancy of the solids in the solution, they remain cloudy or hazy for an indefinite period of time. *Buoyancy*, a measure of the interactions between the particles and the solution, helps keep the solids from precipitating and forming a sediment. In other words, if a colloid exists, it is very difficult to remove it from the beer.

In some cases, the solid particles attract other solids to themselves. This additional binding of solids causes the size of the particles to increase. Eventually, the size rises above one micrometer. When that happens, the solid's buoyancy often isn't enough to keep it suspended in the solution. A precipitate forms that slowly settles out of the solution.

11.2.2 Formation of Colloids in Beer

Colloids form in beer because of the presence of a series of compounds. The first are the different proteins that exist in beer. The majority of the proteins arises from the malt during the mashing process. A small amount comes from the hops and yeast that are added later in the process.

Much of the mass of the proteins is removed during the whirlpool and during settling that occurs during fermentation and cold maturation. However, a small amount can remain as both soluble proteins and suspended solids in the finished beer.

The second important compound that results in colloid formation is the polyphenol. These compounds arise to a small extent from the husk material on the malt during the mashing process. Additional polyphenols show up during the boiling process as hop material is added to the beer. Some of these are removed during the whirlpool, but much of the polyphenols remain dissolved.

The last important compound that aids in the formation of a colloid are the dissolved metals, such as iron and copper, that exist in the brewing water. Even very small quantities of these metal ions can be enough to have a huge impact on the formation of a colloid.

The colloid then forms with the metals coordinate (bind) to the polyphenols and the proteins. The interaction is strong enough that the protein-polyphenol becomes large enough to be insoluble in the beer. As long as the size of the complex is small enough, it stays suspended as a colloid. But when that size grows too large, it will start to precipitate.

The temperature plays a big role in the formation of colloids too. As the temperature decreases, compounds become less soluble in the solution. If there are soluble compounds that are right at the border between soluble and insoluble (i.e., their size is right around one micrometer), they can become a colloid. The result is that a clear beer at room temperature will turn cloudy when the temperature is lowered (Fig. 11.1). This is known as *chill haze*. And when the beer is warmed up again, the cloudiness disappears as the solids re-dissolve in the beer. We initially introduced haze in Chap. 10.

Chill haze is particularly noted when there are lots of proteins and polyphenols dissolved in the beer. Any exposure to oxygen causes the size of the polyphenols to increase. This occurs as the polyphenols get oxidized. Then, they react with each other to dimerize. The dimers continue to add additional units of polyphenols with each oxidation. These large polyphenols bind to proteins to make even larger complexes. And when the complexes get large enough, they can form chill haze.

Over time, the complexes that result in chill haze can get larger and larger. When they do, the particles they form get larger than the one micrometer size. A colloid forms that doesn't redissolve in the beer when it is warmed up. In other words, if not addressed, a chill haze will eventually turn into a *permanent haze*. And if the

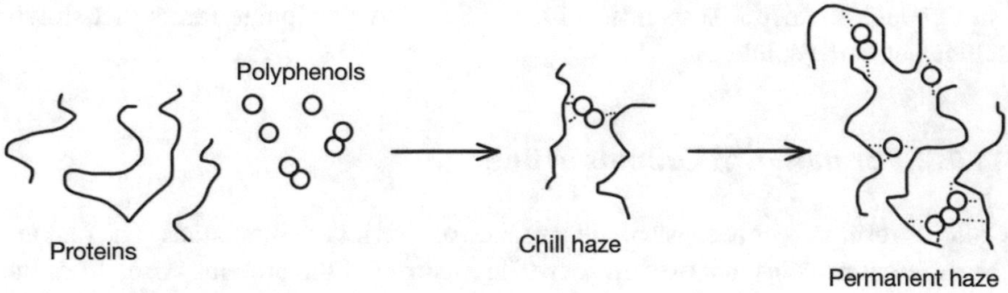

Fig. 11.1 Haze formation in beer. Proteins and polyphenols can combine to make a colloid at lower temperatures known as chill haze. As time, temperature, and oxygen impact the beer, the chill haze colloid gets larger and forms permanent haze

permanent haze is allowed to remain, it will continue to grow until it produces a precipitate. That wonderfully bright beer now is cloudy with sediment at the bottom.

In addition to the chill and permanent haze that results from complexes of proteins, polyphenols, metal ions, and oxygen, there are other compounds that can result in a hazy beer. These include compounds that aren't very soluble in water such as calcium oxalate. It also includes individual cells of yeast and bacteria. Solid calcium oxalate that precipitates, also known as beer stone, arises from excess calcium interacting with small amounts of oxalate present in the malt. This solid forms a very hard crust that can coat the surfaces of the vessels. If the yeast hasn't been removed by flocculating, their presence in the beer can result in a hazy product. And bacteria can do the same thing. Large populations of some bacteria in the beer can also result in a haze.

11.2.3 Turbidity Measurements

Brewers are often concerned with the clarity of their beer. This is especially true for brands that are supposed to be bright. And to ensure that the brand is actually bright, breweries employ a laboratory or install devices in-line to measure the haziness of the beer. These analyses typically measure the turbidity of the beer.

Turbidity is just another word for the cloudiness of the beer. A turbid sample has solid particles suspended within it. Those particles deflect or reflect light as it passes through the sample. Because the light is scattered, the sample appears opaque or cloudy. Milk is a good example of a colloid. It is a water solution that contains large proteins and other compounds that do not settle from the solution. Instead, they are suspended in solution and scatter light to such an extent that the solution appears white.

The measurement of turbidity can be accomplished using one of two different methods. In the first method, known as an absorption method, a beam of light is directed through a sample at a wavelength that is not absorbed by any compounds in the beer. The amount of light that goes through the sample is inversely related to the amount of turbidity.

The second method, known as the indirect detection method, involves the use of an instrument that is similar to the absorption UV-vis spectrometer. Because a turbid sample will scatter light, the indirect detection method is set up to measure that scatter. In other words, the instrument's detector is not placed in the direct line of the beam of light. Then, the detector can measure the amount of light that is scattered. The detector response in this type of instrument is directly related to the amount of turbidity.

Both methods are used in the brewing industry. The first can be done using the UV-vis spectrometer that is used for many other measurements in the laboratory. This is beneficial because the brewery only needs to purchase one instrument to accomplish many different analyses. Unfortunately, the absorption method is not as accurate when measuring slightly-to-moderately turbid samples. It does, however, work very well when the sample is moderately to very turbid. Brewers know this, and for those brands that are designed to be bright, the indirect detection measurement using its specialized instrument provides the best measurement.

Measuring the turbidity is very useful when the brewery wants to quantify the level of cloudiness in a beer. For example, if the brewer suspects a potential chill haze problem with a brand, measuring the turbidity can be useful. If a potential issue with yeast flocculation is noted, the turbidity of the beer during warm maturation could be used to confirm this fact. In essence, the turbidity of a beer is not just a measure of the cloudiness, it can be used to hunt down and solve issues with the brewing process.

11.2.4 Shelf Life

The biggest issue that a brewer faces is the shelf life of the product. This is a measure of how long the beer can survive before the flavor or the cloudiness of the beer changes to a point where it doesn't represent the beer brand's desired quality. The key to this is that beer is a very fragile product. It is made in a highly reducing environment (the yeast removes essentially all oxygen during fermentation), so that the chemical compounds in the beer are very easy to oxidize. In essence, any amount of exposure to oxygen will have a detrimental impact on the finished product.

The flavor of the beer does change over time. Natural aging occurs where the compounds in the beer react with each other and form new compounds. These new compounds can impart a flavor to the beer, and over time, this becomes notable to the consumer. Unfortunately, there are other things that can significantly impact the aging process. For example, if the temperature is not held at the appropriate temperature, the aging process speeds up dramatically. As a rule of thumb, for every ten-degree increase, the rate of aging doubles. The presence of yeast in the finished product can also dramatically increase the rate of aging.

The cloudiness of the beer can also change over time. This occurs as the proteins and polyphenols slowly combine to form haze. Initially, this will show up as chill haze, but over time, it will turn into permanent haze and then into sediment. While these do not tend to impact the flavor of the beer, they do impact the appearance of the finished product and detract from the consumer's enjoyment. Just like with the flavor, the temperature of the beer can significantly impact the colloidal stability of the product. For every ten degree increase in temperature, the rate of haze formation doubles.

In addition to temperature, the most important factor that has an impact on the shelf life of the beer is oxygen. The highly reducing environment of the beer means that any oxygen exposure results in the oxidation of chemical compounds. This impacts the flavor and greatly speeds the rate at which the flavor changes. Even if the oxidation doesn't form the notable cardboard or stale flavors associated with oxygen exposure, the flavor can still change quickly. Equally as important is the fact that the polyphenols quickly react with oxygen and form larger polyphenols. Thus, the rate of haze formation increases dramatically.

The best way to ensure that a beer has the longest shelf life possible is to make sure the beer has a colloidal stability to last. If the beer forms a colloid before the shelf life date, a consumer could pour a hazy beer when the specifications indicate the beer should be bright. And the consumer ends up with a poor opinion of the quality of the product. By controlling the temperature and oxygen levels in the finished product, the brewer can significantly reduce changes to the beer's flavor stability and colloidal stability. This results in the longest shelf life possible and ensures the consumer's perception of quality.

CHECKPOINT 11.1
Other than the ones listed in the chapter, can you name another colloid?
In your own words, describe how haze forms in beer after cold conditioning.

11.3   Clarification

When the brewer is trying to produce a beer brand with the greatest colloidal stability, they focus on clarifying the beer as best as possible. Clarification is the process by which solids in the beer are removed and the precursors to those solids are controlled. Clarification often refers to the removal of solids using gravity.

Beer clarifies naturally by sedimentation. Sedimentation refers to the process by which a solid that is suspended in the beer falls to the bottom of the vessel. The process can be described by Stokes' Law. George Stokes, in 1851, derived an expression that relates how particles move through a solution. One statement of this law is

$$\omega = \frac{2\left(\rho_p - \rho_b\right)g\,r^2}{9\eta}$$

where ω is the velocity or speed of the particle in m/s as it moves in the solution $(\rho_p - \rho_b)$ is the difference in the densities between the particle and the beer in kg/m^3

g is the gravitation force (g = 9.8 m/s^2)

r is the radius of the particles in meters

η is the dynamic viscosity of the solution in kg/m•s

Stokes' law basically says that there are four things that impact how long it takes for a particle to precipitate from the solution. First, the difference in the densities of the particle and the beer is important. In other words, the fluffy, airy particles sediment very slowly from the beer. The dense, compact particles sediment much faster. Second, the gravity of the system determines how fast something precipitates. As we'll see later, if the gravity can be increased, the speed of sedimentation will increase. Third, as the radius of the particles increases, they tend to sediment quicker. And fourth, as the viscosity of the beer increases, the sedimentation takes longer.

11.3.1 During Boiling

During the boiling process, the proteins and polyphenols in the mixture can be impacted. Vigorous boiling tends to degrade the proteins, resulting in smaller pieces. The result, when the temperature cools during the whirlpool is smaller, less-dense particles that slowly precipitate. Longer and more vigorous boiling results in the formation of smaller and less-dense particles. In other words, boiling seems to work against sedimentation.

Luckily, the whirlpool process does remove a lot of the heavier, denser, and larger particles such as hop and malt material. Large carbohydrates such as starch also precipitate at this stage of the process. However, the whirlpool doesn't remove everything. Smaller suspended solids do get to the next stage. In addition, the precursors to the formation of additional solids pass through to the next step.

11.3.2 During Fermentation

The temperature during fermentation is lower than during the whirlpool. Thus, the particle sizes tend to increase as more and more precursors and particles collect into bigger complexes. Large particles start to precipitate from the beer. These particles also begin to adhere to yeast and as it flocculates, the particles fall out of the beer.

Unfortunately, in the initial stages of fermentation, the active formation of carbon dioxide starts churning the mixture. This stirs the mixture and keeps a lot of the solids suspended. Luckily, though, as the fermentation settles down and the yeast start to flocculate, they pull the solids out of the solution.

It is for this reason that the first batch of yeast that flocculates tends to have lots of these solids mixed in. This is fairly obvious when you look at the color of the yeast that has flocculated. Brewers know this and tend to discard the first portion of yeast that flocculates because of the large amount of sediment that is present.

11.3.3 During Maturation

Suspended solids continue to sediment slowly during warm maturation. But when the beer is transferred to cold maturation, the temperature of the system is moved down a lot. The particles become less soluble and start to aggregate more. In other words, their size increases and they start to sediment.

It is this reason that the cold maturation process often takes place in horizontal tanks. The horizontal tanks are much shallower than the vertical CCVs that are typically used in fermentation. In other words, the distance a particle has to travel to completely sediment is greatly reduced. The small and less-dense particles that sediment slowly don't have to go as far. The result is that the beer gets clearer in a much shorter period of time.

Sedimentation can also be enhanced during cold maturation by the addition of a stabilizing agent. These agents are compounds that actively bind to either proteins or polyphenols. The resulting particle gets bigger and denser. That results in their formation of a solid that sediments faster.

Stabilizing agents include *finings* that specifically bind to yeast and proteins. Examples include isinglass (collagen derived from the swim bladders of fish) and gelatin (collagen derived from animal sources). These compounds are dosed into the beer at just the right amount to bind to the yeast and proteins (both the soluble and suspended solids). The brewer often has to estimate exactly the right amount in order to get all of the solids to settle without leaving any behind and without overdosing the beer with these agents. Note that finings don't bind to polyphenols or other compounds that might result in haze or solids.

Other stabilizing agents include *silica gel* (acidified sodium silicate) that binds to proteins, *tannic acid* (mimics polyphenols) that binds to proteins, and *polyvinyl polypyrrolidinone* (PVPP, a synthetic molecule that mimics proteins) that binds to polyphenols. Based on the solids and precursors to solids that exist in the beer, the brewer can select one or a combination of stabilizing agents. The result is a large amount of precipitate and a much brighter beer.

11.3.4 The Centrifuge

In many cases, the use of a centrifuge is the best way to remove the insoluble compounds in beer. While the centrifuge won't remove any of the precursors that may eventually form a solid, the centrifuge does an excellent job of clarifying the product. The centrifuge works by dramatically increasing the gravitational force on the beer. This directly impacts the speed at which the solids settle.

Because of the utility of the centrifuge, it can be employed in many locations along the process stream. For example, the centrifuge can be used to recover beer trapped in the trub at the bottom of the whirlpool vessel. This would result in a reduction in the losses from each batch of beer and provide a drier version of the solids isolated from the whirlpool (less waste). In another example, the centrifuge could be used immediately after cold maturation to remove all of the unsettled solids and provide a much brighter beer.

The centrifuge has a unique design (see Fig. 11.2). In the design shown, beer enters the device through the top and is directed to the bottom of the bowl where it is spun. Depending upon the speed, forces up to about 20,000 times the force of gravity can be made. The result is that the solids move to the outside of the conical-shaped bowl where they form a sludge. Periodically, the seal at the apex of the conical shape is broken and the solids forced outward from the bowl. The seal is then reformed, and more solids are collected. As the bowl fills up with beer, the only place it has to go is between a series of disks and back upward through the exit port. By the time the beer has exited the system, the solids have been removed.

The advantages to the centrifuge are that it can be used at almost any stage in the brewing process, it has a very low oxygen pickup compared to other processes, and that it is a fairly small device. Another advantage is that the centrifuge can be almost continually operated with minor clean-in-place runs between batches of beer.

The disadvantages include the fact that the centrifuge does impart a sheer force on the beer. Thus, any yeast or proteins in the beer can be broken apart. This can be a problem if the yeast are damaged too much – they can impart an off-flavor to the beer if they lyse (break apart). If the proteins are damaged, the result can be a reduction in the foam that forms on the beer.

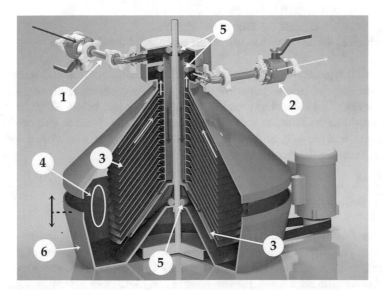

Fig. 11.2 The centrifuge. Hazy beer enters the device at the top (1) and flows downward into the bowl (6). The device spins at a very high rate and the effect of the forces causes the solid particles to collect in the region noted (4). The spinning fins (3) increase the ability of the particles to separate from the beer. The addition of more beer to the system eventually forces the clarified beer up through the fins and out through the top of the device (2). Because of the high rate of spinning, the bearings and seals (5) have to be carefully machined and maintained. Periodically during its operation, the bowl separates and allows the solids to escape

Other disadvantages are relatively high maintenance costs, especially if a part breaks, and a very high noise level. The rapidly spinning bowl is fairly loud. However, the majority of the noise comes from the periodic open and close of the bowl. The loud bang can be shocking to first time users, but the noise is part of the normal operation. The noise is so loud that the manufacturers recommend they are placed in a separate room and that all workers wear hearing protection.

CHECKPOINT 11.2
What are the precursors to the formation of chill haze?
Outline two ways in which excess proteins can be removed from beer.

11.4   Filtration

There are many reasons why the brewer would want to filter their beer. This would be in addition to the natural sedimentation process that occurs in other stages of the brewing process. Thus, at some point after cold conditioning, the brewer evaluates the turbidity of the beer and then determines if the beer needs to be filtered. As we noted in the introduction, the reasons to make the beer bright include:

- The brand is designed to be bright.
- The consumer expects the beer to be bright.
- The brewer wishes to remove all solids and extend the shelf life.

Filtration typically occurs immediately after cold maturation and immediately before sending the beer to pasteurization (if it is flash pasteurized) or packaging. The beer is chilled as cold as possible, typically at or just below 32 °F (0 °C) and then pumped to the filtration unit. Control devices at the filtration unit confirm the temperature and pressure of the beer in order to minimize breakout of carbon dioxide and maximize the settling of solids.

11.4.1 Principles of Filtration

Filters can be broken down into two main classes. In the first class, the device operates on the principle of *sieving*. The holes in the filter indicate the size of the solids that can pass through. Small molecules, such as water, metal ions, and ethanol, pass easily through the filter holes. However, large complexes can get stopped and stuck to the filter. Typical sizes of the particles that might be filtered are shown in Table 11.1. Thus, a filter must be chosen that can hold back the size of the particles that the brewer wishes to remove.

When the beer is ready to be filtered after cold maturation, it is moved into a vessel stationed immediately prior to the filtration device. This *buffer tank* has pressure and temperature controls that can be monitored and potentially adjusted to be

Table 11.1 Particle sizes

Particle	Size	Particle	Size
Yeast	5–10 μm	Bacteria	0.2–2 μm
Haze	~1–2 μm	Metal ions	0.2–0.4 nm
Water	0.27 nm		

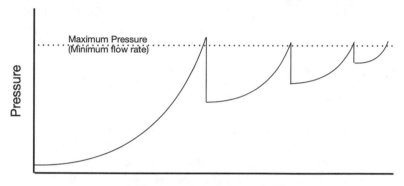

Volume passed through filter

Fig. 11.3 The pressure increases as the volume of the filtered beer increases. The pressure gets so high that it can inhibit the flow of beer through the system. In those cases, a small backflow can be performed that can unclog some of the pores in the system and reduce the overall pressure. This can be repeated until the filter needs to be cleaned

appropriate. The buffer tank is also useful to eliminate or reduce any fluctuations in the flow rate or pressure of the beer entering the filter. Then, when it is needed, the beer is pumped into the filtration device. Back pressure due to pushing the beer through the filter helps to keep the carbonation intact before the filter. The brewer monitors the flow rate through the filter in order to keep the pressure after the filter high enough so that the beer doesn't decarbonate.

Large pumps are needed to push the beer through the filter. And the process increases friction on the beer. It can sometimes result in an increase in the temperature of the system. Thus, a heat exchanger immediately after the filter can be important.

As the beer is passed through the filter, the pressure of the system increases as shown in Fig. 11.3. This is because many of the holes in the filter get plugged with solids. At some point, the pressure gets so high that the risk of a seal breaking or the filter tearing becomes too great. At that point, the process is stopped, the filter cleaned, and the process restarted. In some cases, the filter process can be run backwards to push the solids out of the holes. However, even in those cases, the filter eventually gets plugged so much that cleaning is the only way to fix it.

The other class of device operates as a *depth filter* (Fig. 11.4). In this device, a small amount of a solid material is added to the beer as it is being filtered. The powdered material builds up on a filter. And the beer is filtered as the bed grows. Because the bed continues to build up, the filter doesn't get clogged. The pressure does

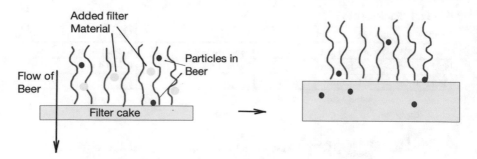

Fig. 11.4 Depth filtration principles. Hazy beer is passed through a filter that is formed by adding powder to the beer. The particles either get trapped in the filter as it forms, or get stuck in the filter because of the very small and winding paths through the powder

continue to rise during the filtration; however, it doesn't do so as quickly as a filter. In fact, the pressure increase is due to the depth of the filter bed. Any particles in the beer are either trapped in the filter bed as it forms or stick to the top of the bed. The beer moves through the filter bed along channels that constantly change as the bed grows.

11.4.2 Filtration Equipment

There are three main types of filters in use in the brewing industry. These include the powder filters, the membrane filters, and the sheet filters.

The powder filters work by the principle of a depth filter. They include the candle, leaf and plate-and-frame filters. The membrane filters operate under the principles of a filter. They include the crossflow filter and the lenticular filter. The sheet filters operate under the principles of a filter, and include not only the sheet filter, but also the lenticular filter (in some cases).

11.4.2.1 The Sheet Filter

The *sheet filter* (see Fig. 11.5) involves a fairly labor-intensive setup. The horizontal device has four bars, two at the top and two at the bottom. A frame is then placed into the device so that it hangs on the top two bars. A filter sheet is then placed next to the frame. Then another frame is placed into the device. This continues until there are alternating frames and sheets across the device. Then, a screw is cranked down slowly until the sheets are compressed tightly between the frames to eliminate or reduce any leaks.

Each frame has an inlet and outlet. The inlets are on one side of the frame and the outlets on the other side. Thus, as the beer moves into the filter, it moves along the system until it enters a frame. It then passes across the sheet and into an adjacent frame, where it enters the exit port of the device.

Often, the sheet filter contains 10 or 20 or more filter sheets. This doesn't mean that the beer is filtered 10 or 20 times. Instead, the beer can "select" one of the

Fig. 11.5 The sheet filter. Filter sheets are added between the yellow and orange frames in the apparatus. The entire device is clamped down onto the sheets to make it water-tight. Beer enters into the system at the top and passes through the sheets on its way to the exit port

frames to enter. After passing through the filter, it enters the opposing frame and exits the device. So, the beer is only filtered once as it passes through the sheet filter. However, the large number of sheets means that a larger quantity of beer can be filtered before the filter fills up with solids and needs to be cleaned.

The advantage to the sheet filter is that the sheets are disposable. In addition, the pore size of the sheets can be selected based upon the level of clarification needed. The disadvantages include the fact that the sheet filter is very labor intensive to set up, perform a pre-run to reduce leaks, and then operate. Luckily cleanup usually involves unscrewing the frames and removing the filters to the waste. The worst disadvantage is that during the setup, a large quantity of beer leaks from the device. This continues until the screws are tightened enough. Even so, under normal operation of the filter, leaks remain and continue to drip beer out of the device. With excellent training and operation techniques, the amount of waste beer can be significantly reduced. However, no matter how it's operated, a small but very real amount of beer is lost during the use of this filter.

11.4.2.2 The Lenticular Filter

The *lenticular filter* is made up of a stainless-steel housing unit that contains a set of filter packets that are stacked around a central tube (Fig. 11.6). The filter packets are made of two layers of filter squished together so that the resulting shape resembles a lentil, hence their name. When beer is to be filtered, it enters through a port at the bottom of the vessel and fills up the entire chamber. It can then pass through either side of any of the filter packets. Once it is in between the filter sheets, it is directed to the central tube where it exits the device.

The lenticular filter uses filter packets that are typically made of membranes. They operate under the principles of a filter and remove particles base on the pore size of the membrane. A large number of filter packets means that the beer can be filtered quickly without clogging the membranes. If needed, the membranes can be

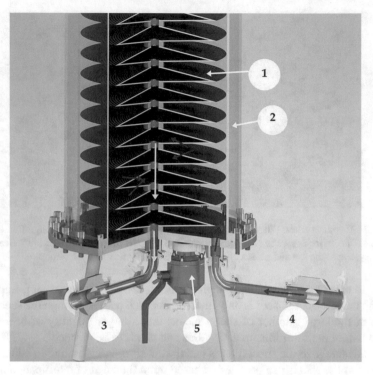

Fig. 11.6 The lenticular filter. The beer to be filtered enters the device at the bottom (4) where it fills the chamber containing the filters (1). The beer passes into the center of each filter where it moves to the central tube and exits the device (3). When the filtration is complete, a valve at the bottom (5) can be opened to help clean the device. The filters can be selected to provide the filtration performance that is desired

de-clogged by backflowing the beer. This gives the filters a longer life before they have to be cleaned. Cleaning is done with hot water or, in some cases with special membranes, cleaned-in-place with warm caustic.

11.4.2.3   Powder Filters

There are multiple versions of the powder filter. Each operates under the same principle of depth filtration. This means that an insoluble powder is added to the beer while it's being filtered. The powder builds up on the membrane and forms a filter bed. That filter bed either traps solids within it as its being formed or stops solids as the beer passes through the filter bed.

This setup requires some additional equipment that can handle the powder. Thus, a powder addition vessel is needed (in addition to the buffer tank). In this vessel, the powder is added with constant stirring to the beer. Then, the beer is pumped into the filter apparatus.

In order to achieve the best filtration, the first powder that is added usually has a relatively large particle size. Thus, this powder can be retained by the membrane and not pass through the pores of the membrane. After the pre-coat is added to the powder addition vessel, the beer is recirculated through the filter until all of the pre-coat has been added and adhered to the membrane. This is easy to tell because the beer that starts out cloudy will eventually become clear as the filter bed sets up.

Then, the filter powder used to clarify the beer is added to the powder addition vessel and the beer is pumped through the filter. As it goes through the filter, the bed increases in size and the beer is filtered of all of the solid particles. The beer is then passed into a buffer tank after the filter and collected there. The pressure and temperature of the beer is monitored in this tank as well to ensure that the carbonation level remains constant.

After the beer has been used in the powder filter, the system is cleaned by backflowing water and then dumping all of the powder out of the device as a slurry. This slurry is sent directly to the waste (in some cases it may go through a centrifuge first to recover any small amounts of beer that were entrapped within the slurry.

The first type of powder filter is very similar to the sheet filter. In this case, the *plate and frame filter* (see Fig. 11.7) is put together just like the sheet filter. Then, as the filtration is operated, the powder is added, and the filter bed is formed. When the filtration is completed, the plate and frame filter is opened, and the powder and sheets are removed and sent to the waste stream. Again, just like in the sheet filter, a small amount of leaking is possible before and during the operation.

Another type of filter is the *leaf filter* (Fig. 11.8). From the outside, this device looks like the lenticular filter. It is comprised of a chamber where the inlet and outlet are attached to the bottom of the vessel. Inside the chamber, there is a central tube with hollow plates spaced out like leaves. The beer and powder are pumped into the chamber and the powder settles onto the leaves and creates the filter bed. When the

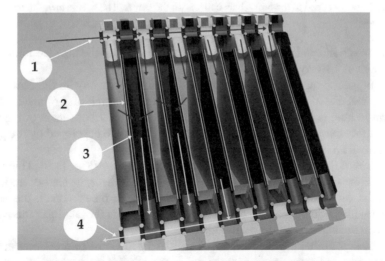

Fig. 11.7 Plate and frame filter. The overall device looks nearly identical to the sheet filter found in Fig. 11.5. However, each frame holds a filter where a filter cake is built up. Then, additional filter powder is added and the filter cake increases. The hazy beer enters through one side of the device (1) and passes through the filter cake (2) into the clarified beer section of the frame (3). Then, it exits the device through a port (4)

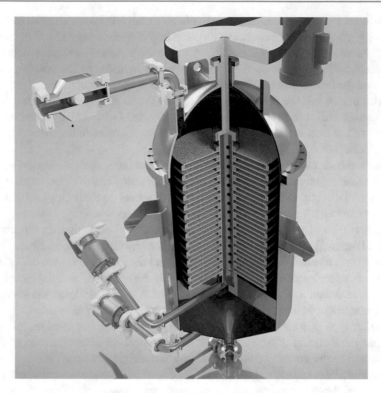

Fig. 11.8 Leaf filter. A cross section of a leaf filter reveals that it is similar to the lenticular filter. A filter powder is added to the beer which collects on the leaves. The beer is then filtered through the filter cake

filtration is completed, the entire device is backflowed and the powder dumped out the bottom as a slurry.

The biggest issue with the leaf filter is that small variations in the flow rate or pressure can dislodge some of the powder and cause it to fall off of the leaf. If that happens, a hole is created in the filter which reduces the effectiveness of the filtration. For this reason, the buffer tank pressure and flow rate developed by the pumping mechanism must be very closely monitored.

The last of the powder filters is known as a *candle filter* (Fig. 11.9). This device has the same outward appearance as the leaf filter, however, instead of horizontal plates, the filtration surfaces are actually long vertical tubes. Again, the beer is pumped with powder into the chamber and then it enters the candles. The powder sets up on the candles and forms the filter bed.

Just as in the other powder filters, a change in the flow rate or pressure can cause the filter cake to loosen or even slough off of the candles. And when that happens, the filtration becomes less effective until the bed sets up again. Unfortunately, if the filter bed is very disrupted, it may not set up so that the pre-coat lays down first. When that happens, the powder can even go through the candles and contaminate

Fig. 11.9 A cross section of a candle filter. Beer enters the device at the bottom (5) and moves upward into the candle section. The candles (4) are made of mesh and catch the filter powder. The beer moves through the filter cake into the top (3) and out through the exit port (1). The pressure of the system is monitored through a port (2). And when the device is to be cleaned, all of the powder can be dumped through the bottom of the vessel (5)

the beer. The brewer is very aware of this and monitors the pressure and flow rates to ensure there is no disruption. Vibrations within the brewery can also be an issue because they can knock the filter cake loose.

11.4.2.4 Crossflow Filters

Given the problems with the other types of filters, the industry has developed a much better filtration device. The *crossflow filter* (Fig. 11.10) is a membrane-based filter that doesn't require powder. Instead, the beer to be filtered is pumped into the device. The beer passes through a series of long porous tubes. It has the option to either stay in the tube and continue moving forward or move through the porous sides of the tube. If it moves across the walls of the tube, it enters another area that exits the device. The solids get stopped by the pores in the walls of the tube.

Because the beer to be filtered can also continue to move through the tube, there is only a minimal pressure that would force the beer through the tube. The beer that makes it through the tubes without crossing the walls gets recycled back to the start of the device. Unlike the designs of the other filter devices, the pressure on the beer in a crossflow filter tends to increase very slightly as the beer is filtered.

Cleaning the crossflow filter is done with a small backflow followed by a caustic rinse. The filter can be cleaned many times without any damage to the filter tubes.

Fig. 11.10 A cross section of a crossflow filter. The overall device usually contains many of these filter setups. Hazy beer enters the device (2) and moves up the filter (4). A small pressure differential allows some of the beer to pass through the filter and enter the filter chamber (5). Then, the filtered beer can exit through the clear beer port (1). If the beer is not filtered, it exits the apparatus (3) where it is recirculated back to the entry port (2)

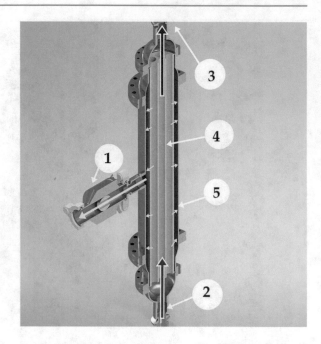

Often a brewer can have multiple filter tube cartridges so that as one is being used, another can be cleaned.

CHECKPOINT 11.3

Outline each of the different filter types and the underlying principles on how they operate.

Which filter requires the most operator time to setup and operate? Explain your reasoning.

11.4.3 Issues with Filtration

The advantages to each filtration device can be used to help select which one is best for a particular brewery or application. There are a wide variety of advantages that have to be considered. However, there isn't just one filter device that is better than all of the rest. In other words, some devices have significant advantages in one application that would not be applied to another use. Things to consider include:

(a) Purchase cost. The sheet filter and lenticular filters are often the least expensive filter devices to purchase.
(b) Operating costs. The operating costs for the sheet filter and lenticular filter are fairly similar. However, the crossflow filter has likely the lowest operating cost.
(c) Space. The lenticular filter and the candle and leaf filters have the smallest footprint and, in some cases, they can be moved to a storage location when not in use.

(d) Longest run before cleaning. The crossflow filter has the longest runs before the filter has to be cleaned.
(e) Versatility. The powder filters have tremendous versatility in that the use of different powders and dosing rates can have a wide range of application. The sheet and lenticular filters are less versatile in that the only filtering options are based upon the filter sheets or filter packs that are available.

Unfortunately, the disadvantages to their use is likely the major reason why a particular filtration device is selected over another. The disadvantages can be broken down into three main categories: product safety, product quality, and worker safety.

11.4.3.1   Product Safety Hazards

The safety of the product can be a strong consideration in choosing a filter. This is especially true with the powder filters. If the pressure of the system increases to a point where the filter membrane fails, it is possible that some of the powder can pass through the filter with the product. The filter powder, depending upon what is used, may pose a health risk to the consumer.

Due to the costs of some of the sheets, membranes, and filter packs, they are often cleaned and reused if possible. Unfortunately, harsh cleaning conditions can result in damage to the filters. So, the brewer may wish to use slightly less harsh cleaners. In this way, it is possible that harmful bacteria can take hold and grow on the filter. Then, when the beer is filtered, those microorganisms transfer to the beer and eventually to the consumer.

11.4.3.2   Product Quality Hazards

One of the more common hazards lies in the area of product quality. The filtration process can add oxygen to the beer. That oxygen can damage the flavor and the colloidal stability of the beer. Thus, when a filter device is set up and made ready for operation, the brewer uses deaerated water. That water is pushed through the system and through the filters in an effort to remove all of the air in the system. Only after that has taken place is the beer introduced to the system. Unfortunately, not all of the oxygen can be removed and some pickup by the beer does take place.

Another harmful addition can occur in filtration. That is the addition of iron. The iron comes from the powders used in powder filtration. Minute traces of iron can be found in many different powders that are commonly used. The manufacturers know this and are constantly improving their procedures to eliminate as much iron as possible. The iron can be detrimental to the flavor of the beer and significantly reduce the shelf life of the finished product.

And, as was mentioned above, it is possible that a powder or filter or membrane isn't sterile before use. The result would be the introduction of microbes to the beer that could damage the flavor or shelf life.

11.4.3.3   Operator Safety Hazards

In most of the filters, following the standard operating procedures leads to a very safe working environment with only a very small risk to worker safety. However, the powder filters do pose a safety hazard. The risk lies in the use of the powder in its dry form.

The powders are made up of very small particles. So small that they create significant dust if they are disturbed. That dust can get into workers' lungs and eyes and cause damage. Most of the powders are made up of *diatomaceous earth*. This is a special type of dirt that comes from the fossils of microscopic sea creatures. The fossils are often not fully intact and have many sharp edges. Thus, when the dust from the diatomaceous earth is inhaled, the sharp edges scratch the inner linings of the lungs. The result is inflammation and scarring of the lungs. And with repeated exposure, this could lead to cancer.

Diatomaceous earth is safe when it is wet and part of a solution or slurry. Unfortunately, it isn't supplied as a slurry from the manufacturer. Instead, it typically comes in bags that have to be opened and poured into a powder slurry vessel where water is added. That means workers can be exposed to the powder at this point. And, if the used slurry is allowed to dry, the problem occurs again. The issue of diatomaceous earth is very real such that in some locations it is classified as hazardous waste. And if that is the case, it will require special handling (and extra cost to dispose of) once it has been used in the brewery.

Chapter Summary

Section 11.1

Consumers and brewers have expectations as to the level of cloudiness of their beer.

The shelf life of a beer is directly related to the clarity of the beer.

Section 11.2

Colloids are solutions that contain suspended solids. They appear opaque or cloudy.

A complex can form between proteins and polyphenols in beer. Oxygen and metal ions enhance the formation of that complex.

If the complex gets too large, it can become insoluble in the beer at low temperatures. This is chill haze.

Chill haze can become permanent haze over time.

The cloudiness of a beer can be measured using either an absorption method or an indirect detection method.

The shelf life of the beer is related to the flavor and colloidal stability of the beer.

Section 11.3

Stokes' law governs how fast a particle can precipitate from a solution.

Naturally, the solids in a beer will sediment. They can be encouraged to sediment quicker with finings, clarification agents, and cold temperatures.

The centrifuge can be used to increase the rate at which sedimentation takes place.

Section 11.4

Filtration of the beer is possible using sheet or membrane filters and depth filters.

Each has its advantages and disadvantages that depend upon the specific brewery and specific application.

Issues with the safety of the filtration process can impact the product safety, product quality, and worker safety.

Questions to Consider

1. What would a brewer do if the consumer expectations for a particular brand were not met by the beer that was purchased?
2. Is a solution of 1.0 g salt in 1.0 L water a colloid? Why or why not? (You may need to consult the internet to assist with the answer.)
3. What are the two important precursors to the formation of haze?
4. A particular brand of beer has a high concentration of metal ions other than calcium in it. What effects would this likely have on the finished beer?
5. Describe the role of oxygen in the formation of haze in beer.
6. What steps should a brewer take to ensure the longest shelf life for a particular brand of beer?
7. As the viscosity increases, what happens to the speed at which a particle precipitates from solution?
8. If the particle density is very similar to the solution density, what effect does this have on the sedimentation rate?
9. List two ways in which a suspended particle can be removed from beer.
10. How does the temperature of the beer affect the rate of sedimentation?
11. Which sediments faster, a particle in a CCV or a particle in a horizontal tank? Why?
12. Describe how an in-line turbidity meter might work using absorption techniques.
13. Describe three things that can be used to help remove proteins from beer.
14. How does a centrifuge result in removing suspended solids from a particular beer brand?
15. What are the two principles of filtration and how does each work to remove suspended solids?
16. Describe how a sheet filter works in your own words.
17. What is one of the biggest issues with the use of a leaf filter or a candle filter?
18. Which filter system has the slowest increase in back pressure? Why?
19. What are the dangers of the use of a powder filter?
20. Which filter system would be most appropriate for a small craft brewer? Explain your choice.
21. What are the issues associated with the use of diatomaceous earth?

Laboratory Exercises

Filtration

This experiment is designed to illustrate the differences in different filtration types and how they work.

Equipment Needed

Erlenmeyer flasks (5125 mL flasks)
Powder funnel
Sand (at least 30 g per student)
Filter paper
Cotton

Experiment

A colloidal solution is obtained by using either a commercial hefeweizen or through the following process:

Add a packet of yeast to 500 mL of malt solution made with dry malt extract to result in a gravity of 1.035–1.045 g/mL. The solution is then stirred for 6 to 12 hours at room temperature and then used immediately.

Each student is given 200 mL of the colloidal solution.

The solution is divided into 4 parts and each 50 mL treated separately. Each is poured into a funnel slowly until at least 30 mL have passed through the filters:

(a) Solution 1 is poured slowly through a funnel until 30 mL is obtained.
(b) Solution 2 is filtered through a funnel lined with a piece of filter paper.
(c) Solution 3 is filtered through a funnel containing a small plug of cotton.
(d) Solution 4 is filtered through a funnel containing a small plug of cotton and 30 grams of sand.

After filtration of the three samples, the results are compared. Specifically, the solutions are lined up in order of their turbidity with the most turbid as the first sample and the least turbid as the last sample.

Which of the solutions was filtered the best?

Which of the solutions was filtered using the principles of sieving? ....of depth filtration?

Using the results of this experiment, which of the filtration processes used in a brewery would likely provide the best result for a filtered beer? Explain your answer.

Packaging 12

12.1 Introduction

At this stage in the process of producing beer, the product is ready for serving to customers. It has gone through the steps from grain to glass, almost. The beer sits in the conditioning tank and just needs a forum to enjoy. In some cases, the brewer wants to serve the beer directly to the customer. This is possible if the conditioning tank is connected to the tap AND if the customer can come to the brewery. But not all customers can.

In this chapter, we'll explore the ways in which the brewer can distribute their craft to the customer. In addition to the taproom or bar at the brewery, the brewer can package their product in what is known as *small pack* or *large pack*. Small pack includes the single serving sized cans or bottles. Large pack refers to multi-serve containers such as kegs. Let's start our exploration with a look back into carbonation.

12.2 Carbonation and Other Gases

In Chap. 10, we uncovered the use of carbon dioxide to carbonate beer in the maturation or conditioning tank. We discovered the process by which brewers calculate the amount of carbon dioxide dissolved in their beer. We even found out the way to determine the number of grams of CO_2 that are dissolved in a given volume of beer.

12.2.1 Pressure Loss in Transferring Liquids

The volumes of CO_2 that are dissolved in beer in the conditioning tank must be retained as the beer moves into the package. Loss of the carbonation along the way must be monitored and avoided at all costs. This means that the beer must be

© The Author(s), under exclusive license to Springer Nature Switzerland AG 2021
M. Mosher, K. Trantham, *Brewing Science: A Multidisciplinary Approach*,
https://doi.org/10.1007/978-3-030-73419-0_12

protected from any reduction in the pressure of the beer due to transferring it from the conditioning tank to the packaging line.

Initially, it may seem that all the brewer needs to do is to maintain the pressure of the maturation tank during the transfer. In other words, as the liquid flows along a horizontal pipe, the pressure of the liquid should remain constant. But this is not the case. As beer, or any liquid for that matter, is pumped from one location to another, it experiences friction. That friction can slow the flow of the liquid. And, as the liquid slows, the pressure of the liquid reduces. If that pressure is reduced to levels that won't support the level of carbonation that exists, the beer will degas.

Friction exists in the piping in many different ways. Henri Darcy studied this effect on the pressure of a liquid as it moves along a pipe. He found that the change in pressure (ΔP) could be calculated by an equation we now call the Darcy Equation (Eq. 12.1). This equation relates the loss of pressure due to the length of a pipe, its flow rate, and the speed of the fluid flow. Note that this is NOT the same as Darcy's Law:

$$\Delta P_{pipe} = 4\phi \frac{L}{d} \rho u^2$$

(12.1)

where ΔP is the pressure change due to friction in Pascals (Pa)
ϕ (phi) is the friction factor (unitless)
L is the length of the pipe in m
d is the diameter of the pipe in m
ρ is the density of the liquid in kg/m^3
u is the mean velocity of the liquid in m/s

The friction factor (ϕ) in this equation is not a constant, but instead is related to a value known as the Reynolds' number (Re). Reynolds' number is a value that relates the forces that move the liquid to the forces related to the viscosity of the liquid. In other words, the Reynolds' number can help determine if the liquid is experiencing laminar flow or turbulent flow. Laminar flow occurs when the liquid flows without mixing laterally. Turbulent flow is just the opposite. Turbulence, or mixing, occurs laterally in a turbulent system. As we would expect, liquids that flow laminarly do not experience much impact from the roughness of the inside of the pipe. Turbulently flowing liquids are greatly impacted from the pipe roughness. In fact, the friction factor for liquids flowing in a pipe can be determined through the use of a Pipe Friction Chart that plots the Reynold's number versus the pipe roughness.

Bends in the pipe, obstructions to the flow, and valves also impact the pressure of the fluid as it flows. The pressure drop can be quite large, even if the length of the pipe isn't very long. Adding each of the friction losses for the individual obstructions allows us to calculate the pressure drop due to these obstructions. This is done using Eq. 12.2:

$$\Delta P_{\text{fittings}} = \Sigma \left(\frac{k \rho u^2}{2} \right)$$ (12.2)

where ΔP is the pressure drop in Pa
k is the constant of the particular obstruction (see Table 12.1)
ρ is the density of the liquid in kg/m³
u is the mean velocity of the liquid in m/s

Table 12.1 lists some approximate constants that indicate the measure of the friction of the particular obstruction. Values less than 1 suggest that the bend, valve, or obstruction has little impact on the pressure change in the flowing liquid. Values greater than 1 suggest a device that has a very large impact on the pressure of the system.

The total pressure loss of the system is then the sum of the pressure loss for the pipe run plus the pressure loss due to the fittings or obstructions:

$$\Delta P_{\text{total}} = \Delta P_{\text{pipe}} + \Delta P_{\text{fittings}}$$ (12.3)

Let's consider an example to determine the pressure differential along a pipe run. Assume beer with a density of 1010 kg/m3 is moving along a 50 m pipe ($\phi = 0.0038$) at a velocity of 2.0 m/s and that the pipe is 5 cm in diameter. This is shown in more detail in Fig. 12.1.

Inserting the values that we are given in the problem, and then solving the equation, provides the frictional losses due to the length of the pipe alone:

$$\Delta P_{\text{pipe}} = 4 \phi \frac{L}{d} \rho u^2$$

$$\Delta P = 4 (0.0038) \frac{50\,m}{0.050\,m} \left(1010 \frac{kg}{m^3} \right) \left(2.0 \frac{m}{s} \right)^2$$

$$\Delta P_{\text{pipe}} = 61408\,Pa = \mathbf{61.408\,kPa}$$

Table 12.1 Pipework frictional constants. These values are approximate and may be different for a particular system. In most cases a range of values is possible based on a number of different factors

Pipework or obstruction	k (unitless)	Pipework or obstruction	k (unitless)
Valves		Fittings	
Ball valve – open	0.05	90° square	1.3
Ball valve – 2/3 open	5.5	90° rounded	0.75
Ball valve – 1/3 open	200	45°	0.35
Butterfly valve – open	0.2	T – straight run	0.4
Butterfly valve – ½ open	15	T – used as elbow	1
Globe valve – open	10	Coupling	0.04
Gate valve - open	0.15		
Gate valve – ½ open	2.1		

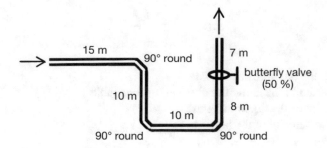

Fig. 12.1 Example pipe run. There are three 90° rounded bends and one butterfly valve that is half way open

The frictional losses due to the fittings along the run are determined by adding each of the losses due to the fittings and the half-opened valve. Note that we have three 90° rounded bends and one butterfly valve. Solving the equation:

$$\Delta P_{fittings} = \Sigma \left(\frac{k \rho u^2}{2} \right)$$

$$\Delta P = 3 \left[\frac{0.75 \times 1010 \frac{kg}{m^3} \times \left(2.0 \frac{m}{s} \right)^2}{2} \right] + 1 \left[\frac{15 \times 1010 \frac{kg}{m^3} \times \left(2.0 \frac{m}{s} \right)^2}{2} \right]$$

$$\Delta P = 3 \left(1515 \, Pa \right) + 1 \left(30300 \, Pa \right)$$

$$\Delta P_{fittings} = 4545 + 30300 = 34845 \, Pa = 34.845 \, kPa$$

The total pressure loss along the pipe run is then the sum of these two frictional losses:

$$\Delta P_{total} = 61.4 \, kPa + 34.8 \, kPa = 96.2 \, kPa$$

This is a significant loss of pressure in the liquid over the 50 m run. Keep in mind that this is a very long run in a microbrewery, but a fairly short run of pipe in a macrobrewery.

12.2.2 Other Gases Used in "Carbonation"

In addition to carbon dioxide, the brewer may decide to serve the beer using a different gas. In order for this to work, the gas must have a measurable solubility in the beer. For example, an April Fool's Day joke has been passed around suggesting that helium gas could be used to "carbonate" a beer. The joke is that you could drink the beer and when you talked afterward, your voice would become high-pitched from the helium. While it may be possible to force helium to dissolve in the beer, once the pressure was removed from the beer (i.e., the bottle was opened), the beer would completely, instantly, and violently degas. This would occur because helium has essentially no solubility in the beer. If someone were actually able to do this, the real joke would be the tremendous fountain of beer that would gush from the bottle as the cap was removed. Moreover, when you drink a beer, you tend to only breathe in a very small amount of the gas in the beer (most of the CO_2 in the beer is lost as bubbles to the atmosphere; the majority of the rest of the CO_2 ends up in your tummy.) So, the amount of helium a helium beer drinker would breathe would likely not be enough to change the pitch of their voice.

Which gases are soluble in water? The solubility is based on the ability of the gas to interact with water as governed by Henry's solubility constant (Table 12.2). This constant indicates the ratio of gas that dissolves in a liquid versus the partial pressure of the gas. As the constant gets larger, the gas becomes more and more soluble in water or beer. At the molecular level, the more polar the gas is, the more it interacts with water. The more interaction, the more the gas tends to dissolve in water. Too much interaction and the gas will be primarily dissolved in water and not form bubbles when the pressure is lowered. A nonpolar gas interacts very little with the water and does not dissolve. Too little interaction and it will degas too quickly to be useful to the brewer.

Nitrogen gas (N_2) is much less soluble in water than carbon dioxide (CO_2) as can be shown by examining the values of the Henry solubility constant from Table 12.2; N_2 (0.016) versus CO_2 (0.83). Nitrogen's limited solubility means that it will degas quite well as the pressure of the beer is reduced and the beer is served. This leaves the beer much less "bubbly" after all of the nitrogen has left the beer because very little nitrogen will remain in the beer. In addition, if nitrogen alone is used to "carbonate" the beer, the formation of carbonic acid that provides a tart, crisp note to the beer will be lacking. The result would be a flat beer with a very nice head.

Nitrogen gas, as it forms bubbles in the beer, produces very tiny bubbles. This is likely due to the limited interaction between nitrogen molecules. These tiny bubbles tend not to combine into larger bubbles when they form a head on the beer. The result is a creamier foam.

In practice, brewers tend to use a mixture of nitrogen gas and carbon dioxide gas. The ratio is about 25% nitrogen to 75% carbon dioxide. Both gases dissolve in the beer to some extent. The result allows carbon dioxide the opportunity to interact with the water in the beer and form carbonic acid. This still provides the tart flavor

Table 12.2 Henry's solubility constants for selected gases

Gas	H (unitless)	Gas	H (unitless)
H_2	0.019	N_2	0.016
O_2	0.032	CO_2	0.83
NH_3	1400	N_2O	0.61
H_2S	2.4	SO_2	29
He	0.0090	CH_4	0.033

Calculated using data from Sander, R., *Atmos. Chem. Phys.*, **2015**, *15*, 4399–4981

to the beer. The nitrogen gas rapidly forms the tiny bubbles and degases much of the beer when it is poured, but with enough CO_2 still in the beer, it doesn't go flat.

One fascinating thing about gases in beer is the mechanism of the formation of bubbles. Research is still being conducted to try to uncover the details of the mechanism, but recent results suggest that bubbles don't form spontaneously on their own. They can, but not at the pressures used to carbonate beer. Something has to initiate the formation of the bubble. This can be microscopic imperfections in the glass, dust, sediment, or a whole host of other things. The particles or imperfections act as nucleation sites where the gas can form a microscopic bubble. The bubble then clings to the site as it grows. If the gas is CO_2, the interaction is relatively strong, and the bubble has to grow large enough in order to release from the nucleation site and drift up through the beer. If the bubble is N_2, the interaction of the bubble of gas with the nucleation site is fairly weak. This means that the bubble releases when it is much smaller than if it were CO_2.

In addition, the smaller bubbles of nitrogen gas tend not to interact with neighboring bubbles as much as carbon dioxide. Thus, the bubbles tend not to grow into larger bubbles. When only CO_2 is the gas in the beer, adjacent bubbles can interact with each other. The chance exists that these bubbles will merge into larger ones. And, the amount of surface area of the tiny bubbles means that a tremendous amount of beer is entrapped and suspended in the foam. The foam is very thick.

In actuality, when two bubbles merge, they do so by a process known as *disproportionation*. In this process, one of the bubbles gets larger while the other gets smaller. The reason for this mechanism of merger can be explored as shown in Fig. 12.2. Imagine that we've blown up each balloon and attached it to either end of the hose. In our example, we've placed a hose clamp to isolate each balloon from each other. Let's assume one of the balloons is slightly larger than the other. When the hose clamp is removed, the smaller balloon gets smaller and the larger balloon gets larger. For the balloons this occurs because the small balloon has more tension in the plastic of the balloon. The more "stretched out" larger balloon is easier to stretch even more and gets larger. The latex walls of the smaller balloon are more taut and much harder to stretch.

The analogy of the balloons is directly related to what happens to the bubbles in the beer. The *Laplace pressure* drives the process. The Laplace pressure is the difference in pressure outside of the bubble and the inside of the bubble. It is very dependent upon the surface tension of the bubble. It turns out that the Laplace pressure is inversely related to the radius of the bubble. In other words, larger bubbles

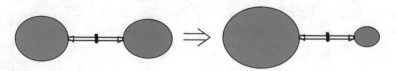

Fig. 12.2 Disproportionation analogy using balloons. On the left is the setup before the hose clamp is removed and the result is on the right after it is removed

have a smaller Laplace pressure and tend to grow in size when they merge with other bubbles.

One property of nitrogenated (i.e., 25% N_2 and 75% CO_2) beers is that the perceived bitterness of the beer is greatly reduced. Research still needs to be completed to understand the mechanism of this effect. One theory involves the degassing that occurs when the bubbles rapidly form. This greatly reduces the carbonation level of the beer, its tartness, and its acidity. The reduction in these three factors will have an impact on the perceived bitterness of the beer. Evidence for this can be found by comparing the flavor of a normally carbonated IPA that has been degassed by stirring for 10 min versus one that is freshly poured from the bottle.

CHECKPOINT 12.1

What is the pressure drop in a wort ($\rho = 1055$ kg/m^3) that is pumped through a 10 m length of 4 cm diameter pipe ($\phi = 0.0042$) with two 90° square bends and two 45° bends and ends in a wide-open ball valve? Assume the velocity of the beer is 1.0 m/s.

Would you expect a beer that it is pressurized with laughing gas (N_2O) to have a head similar to a nitrogenated or to a carbonated beer?

12.3 Packaging

Once the finished beer is ready to be shipped outside of the brewery, it is piped to the packaging line. The packaging line is a system that prepares the package (or "pack" for short), fills it, and seals it. In a brewery, this is the noisy machinery that customers love to watch. It is fascinating to see all of the intricate movements that the bottles or cans make as they move along the line.

In this section, we will explore the two options for packaging beer – small pack and large pack. Small pack refers to the single-serve packages such as bottles and cans. Large pack refers to kegs. The general principles of the packaging line are the same for both types of packages, irrespective of their size. In essence, the package must:

- Protect the beer from the environment
- Protect the beer so it doesn't harm the customer
- Contain a legally measured volume
- Dispense the product easily

Packaging also serves the function of advertising to the customer. This includes a number of key features such as looking good on the shelf, providing useful information to the customer (such as the style of beer, the amount within the container, etc.), and drawing in customers through advertising. While we often overlook this part of the package, it is very important to the brewer. More often than not, this can be the only time the brewer has the opportunity to entice the customer to try the beer and become a regular consumer. So, the package is very well thought out and very well planned.

Not only is the label important (the look of the label, the artwork, the dimensions of the name of the beer), but the shape of the package, the color of the package, and even the name of the beer become extremely important. With the sheer number of different beers in the cooler at your local liquor or grocery store, standing out from the crowd is becoming more and more important. This is especially true in those stores with limited shelf space. The package must catch the eye of the consumer in order for the beer to sell. Then, once the beer is sold, the contents must appeal to the customer for them to consider a repurchase later.

12.3.1 Small Pack

In the United States, *small pack* is likely the main mode by which the customer consumes a particular beer. While the influx of new microbreweries continues to grow, there is still considerably more beer consumed outside of the taproom, bar, or tavern. Choices for the brewer to package their beer are everywhere.

12.3.1.1 Bottles
The most common of the small pack, especially for the craft beer industry in the United States are bottles. Bottles are glass containers that come in nearly any shape, size, or color. The standard bottle is shown in Fig. 12.3. The main areas of the bottle include the finish, the neck, the shoulder, the body, and the base. Labels that can be added to the bottle include a neck label, a body label, a back label, and a footer label (placed just above the base).

Bottles are manufactured in a very energy-intensive process in a furnace at approximately 1500 °C. Silica (SiO_2) is first mixed with sodium carbonate (Na_2CO_3, also known as soda ash), calcium carbonate ($CaCO_3$, limestone), and colorizing agents. Cullet (broken glass) is often added to the mix in as much as 90% of the total. This addition can significantly reduce the temperature where the glass melts, making the preparation of new glass more efficient.

The color of the glass is determined by the trace amounts of metals that are added to the mixture. For example, if nothing is added, the glass is clear. If iron is added, the glass turns green and if chromium is added, the glass turns brown. Addition of cobalt causes the glass to turn blue. The typical recipe for a batch of glass is shown in Table 12.3.

The powdered ingredients are blended and then fed into a furnace that is heated to around 1500 °C (Fig. 12.4). This mix floats on top of the existing molten glass in the furnace. As it melts and combines with the existing glass, it moves along the furnace until it reaches the *bridge*. There it must pass under the bridge through the *throat* and into the *working end* of the furnace. This pass under the bridge ensures that only the molten, mixed glass reaches the working end of the furnace. The molten glass then enters the *conditioning channel*, a shallow portion of the furnace

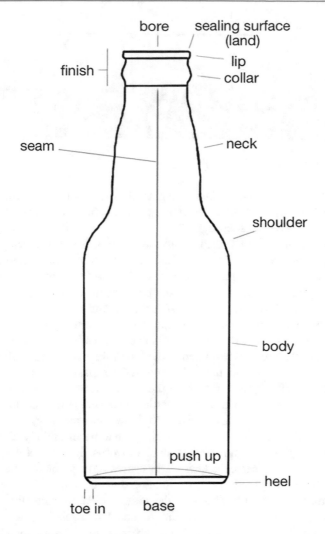

Fig. 12.3 Parts of a bottle

Table 12.3 Components that make up glass

Component	Percent
Silica (SiO_2)	70%
Sodium carbonate (Na_2CO_3)	15%
Calcium carbonate ($CaCO_3$)	10%
Colorant	5%

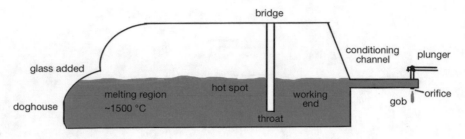

Fig. 12.4 Glass furnace

where the glass begins to cool. The liquid glass flows into an *orifice* at the end of the channel and a *plunger* pushes a given amount of glass through the orifice. Shears cut the *gob* of glass as it passes through.

The gob then drops into a mold where the bottle begins to take shape. First the mold is sealed and pressurized on top of the gob. This forces the molten glass to the bottom of the mold. A wand at the bottom of the mold then blows air up into the center of the gob. This forms the top of the bottle as the molten glass has filled the bottom of the mold. The push of the air into the gob inflates the molten glass, cooling it enough that it begins to hold its shape. The glass at this point is a half-formed bottle known as a *parison*. Before the parison collapses back into a ball, the mold then inverts and the pressure reapplied to the top of the mold. This pushes the glass outward to fill the mold. This blast of air causes the glass to cool even more. The temperature is finally at the point where the glass doesn't flow quickly at all anymore. The mold is then opened, and the bottle moves on to the finishing stages. The overall process is known as *blow molding* (or blow-blow molding).

The greatest disadvantage to blow molding is that the air initially blown into the gob doesn't always form symmetrically. In other words, the walls of the glass bottle may not end up with the same thickness. This causes some problems with packaging later especially if the bottle is weakened on one side due to the thinner walls.

The alternative to the blow mold is the *press-and-blow mold* (Fig. 12.5). The process was developed in Germany and refined by Owens-Illinois, one of the world's leading glass manufacturers. In this process the gob is dropped into a mold (Fig. 12.5a) and the mold is sealed (Fig. 12.5b). Then, an air injector is raised into the gob and air pressure is applied forcing the gob to take the shape of the mold (Fig. 12.5c). The parison is then removed (Fig. 12.5d), inverted, and placed into another mold. A seal is placed on top of the mold that also helps form the finish (Fig. 12.5e). Air is then pushed into this second mold forming the bottle (Fig. 12.5f). Immediately after forming (Fig. 12.5g), the bottle is put into an oven known as a lehr (pronounced "leer"). Inside the lehr, the bottles are reheated and then cooled slowly under controlled conditions in a process known as *annealing*, so that the glass doesn't crack as it cools.

The final step in the process for manufacturing a glass bottle is to spray the bottles with a lubricant to ensure that the bottles don't get scuffed as they are manipulated during the packaging process. In addition, the bottles are sent to a measuring

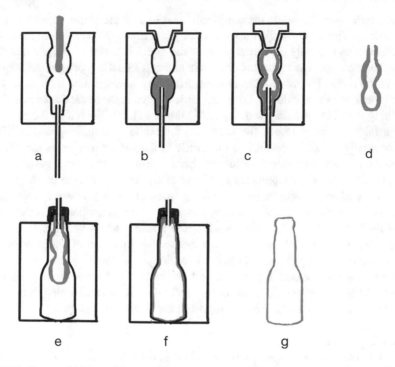

Fig. 12.5 Press and blow molding process

machine to ensure that each part of the bottle is within specifications, free of bubbles, and ready to be used. Any rejects are simply ground up and added back into the glass furnace.

Once at the brewery, the bottles join up with the finished beer at the packaging line. Most breweries that bottle have an automated process that involves machinery and robots that unpack the bottles from the shipping packaging. The bottles are then inverted and washed to remove any dust or lubricant that may have inadvertently gotten inside the bottle. The brewer can also spray the bottles with a sanitizer to ensure that all microbes are killed before the beer is added. However, if the bottles are sprayed with hot water (~80–85 °C), the heat of the water is often enough to sanitize the bottles. The action of the spray inside the bottles is also important as it can help to dislodge any dirt.

The bottles are then lined up and placed one after the other into the filler. Here the bottles are attached to the filler arm that comes down and seats on the sealing surface of the bottle. A vacuum is applied to the bottle and then the bottle is filled with CO_2. This is repeated a second time such that the gas in the bottle is about 99% CO_2. Alternatively, the bottle can be nitrogen dosed by squirting a small amount of liquid nitrogen into the bottle. The liquid nitrogen instantly warms up and evaporates into gas that swells and pushes all of the air out of the bottle. This method results in filling the bottle with about 99.9% N_2.

Beer then enters the bottle through a tube and fills up the bottle. The tube can be short or long. The long tube filler does a better job at filling the bottle from the bottom up. This means that only the top surface of the liquid is exposed to the atmosphere inside the bottle and results in a lower uptake of any residual oxygen gas that happens to be in the bottle. For many other reasons, however, the short tube filler tends to be found more often in the industry. With very little oxygen in the bottle due to evacuation and filling with CO_2 or dosing with liquid N_2, the short tube filler makes the case.

The filler continues to add the beer into the bottle by counter-pressure filling. This occurs by simply pressurizing the bottle to the same pressure as the finished beer and then opening a valve to allow the beer to flow into the bottle. A relief valve opens to allow the excess pressure of filling to continue to fill the bottle. It stops adding beer when the beer begins to exit the filler, or when a sensor is touched by the liquid. Then, the filler arm is removed as a puff of air is applied. This causes the top of the beer to foam. Alternatively, a spray arm can add a very small squirt of sterile water to the top of the beer in the bottle. The injection of water causes the beer to foam up, filling the top of the bottle entirely with foam.

Before the foam settles, a cap is placed on the bottle. A sleeve then pushes the cap into the bottle, sealing it. At this point, the bottle is filled and sealed. It moves to an inspection area where the level of liquid in the bottle is measured.

12.3.1.2 Cans

Aluminum cans are becoming more prevalent in the United States as a beer package. They offer some useful advantages over bottles:

- Completely impervious to light
- More impervious to oxygen than a bottle
- Lighter and easier to transport
- Less expensive

Cans are manufactured from a sheet of aluminum in a roll. As the roll is fed into the machinery, it is sprayed with oils to aid in lubricating the machines and protecting the aluminum from uneven stretching and cracking. The first machine the sheet aluminum enters is the cupping press. Here, a press stamps out circles about 14 cm in diameter and then bends the aluminum into the shape of a cup. These cups are thicker than the finished package and much shorter. However, additional machines will stretch them out into the shape of the finished aluminum can (Fig. 12.6).

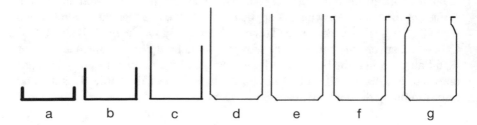

Fig. 12.6 Stages of development of the aluminum can

The cups (Figure 12.6a) are then fed into the body machine. The cup is fitted onto a plunger that presses the cup through a ring. The cup stretches as it passes through the ring, elongating and thinning the cup. This occurs a number of additional times, each time the ring is slightly smaller. The finished product is a straight can that is open at one end ((Figure 12.6b, c, d). The bottom of the can takes the shape of the plunger and looks just like the bottom of the finished can. The final thickness of the can ends up at about 115–120 μm.

Early aluminum cans were much thicker than the modern ones. Trial and error led to the development of "lighweighting" the cans by making them thinner. Today computer-aided design has allowed manufacturers to shrink the thickness of the can to the point where the can is significantly lighter (about 13 grams compared to the original 80 grams). This saves money on transportation and cost for the can.

The elongated can is then trimmed to the correct length (Figure 12.6e) and sent to the washer. The cans move along a conveyor upside down and are sprayed with hot (60 °C) hydrofluoric acid (HF). This acid cleans the interior of the can and the exterior of the can, eliminating all contaminants, oils, greases, and other soils. The cans are rinsed, still upside down, with hot (60 °C) distilled water and then dried.

The cans then move under a roller that applies a coat of varnish to the bottom of the can. This is applied to aid in the movement of the cans along a conveyor when they are turned right side up. A label is painted onto the can at this point. Typically, this is done in a rotatory machine where multiple rollers apply different patterns of ink to the surface of the can. Typical paint application can occur at the rate of over 1000 cans per minute. After all of the different colors have been applied, a coat of varnish is applied to the exterior of the can. This coat is placed over the paint to protect it. Heating the can in an oven dries the paint and varnish and hardens the exterior markings on the can.

Alternatives to painting a can are possible. While stickers or other glue-affixed labels are possible, they are typically not applied to the can. The cans are too flexible to ensure that the label doesn't peel off or fall off at some point later on. Instead, sleeves of plastic that have been printed with the label can be placed onto the can and then heated. The shrink-wrap goes on easily and can even be applied after the can is filled (which is the preferred time).

Next, a water-based varnish is sprayed on the inside of the can. This varnish is put in place to act as a barrier between the aluminum and the contents of the can. Because beer is somewhat acidic, it will react with aluminum slowly. If left in contact with the aluminum, the beer would eat its way through the can, causing it to rupture from the pressure.

In the final two steps, the can enters the *necker* – a machine that slowly rolls the can and forms the contraction at the top of the can known as the neck (Figure 12.6f). The process of forming the neck occurs in increments (as many as a dozen or so) so that the aluminum isn't torn or damaged in the formation of the neck. The *flanger* is the final stop for the can (Figure 12.6g). Here it spins around rollers that crimp the top of the can and bend it at a 90° angle. This angle and the amount of the flange at the top is very important in the crimping step after the can is filled.

At the brewery, the cans undergo a very similar treatment to that experienced by the bottles. They are inverted and washed, then righted and enter the filler. The filler arm and can seal along the flange at the top of the can and the beer is placed in the can. The major difference here is that while the bottle can be evacuated repeatedly in order to replace the atmosphere inside the bottle with CO_2, pulling a vacuum on a can would instantly crush it. If the brewer wishes to replace the atmosphere inside the can prior to filling it, they can add a nitrogen doser to the packaging line just before the cans are filler.

Once full of beer, the cans move along the conveyor belt and a cap is placed on top as a jet of CO_2 disturbs the top of the beer. This causes the beer to foam slightly so that the end cap of the can floats on the foam, indicating that no oxygen is present inside the can once it is seamed. Then, the cap and can are joined in the seamer. The seamer spins the can while a device called a *chuck* rolls the end of the cap under the flange of the can. A second chuck then presses the seam together making a tight seal (Fig. 12.7).

Disadvantages to canning instead of bottling exist. Beer in a can is a little less carbonated than beer in a bottle due to the way that the cans are filled. In other words, bottles are counter-pressure filled so that the pressure of the finished beer is not different from the pressure in the filler tank. Because cans are not filled this way (i.e., beer is simply poured into the can), the carbonation decreases slightly after sealing until the pressure in the can rises to meet the pressure of the carbon dioxide that is dissolved in the beer. This can be adjusted by slightly over-carbonating the beer at the filler or maturation tank prior to canning. The other issue with cans is related to the way they are filled. It is very difficult to eliminate all of the oxygen from the can during the filling and sealing process. Thus, higher levels of oxygen can be entrapped inside the can as it is sealed. The levels are still quite low and often

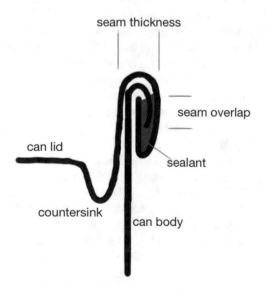

Fig. 12.7 Close-up of a can seal. Note the overlap and sealant help make the can air- and liquid-tight

have only a minimal impact on the beer. However, some brewers strongly abhor the idea of oxygen, at any level, in their finished product.

12.3.1.3 Plastic

Plastic beer bottles are starting to have some impact on the industry. Plastic does have some advantages over the other small pack, but it also has some disadvantages as well. The main advantages are:

- Lightweight – about 25 g per bottle, but still heavier than aluminum
- Recyclable – but so are glass and aluminum
- Inexpensive – actually more expensive when you consider the requirement to use multilayer bottles (see below)
- Noise reduction – significant noise reduction compared to aluminum and glass during the fill process
- Insulated – the plastic does offer a little insulation for temperature

The disadvantages probably outweigh the advantages. Especially if we consider the arguments for each of the advantages:

- Slower filling speed – reduction in production levels
- Permeable – gases can permeate through plastic
- Not thermally stable – requires modifications to pasteurization process because the plastic may deform under heat

While some manufacturers and brewers are moving to the plastic bottle because of its versatility in design, the plastic bottle has yet to hit the market with force. Addressing the issues of permeability and pasteurization are the hot topics of the day, and their solutions could see this form of small pack everywhere in the marketplace.

> **CHECKPOINT 12.2**
> Outline the steps required to bottle a beer, starting with silica and ending with the finished product.
> Name two advantages and two disadvantages to the use of the glass bottle.

12.3.2 Large Pack

Large pack refers to the multi-serve vessels that are used to distribute beer outside of the brewery. Go to any bar or tavern in the United States and you'll find a series of beers on tap. In other words, the bar has a number of kegs of beer from which to serve a beer. For the consumer, the best price for a pint of ale is had out of the keg. For the brewer, the lowest profit margin is on the keg. But only so many people can

Fig. 12.8 Cutaway view
of a keg

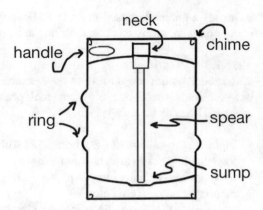

visit the taproom at the brewery, so this is the way to distribute the product to
the masses.

Kegs originally were manufactured from aluminum. However, due to the malle-
ability and ductility of aluminum, those kegs tend to be difficult to find anymore. It
didn't help the situation that the aluminum keg was easily taken to the recycling
center – a fate of many kegs that were stolen. Modern kegs are made from 304 stainless steel. The keg itself is made in two parts by cutting circles of stainless steel and
then forming a top half and bottom half using a technique similar to making an
aluminum can. The final thickness of the stainless steel ends up at about 1.5 mm.
The two halves are then TIG (tungsten inert gas) welded around the middle of the
keg to make the final container. The neck and chimes are then welded onto the container to give the overall shape of the keg (Fig. 12.8).

Like aluminum kegs, stainless steel kegs can expand. Over time, the keg does
grow slightly in size due to heating while under pressure or freezing of the product
inside. Heating a keg can be very disastrous. As the keg gets warm, the contents
inside also heat up. The beer expands in volume as it gets warmer, and if there is not
enough headspace in the keg to handle the increased volume of warm beer, the keg
can actually rupture (at least 0.8 L of headspace is required to prevent this if the keg
is heated to 70 °C). The main issue is that kegs get warm accidentally. If a keg gets
left in a car out in the sun, the keg is placed directly in the sun, or the keg might even
be left under a plastic tarp prior to an event, the temperature of the keg can increase.
If the keg is placed in the freezer, one also needs to be careful. The cold temperatures might be cold enough to cause the beer to freeze inside the keg. Water, the
major constituent in beer, expands as it approaches the freezing point. Thus, the
frozen beer occupies much more space than cold beer. The keg could rupture or at
the very least deform from the pressure.

Kegs are pressure tested before they are finished in the manufacturing process.
The standard pressure test ensures that the keg will not rupture below 90 psi. Since
the standard operating pressures of the keg rarely get as high as 30 psi, this provides
a large safety range for the keg.

In addition to the body of the keg, a *spear* is added. The spear attaches to the neck of the keg and provides a way to admit carbon dioxide and withdraw beer from the bottom of the keg. Because kegs are used worldwide, and because each region uses a variety of keg sizes and styles, the spear is also just as variable. However, every spear has four basic parts: body, stem, seal, and spring. The *body* is the point of attachment of the spear to the keg. Due to thefts, the attachments often involve locking rings or threaded ends that require special tools to dismantle. The body is also where a coupler is attached to connect the keg to a tap, filler, or cleaner. The *stem* is a long tube that reaches almost to the bottom of the keg. Often, the keg has an indentation at the bottom called a *sump* where the stem reaches. This is placed so that the majority of the product in the keg can be removed via the stem. A *seal* seats against the body that keeps the keg closed until the coupler is attached. And a *spring* holds the seal in place. When the coupler is attached, it pushes down on the spring, and moves the seal out of the way. This allows gas to enter the keg outside of the spear, and liquid to be forced up the spear and through the coupler to the tap.

As we noted, there are many different spear arrangements based on where you are in the world. Here in the United States, the most common spear is known as a "D-system" valve. This spear has an indentation in the body where the coupler seats and locks into place before the spring is pushed down. Other common arrangements include the "A-system" valves that are flat. These require that the coupler attaches to the outside of the spear body.

Kegs in the brewery take longer to fill than small pack but follow a very similar process. First the kegs are placed onto a conveyor belt by either a robot or manually. The kegs are then inverted and passed into an external washer. The external washer sprays caustic and rinse water successively on the kegs and ensures that the outside of the keg is as clean as possible. Then, the kegs move on to the cleaner/filler. In some cases, a special machine is located before the cleaner/filler that checks to make sure that the spear is properly attached to the keg.

In the cleaner filler, the inverted keg is coupled to the machine and the spear engaged. In cycles, a blast of water is injected into the keg to rinse out any remaining beer or other material from the interior of the keg. This is then pushed out using air. Spraying caustic or another cleaner through the stem washes the keg. Because the keg is upside down, the caustic rushes to the end of the stem and sprays against the bottom of the keg. The caustic then collects at the top of the keg and exits through the gas port. The keg is then rinsed with water, and then sanitized by either spraying sanitizer through the keg or by rinsing the keg with water that is at least 80–85 °C. This hot water rinse can ensure that all microbes are killed inside the keg.

Finally, the keg is turned over and carbon dioxide is blown through the stem. The CO_2 exits through the top of the spear until all of the air and oxygen have been removed. Restricting the flow of the gas exiting the keg increases the pressure of the CO_2. When the pressure is approximately the same as the pressure of the beer in the conditioning tank, beer is allowed to flow into the keg down the stem. In this way, the beer fills the keg from the bottom up, pushing out the excess gases in the way. A scale or flow meter is used to ensure that the exact amount of beer is added into the keg.

The keg is then detached from the machine and conveyed to the next station that cleans the body of the spear and applies a label around the neck of the keg. The station also attaches a cap to the neck to signal that the keg is full. They are then placed into the refrigerator and await their transportation to the bar or tavern.

Some keg fillers operate completely in the inverted position. The keg is washed in an inverted position as before but left inverted as it is filled through the gas port. This means that the beer still fills from the bottom up as the gas exits through the stem of the spear. The keg is still filled until the volume of beer in the keg matches the correct fill level. Again, this can be measured by many different methods.

12.4 Pasteurization

What is pasteurization? This is a process that was first discovered by French chemist Louis Pasteur in 1864 while he was exploring why wine tended to sour with age. By this time, he had already determined that yeast were the reason why the sugars were converted into alcohol and carbon dioxide, and that oxygen was not needed for yeast to survive. It was in the 1870s that he began working on fermentation problems in the brewing industry. His hope was to provide information that would help the French brewing industry surpass the German industry (who at the time had a very well-developed brewing industry and produced some of the best beers in the world.)

What Pasteur found was that most of the yeast used to make beer in France was actually not pure yeast, but instead a mixture of bacteria, mold, yeast, and other fungi. He strongly advocated for heating the beer after it was made to eliminate these organisms and allow the beer flavor to be stable. He also suggested that tartaric acid be used to purify the yeast prior to its use. Unfortunately, this didn't work very well.

It was Emil Hansen, a Danish chemist, who carried Pasteur's work further. Hansen worked at the Carlsberg Laboratory (a research company spun off from the Carlsberg Brewery by its founder Jacob C. Jacobsen) on the problem of impure yeast. He determined that it was possible to separate a specific colony of yeast into a pure strain that would cleanly ferment wort into beer. This first strain of yeast he named *Saccharomyces carlsbergensis* after the laboratory where he worked. It wasn't until the mid-1980s that it was determined that this is the same species of yeast as *Saccharomyces pastorianus*, the lager yeast named after Louis Pasteur that was identified almost 15 years earlier than Hansen's work.

It was clear even then that Pasteur's contributions to the brewing industry were incredibly valuable. By far one of Pasteur's most important discoveries was pasteurization. This is the process that eliminates almost all microbes. It has been applied to milk, wine, juices, and beer. The goal of this process is to accomplish the task of eliminating microbes without changing the flavor of the product.

There are many different levels of elimination of microbes that must be considered. This is the same argument as between the words clean, disinfect, sanitize, and sterile. It typically results in the definition of the degree of elimination of microbes.

For example, if something is made sanitary, it results in a 99.999% of all microbes. This is typically the level of cleanliness needed in the brewery. To make something sterile, it needs to result in 99.99999999% reduction of the microbes. This can be difficult to obtain.

How is the degree of pasteurization measured in the industry? The amount of "pasteurization" depends on the temperature and length of time. A pasteurization unit (PU) is one way to measure how much "power" is applied to the elimination of microbes in beer. Specifically, 1 PU is defined as the amount of "power" applied in a product heated to 60 °C for one minute. First defined by Del Vecchio in 1951, the PU is calculated from:

$$PU = t_{(minutes)} \times 1.389^{\left(T-60^\circ C\right)}$$
(12.4)

where t is the hold time in minutes, and T is the temperature of the beer reported in °C.

This equation works to calculate the number of PU that are applied to a beer when it is heated at any temperature. Note from the equation that this implies a tenfold increase in effect for every 7 °C increase in temperature. In addition, temperatures less than 60 °C generate very small PU values. In other words, holding a beer at a temperature less than 60 °C will require a significant amount of time to obtain any measure of PU value.

Let's assume the brewer wishes to achieve 30 PU. To obtain this many pasteurization units, the beer would need to spend 30 minutes at 60 °C. However, because of the equation, the degree of pasteurization is not linear with temperature. So, elevated temperatures are more effective at killing microorganisms. And, at those higher temperatures, the time to achieve the same PU will be reduced. As an example, let's rearrange Eq. 12.4 to solve for the time to achieve 30 PU at various elevated temperatures:

$$\frac{30PU}{1.389^{\left(T-60^\circ C\right)}} = t_{(minutes)}$$
(12.5)

Table 12.4 does just this. It shows the hold times for a beer at various temperatures in order to get the same 30 pasteurization units. Note that increasing the temperature rapidly decreases the time required for the same effect. While it seems that the best option would be to choose the greatest temperature, this is not the case. Increased temperatures have the potential to change the flavor of the beer. (What

Table 12.4 Hold times at different temperatures to achieve 30 PU

T (°C)	T (°F)	t (minutes)	t (seconds)
60	140	30.00	1800.0
65	149	5.80	348.1
70	158	1.12	67.3
75	167	0.22	13.0
80	176	0.04	2.5

Table 12.5 Pasteurization
unit relationship to provide
similar reductions in
cell counts

Organism	PU
Saccharomyces cerevisiae	1
Pediococcus	1
Lactobacillus	5
Wild yeasts	10

chemical reactions would you predict to occur and how would they change the flavor of a beer?)

Other values based on Eq. 12.4 are commonly used in the discussion of the pasteurization of beer. One predominant value is the decimal reduction time (D). This
is the amount of time required, at a given temperature, to kill 90% of the specific
organism being studied. This value is very specific, so can only be compared
between organisms under very similar conditions. For example, in one study by
Tsang and Ingeldew, *S. carlsbergensis* was found to have a D = 0.004 min,
Pediococcus acidilactici had a D = 0.867, and *Lactobaccilus delbrukii* had a
D = 0.091. Thus, *Pediococcus* is more thermally stable than other microbes.

The amount of PU needed to effectively eliminate microbes in beer will depend
on the type of microorganisms we are trying to kill (as indicated as well by the decimal reduction time. For example, wild yeasts in are more resistant to heat than the
standard Brewer's yeasts (*S. cerevisiae, S. pastorianus*). Therefore, the beer contaminated by wild yeasts will require more PU to make the product stable. Table 12.5
lists some common spoilage organisms in beer and how many PUs are required to
have the same killing power as 1 PU on brewer's yeast. Note that Table 12.5 doesn't
list enzymes. Flavor-damaging enzymes can also survive into the finished beer. To
make sure that the beer is truly flavor stable as long as possible, it is imperative that
any enzymes are destroyed in pasteurization. They tend to be quite thermally stable.

How effective is pasteurization? Applying 20 PU to a beer during pasteurization
indicates that only 1 in 10 billion microbes has a chance of surviving. At 30 PU, this
drops to only 1 in 10^{15}. The survivor, if one did survive, would be hard to find if you
looked through an entire bottle of beer. Luckily, brewer's yeast produce invertase
and the activity of this enzyme can be used to determine the effectiveness of the
pasteurization. Typically, this is done by adding invertase to a beer and measuring
its activity prior to the pasteurization and after the pasteurization. Then, the reduction of the activity is a measure of how effective the treatment is on the product.

CHECKPOINT 12.3

Some brewers only provide 15 PU to their product before shipping. How long
would it take to obtain this amount of pasteurization if the hold temperature was 65 °C?

12.4.1 Tunnel Pasteurization

There are two common ways in which beer is pasteurized. The first method is called *tunnel pasteurization*. In this approach, the beer is first packaged in bottles or cans and sealed. Then, the sealed packages are sent through a "tunnel" where they are sprayed with recirculating heated water. The product's temperature is slowly raised from bottling temperature to the desired pasteurization temperature and held there for the desired rest time. Then, the temperature is lowered back to bottling temperature by spraying the packages with cooler water.

An example tunnel pasteurization configuration is shown in Fig. 12.9. In this simple design, there are five regions where the product is either heated or cooled, and one region where the temperature remains constant. As the bottles or cans enter region 1 on a conveyor, they are slowly heated by water that was used to cool the packages leaving region 5. By coupling the entrance and exit water recirculation in this manner, the amount of energy required from outside of the system is minimized. Heat energy is added or removed as necessary through heat exchangers in order to keep the temperature of region 1 and 5 constant. Regions 2 and 4 are likewise coupled, again raising the temperature of the packages as they travel through region 2 and cooling through region 4. Region 3 in Fig. 12.9 is the main pasteurization region. It is here where the packaged beer is at our final, elevated pasteurization temperature and will spend most of its time through the tunnel.

Tunnel pasteurization does not work well with larger packages, such as kegs. With the increased volume of beer in the sealed container and the materials of the package itself, it would take an unreasonable amount of time and energy to ensure that the packaged beer is thoroughly heated. If pasteurization is needed for large pack, another method needs to be utilized by the brewer to effectively remove bacterial contamination from the finished beer.

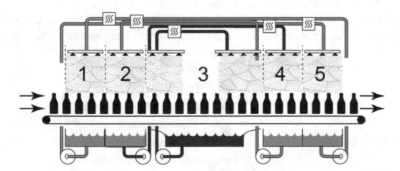

Fig. 12.9 Tunnel pasteurization system. Note that region 3 is at the final pasteurization temperature, and that the speed of the conveyor determines the length of time that the product spends at this final temperature

12.4.2 Flash Pasteurization

A second method is called *flash pasteurization*. In this approach, the beer is heated to pasteurization temperatures, held at that temperature for the desired amount of time, and then cooled *before* being placed in a package. This approach can be applied to virtually any packaging vessel: bottles, cans, or kegs. The approach requires the packaging vessel to be aseptically clean prior to filling. In addition, it also requires that the piping, tubing, and packaging apparatus itself be sterile. It would do no good at all to have a contaminated feed line from the pasteurizer to the bottle.

A generalized flash pasteurization method is shown in Fig. 12.10. The apparatus essentially consists of a plate chiller that is converted to allow one region to heat the beer, a second region to hold the hot beer, and a third region to cool the beer back to bottling temperature. The first step involves pumping the finished beer into a heat exchanger that warms the beer to the pasteurization temperature. The warm beer travels through a heated array of pipe ensuring that the beer spends an appropriate amount of time at pasteurization temperatures. Then the beer is cooled right before being sealed in an aseptically clean vessel.

Since the degree of pasteurization depends on the time the beer is held at an elevated temperature, we modify Eq. 12.4:

$$PU = \left(\frac{V}{Q}\right) \times 1.389^{(T-60^\circ C)} \tag{12.6}$$

where V is the volume of holding tube at the elevated temperature (in m^3), and Q is the volumetric flow rate (in m^3/min).

The units of V/Q are then m^3 / (m^3/min) which reduces to min. This must be in minutes to calculate PU using this equation.

As an example, let's assume that our pasteurization temperature is 65 °C. Table 12.4 indicates that we must leave the beer at this temperature for 5.5 minutes, or 348 seconds to achieve 30 PU. What we would like to calculate is the length of piping needed so that the beer stays at this temperature for the required

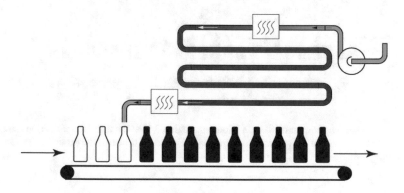

Fig. 12.10 Example configuration for flash pasteurization

time. First, let's assume that our bottle filler is capable of filling 6 standard 12 oz. bottles every second. This implies we need a volumetric flow rate of

$$Q = 6 \times \frac{12oz}{s} \times \frac{29.6 \text{cm}^3}{1oz} = \frac{2131 \text{cm}^3}{s} \tag{12.7}$$

Note that are units aren't in m^3/min, but we'll address this discrepancy as we move forward. Since the demand for 30 pasteurization units requires a time of

$$\left[\frac{V}{Q} \right] = 348s, \tag{12.8}$$

all we need to do is find the volume of the piping. Plugging the answer to Eq. 12.7 into Eq. 12.8, we get

$$\left[\frac{V}{2131 \text{cm}^3 / s} \right] = 348 s \tag{12.9}$$
$$V = 741,588 \text{cm}^3,$$

Assuming the pipe is a cylinder, Eq. 12.9 implies that

$$V = \pi r^2 l = 741,588 \text{cm}^3 \tag{12.10}$$

where π is a constant equal to 3.14159
r is the radius of the pipe
l is the length of the pipe

If we assume that the pipe is 5 cm (2 inch) diameter pipe, this implies that the length must be 9442 cm long (94.42 m). If we raise the pasteurization temperature to 70 °C, the time that the beer needs to spend in the pipe drops to 67.3 seconds, so the length is now about 18.26 meters. Again, there is a trade-off; the higher temperature will reduce our equipment needs, but one runs the risk of changing the flavor of the beer.

If we wanted only 10 PU applied to the beer for pasteurization, the length of tubing (at the same volumetric flow rate) gets even shorter.

One significant issue is important to consider with the use of a flash pasteurizer. Because the beer will be heated well above the finished beer temperature, we must consider that effect on the carbonation of the product. The same effect is seen in the tunnel pasteurizer, but because the bottles or cans are sealed, no carbonation is lost. The flash pasteurizer, on the other hand, can cause the beer to degas as it moves into the holding region. This can be a serious issue (loss of carbonation). So, the brewer must pay particular attention to the pressure of the system and increase it significantly as it goes through the flash pasteurizer to ensure that the beer doesn't degas.

The heat exchangers themselves are very similar to the plate chillers that we explored in an earlier chapter. These devices use hot water to warm the beer to the appropriate temperature. They can be counter-flow or concurrent flow and the volumetric flow rate becomes the determining factor in how much heat is applied to the beer. The typical design is as a counter-flow exchanger. Often, the water (or water-glycol mixture) is recycled from the chiller portion of the heat exchanger into the heater portion. Additional heat can be applied to ensure that the heater reaches the desired temperature. And then the water or water-glycol mixture is cooled down to allow the chiller to reach the appropriate temperature. This recycling saves tremendous amounts of energy for the brewery.

12.4.3 Other Methods of Pasteurization

Other types of devices exist to effectively result in pasteurizing beer. One such device involves the use of a filter.

A plate and frame filter is a device where a filter, typically a sheet of porous plastic with very small holes, is placed in a frame and then sealed. The beer is pumped into the frame and forced through the filter. Depending upon the size of the holes in the filter, the beer can be significantly cleaned of bacteria, yeast, and other particles.

Often the beer is passed through multiple filters that step down in size gradually until the holes are small enough so that no bacteria or yeast can pass through. After the beer is transferred, cleaning water is passed backward through the filter to remove the particles that were captured. Alternatively, the frames can be opened and the filters discarded.

This is a very effective way to remove any particles from the beer. The biggest advantage is that microbes can be removed almost completely from the beer without the use of flavor-damaging heat. However, like the flash pasteurizer, contamination can be a significant concern post-filter. In addition, forcing the beer through small holes results in a decrease in pressure in the beer that can be significant enough to degas the beer. The pressure must be increased as the beer flows through the plate and frame filter.

Chapter Summary

Section 12.2
>   Friction from pipes can reduce the pressure of a moving liquid. This can be determined using Darcy's equation.
>   Nitrogen can be used to provide the bubbles in beer.
>   Bubbles in beer become larger through the process of disproportionation.

Section 12.3
>   Small pack refers to single serving containers. This includes the bottle and the can.
>   Large pack refers to multiple serving containers such as kegs.

Section 12.4
>   Beer can have a longer shelf life if it is pasteurized.
>   Tunnel pasteurization and flash pasteurization can be used to eliminate microbes from the finished product.

Questions to Consider

1. How many pasteurization units would you expect from holding a beer sample at 25 °C for 6 hours? Explain your answer.
2. What is the pressure drop in a 60 m pipe ($\phi = 0.0038$) that is 5 cm (2.5 in) in diameter, if the flow rate is 0 m/s? Assume the density of the liquid is 1200 kg/m^3. What if the flow rate was 1.0 m/s? …0.2.0 m/s?
3. A liquid with a density of 1100 kg/m^3 is pumped at 1.25 m/s along a 15 m pipe ($\phi = 0.0042$) that has a diameter of 10 cm (5 in). What is the pressure drop in pascals?
4. A carbonated beer at 110 kPa is pumped from the conditioning tank to the packaging line along a 30 m straight pipe (5 cm; $\phi = 0.0029$). If the density of the beer is 1010 kg/m^3, what is the maximum velcocity allowed such that the beer doesn't decarbonate? Assume the beer will degas if the pressure falls below 100 kPa. What happens to the pressure if a 90° rounded bend is placed in the line?
5. A pipe run of 10 m (5 cm diameter, $\phi = 0.0032$) contains two 45° bends and two butterfly valves that are half-way open. If the liquid has a density of 1045 kg/m^3, what is the pressure drop across the pipe run?
6. Use the Internet to find a video that shows how aluminum cans are made. Outline the steps at each of the stages and compare that to the information in this chapter.
7. Use the Internet to find a video showing how bottles are made. How are designs and patterns on the glass put there?
8. Bottles tend to have a seam running down their length. From where does that seam arise?
9. Provide a list of pros and cons (two each) for each of the small packages that we discussed in this chapter.
10. A customer purchased a keg but didn't use all of the beer in it. How is that beer removed when the keg is returned to the brewery.
11. Diagram the setup that would allow a brewery to use two wort chillers to flash pasteurize their beer. Show how the cooling or heating water could be recycled in this system.
12. Why is ammonia not a good gas to use to pressurize beer?
13. Which would likely produce bigger bubbles if it were used to pressurize a beer, O_2 or CH_4? Why?
14. Neither of the gases in question 13 is suitable for pressurizing beer. Explain why?
15. Would there be an issue if the brewer decided that they only wanted to produce 3 PU when pasteurizing their beer?
16. Cans are typically cleaned with HF in the manufacturing process. Use the Internet to look up HF and its safety information. Is this acid safe for consumption?
17. From the information in the text about aluminum cans, extrapolate some pros and cons that you might expect for an aluminum keg.

18. Which is more restrictive to the flow of a liquid, a ball valve or a gate valve? Explain your answer.
19. Two identical pipe runs are compared. One is 5 cm pipe, the other is 3 cm pipe. Which would have the greater pressure drop?
20. Outline the pros and cons (list two of each) for a plate and frame filter as compared to a flash pasteurizer.
21. A brewer has a flash pasteurizer that has a 3-meter length of pipe (d = 1.5 in) to use at the hold temperature. How fast in m/s could the beer ($\rho = 1.005$ g/mL) flow through the pasteurizer if she wanted 68 °C as the hold temperature and 1.5 minutes of time?
22. What would be the pressure drop in question 21 across that pipe length assuming that there were six 90° square turns? Assume, as well, that the pipe is relatively smooth ($\phi = 0.0028$).
23. A brewer needs to fill 24 16 oz bottles every 10 seconds in the filling line. If the pipe coming to the packaging room is 5 cm in diameter, what is the velocity of the beer in that line?
24. Provide a list of at least three ways in which a brewer would know that a keg was filled to the appropriate level.
25. Prepare a drawing that explains the steps in a blow mold. Consider Fig. 12.5 as an example of the resulting drawing.

Laboratory Exercises

Thermal Expansion of Water

This experiment is designed to illustrate the relationship between the volume of water and its temperature. From this, inferences about headspace in packaging can be made.

Equipment Needed

Graduated cylinder (100 mL)
5 beakers (250 mL)
Water
Hot water bath – adjustable
Balance

Experiment
Caution. Hot glassware looks very similar to cold glassware. Wear heat-resistant gloves when performing this experiment.

   Use the hot water bath to warm approximately 1 L of water. As the water heats, use one of the beakers as a ladle and pour 100 mL samples into the graduated cylinder. Complete the task quickly so that the water maintains its temperature. Then pour the water from the graduated cylinder into a tared beaker. Obtain water samples at room temperature, 40 °C, 60 °C, and 80 °C.

Allow the four samples of water to cool to room temperature by leaving them on the bench, and then placing the beaker in a cool water bath. Once all four are at the same temperature, obtain the mass of each beaker. Don't forget to dry off the beakers after you have taken them out of the bath.

Plot the density of the water (all were 100 mL) versus the temperature of the water. Is there a relationship? Is it linear?

Use this information to determine how much headspace would be required (in mL) in a 12 oz. bottle that passes through a tunnel pasteurizer. Assume the Pasteurizer holds the beer at 75 °C for 2 min. What would happen if the calculated amount of headspace did not exist?

Quality Assurance and Quality Control 13

13.1 What Is Quality?

We briefly touched on the definition of this word in Chap. 4. Let's look at the subject again, but in a more detailed way. As we saw in Chap. 4, quality is a loaded word. It means different things to different people. A high-quality beer has one meaning to the customer and a different meaning to the brewer. To the customer, the measure of quality is likely linked to their level of satisfaction with the beer. This satisfaction comes from their perception of the taste, how they feel the beer looks in the glass, and, believe it or not, how much money they pay for the beer. In fact, there are many factors that lead into the customer's decision about the quality of a particular beer. And, a beer that one person believes is high quality may be considered quite the opposite by another person. The expert taster and beer judge may consider a beer to be high quality if it conforms adequately to a particular beer style. If it is out-of-style, the beer judge may consider the beer as low quality. Even if everyone believed a beer was high quality, the brewer may disagree.

That disagreement arises because quality has a different meaning to the brewer. To the brewer, the cellarman, the packaging specialist, and everyone else working in the brewery, quality may imply adherence to a particular style or even agreement with the design specifications. A brew that falls outside of those specifications or style guidelines may be rejected by the brewer and discarded. It may be perfectly fine to drink, but the brewer may consider it to be low quality. It may be tasteful enough to sell in the taproom, but because it is classed as a low-quality product, the brewer doesn't do so.

The discrepancy between quality to the consumer and quality to the brewer gives rise to a quandary. Who is right? Is everyone correct? The answer is based in the meaning of the word. Every individual determines the quality of a beer using his or her own metrics. So a high-quality beer to the brewer may not be high quality to others. Just because the beer judge classifies a homebrewed ale as high quality, drinkers may completely disagree.

© The Author(s), under exclusive license to Springer Nature Switzerland AG 2021
M. Mosher, K. Trantham, *Brewing Science: A Multidisciplinary Approach*,
https://doi.org/10.1007/978-3-030-73419-0_13

But then, how do we determine if a beer is high quality or low quality? It's obvious that the perception of quality lies with the individual, so determining the quality of the beer would be nearly impossible (unless, of course, everyone agrees). Instead of worrying about the quality of the beer, the brewer focuses on creating a product that will sell to the customer. If there is a demand for the beer, the brewer has made a good product. If the demand is lackluster or nonexistent, the brewer doesn't repeat that beer.

Not all breweries look at quality this way. Some are fixated on the production of beers that match their specifications for a particular style. Others focus on the design parameters and require their beers meet them every batch. Still others ignore all of this and just make beer, likely figuring that if they make it, people will drink it. Most of these breweries struggle to stay in business, or close their doors permanently.

So while high quality is something that every brewer tries to make, customers are the ones that make the decision. Focusing on making high-quality beers, then, requires that the brewer includes the customer in the equation. Brewers still need to ensure that the current batch of beer is the same as previous batches. The focus is on consistency. Consistently making a beer requires systematic approaches to the control of the parameters of the beer. Everyone in the brewery must be following that plan. Measurements must be taken and decisions on the next steps made in order to maintain that consistency. However, in the end, it all boils down to the decision of the customer. In other words, a brewery could manufacture a beer very consistently such that every parameter of the beer was within the specifications, but it could still not sell because the customer has decided that the beer isn't high quality.

13.2   Quality Control

To ensure that a particular beer is the same as others, the brewery often institutes a plan that measures key features of the beer as it goes through the brewing process. Simply measuring a parameter and then writing it down won't work. Even if the best analytical techniques are used and the value of the parameter is accurate and precise to the greatest number of significant figures, the process isn't complete. The plan must include a way to use the measured values of the beer's parameters to adjust, evaluate, and inform the brewer about the state of affairs on a particular batch. There is no point in measuring a particular parameter if the end result is to have a record of it.

In addition, as we'll see in the next section, measuring those parameters must have a relational feedback into a process. A plan to maintain consistency, any quality control process, works well only if it is part of the bigger picture. For example, let's assume a brewer measures and keeps track of the mash temperature in every batch. Adjustments can be made to the temperature of a particular mash in order to maintain consistency in the batch versus other batches. However, unless feedback into the brewery exists where those values are evaluated by the entire brewery, the variability in mash temperature may become commonplace. That would imply that the brewer would need to measure and adjust the mash temperature for every batch.

Constant adjustment of the batches doesn't always put the particular batch back on track to be consistent with earlier batches. For example, let's say the brewer notes that the current mash-in resulted in a temperature of 60 °C (140 °F) when it was supposed to be 67 °C (153 °F). The time it takes to raise the mash temperature to the specified 67 °C means that the current batch has rested at b-amylase's optimum temperature. This rest, even briefly (assume it takes 10 min) while the temperature was raised to the specified temperature, has adjusted the ratio of fermentable to unfermentable sugars in the resulting wort. This could result in significant variations that would need adjustment at every stage after the mash. However, with feedback to the entire brewery, it might be found that an earlier step is not being monitored and is the reason for the variability.

Quality control is necessary. It is the working end of the process, where different control parameters are monitored and measured. It is the physical measurement of properties of the product as it goes through the brewing process. But without the integration of those measurements into a larger plan, quality control becomes more of a knee-jerk reaction to what is going on. While breweries that perform only quality control can react and make adjustments to values that fall outside of the specifications for a particular brew, the overall consistency that all brewers try to attain will fail to be adequately achieved.

13.2.1 Methods in Quality Control

Quality control methods require not only measuring the parameter of a particular process, but also recording that measurement for use in the quality assurance portion of the management of the brewery. Charting is one of those methods that can be employed and used on the brewery floor for rapid "go-no go" determinations of product quality. Charting is the process of adding measurements to a graph and noticing the result of the placement of the measurement.

For example, a chart could be constructed that illustrated the mash-in temperature for a particular beer. Figure 13.1 illustrates an example of a chart that could be used. The chart contains maximum and minimum permissible values of the measurement. It also contains a line drawn in the middle of the chart that indicates the quality standard for that parameter. Each measurement is placed on the chart as the next entry.

Note that this chart indicates that over the last 10 batches, none of the mash-in temperatures have been above the maximum permissible or below the minimum permissible value. This would immediately indicate that in each batch, there was no need to adjust the temperature of the batch.

Note as well that selection of the maximum and minimum permissible values must be well thought out. The brewer could have indicated the maximum temperature as 82 °C (180 °F), but this would be unusable in the brewery. Why? Because if the mash-in occurs at 81 °C (179 °F), still below the maximum, all of the enzymes would be denatured, and no mashing would take place. Similarly, if the brewer required the maximum temperature as 67.278 °C (153.10 °F) and the minimum

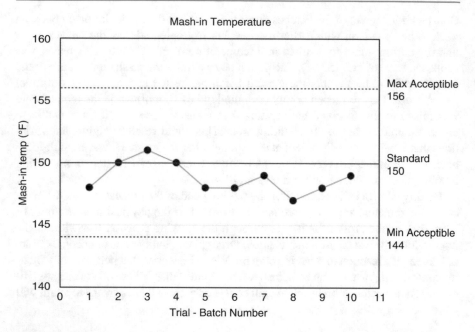

Fig. 13.1 Charting example. Here the example is the plot of mash-in temperatures for a porter, obtained during each batch made in the brewery

permissible temperature as 67.272 °C (153.09 °F), it may be extremely difficult or even impossible to accurately measure the temperature of the entire mash to that level of significance.

Another charting example that is more useful is the *CUSUM chart*. CUSUM stands for CUmulative SUM. In this chart, the difference of the measurement and the target quality standard is added to the previous difference. The chart still contains maximum deviations from the quality standard, and these still act as action levels. But, the difference in this chart is that trends in the data appear. Figure 13.2 shows exactly the same data as does Fig. 13.1, but here, we notice a problem indicating action. In other words, the temperature of the mash-in has been consistently running lower than desired. This appears easily on the CUSUM chart indicating an issue somewhere in the overall process.

The trends found in a CUSUM chart are very useful in determining issues in the overall process. If an issue exists, it will show up as a trend going either up or down on the chart. That provides a very good signal to the brewer that something is not right and needs to be adjusted.

CHECKPOINT 13.1
Outline the key features of a successful worker safety program.
Define an SOP and explain why it is useful to have in the brewery.

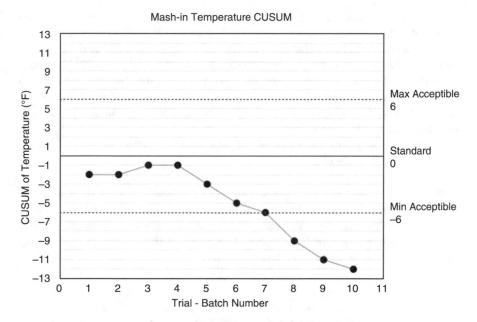

Fig. 13.2 CUSUM chart of mash-in temperatures. Data is taken from Fig. 13.1

13.3 Quality Assurance

As we discovered in the previous section, the quality control program is simply not enough to produce a quality product consistently. That is where quality assurance comes into play. Quality assurance is the plan that implements the quality control program and combines it with good manufacturing practices. Where quality control is the inspection of the process, quality assurance is the process where the results of the inspection feed back into the overall operation of the brewery. Quality assurance works to prevent mistakes in production that effect product quality by considering the operation of the entire system. Quality control, on the other hand, detects and adjusts or rejects an individual process product.

Quality assurance compares the outcomes of the well-performed quality control program to specified quality standards. Deviations from those standards dictates courses of action that may require re-examination of the particular processes in the brewery, re-tooling of equipment, replacement of processes with better ones, or re-evaluation of the process as a whole. Remember that quality assurance is a system-wide examination of the process, whereas quality control is involved only in individual product output.

Quality standards are part of the equation in quality assurance. These are parameters determined from the brewer's specifications and customer preferences. They cannot be set by only one of these two groups of people. They must include the brewer because the brewer is the one that is developing the product and producing it. Only the brewer knows the capabilities of the process and the limitations of what

can and can't be accomplished. The quality standards must include customer preferences. Only the customer can talk to the level of quality of the beer from a purchasing standpoint. And input shouldn't be limited to just one customer group. For
example, a beer that is made in one town should not only seek the customer preference from that town. The brewery should also seek input from nearby towns based
on their distribution pattern. If the brewery plans to increase sales in a different
town, they should first seek customer preferences in that town and develop quality
standards that speak to those preferences.

Approaches to quality assurance reflect the systemic examination of the process.
They include the following.

Stress Testing   The beer that is produced can be evaluated for its shelf life. The
resulting examination should be very detailed and include not only sensory evaluation of the product, but also other analysis such as measurement of the carbonation
level, viscosity, and pH of the product. The result of this extensive testing may
indicate a systemic issue that can be addressed.

Total Quality Management   This process involves a four-pronged approach to
quality where the entire brewery works together to plan, manage, evaluate, and
improve the production profile. The four prongs include:

(a) Definition of the product's quality. Where many quality definitions might
    include input from within the brewery, total quality management relies on the
    consumer to determine the quality of the product.
(b) Buy-in from the entire brewery. While the workers, brewers, and other people
    are charged with making a quality product, the managers of the brewery have
    the responsibility to ensure quality. This ensures that funding and decisions can
    be injected into the production process.
(c) Every process in the brewery must be systematically evaluated. The evaluation
    should be very detailed and cover every aspect of production. In doing so, each
    action along the production line can be adjusted, modified, eliminated, or
    enhanced if it has an impact on the quality of the product.
(d) Improvement in the production process must be done all the time. This implies
    that the processes are constantly being modified to streamline and improve
    quality. This is known as continuous improvement and results in a brewery that
    works hard to not only make beer, but to make it with attention to quality all
    the time.

Statistical Analysis and Control   This approach to quality assurance requires that
the brewery and its employees evaluate all of the quality control data all the time.
This allows the brewery to recognize and eliminate defects in production before
they occur.

ISO 9001 This approach involves using the standards set out by the International Organization for Standardization (ISO). These standards are evaluated, constructed, and improved by a group of over 130 national standards organizations from around the world. A brewery can work, train, and become registered as ISO-9001 compliant. The basic principles of this standard include:

(a) Considering the customer at all times
(b) Following good leadership practices
(c) Engaging the workforce
(d) Focusing on the process rather than the product
(e) Working to improve the production process
(f) Making decisions using evidence
(g) Managing relationships within the brewery

Other quality assurance methods and practices exist. Each of these methods tends to focus on essentially the same things. That is because these items are the most important to ensuring quality in the brewery product. Because each brewery operates differently, the specific quality assurance method that gets implemented may differ between breweries. However, the basic principles in each of these systematic methods are a focus on the customer, working toward always improving, and creating a business that is management focused.

So, does a brewery have to have and follow a quality assurance protocol in order to produce a high quality product? The answer is yes and no. It's no because many brewers can make excellent quality beer without following a quality assurance protocol. However, the answer is yes because most governments are now requiring that breweries follow a general process that ensures their products are not only "high" quality but also safe for human consumption.

13.3.1 Good Brewery Practice

One very useful method that ensures quality in production is the application of good brewery practices. This is essentially the same thing as good manufacturing practices (GMPs) that have been applied to the brewing industry. These guidelines are different depending upon country in which the brewery is located; however, the overall guidelines are similar from country to country.

In general, the brewery must have rules and regulations in place that address each of the areas outlined below. In many places, the specific nature of how these guidelines are implemented is not dictated. Any inspections would be focused on making sure each of these practices is in place.

- Maintain a clean and hygienic workspace.
- Maintain controls so cross-contamination between products is not possible.
- All equipment is validated to perform the work it is designated to do.
- All processes must follow SOPs and any changes evaluated and monitored.

- All workers must be adequately and routinely trained.
- Records must be made and kept of each step in the production process.
- All products must be traceable back to their specific records.
- A system for recall of any product must be in place.
- All consumer complaints must be investigated, and steps taken to prevent recurrence of any safety issue.

More importantly, the implementation of good brewery practice or good manufacturing practice ensures a baseline level to maintaining product safety and product quality. The more detail a brewery puts into their GMPs, and the stricter the brewery is in following their own GMPs, the more control the brewery has over their product. And that's the first step required to producing quality product.

In some cases, a well-documented GMP plan (with appropriate SOPs, records of training, records of production, and records of recalls and complaints) can be used if any legal issues arise. This can save hundreds of thousands of dollars if a lawsuit were ever brought against the brewery. Those records can be used to verify the exact specifications of each batch of beer and confirm or refute any query about the product.

How does a brewery go about setting up and implementing a GMP plan? Typically, a team of managers and workers gathers and evaluates each of the required guidelines. That evaluation reveals all of the steps that must be taken to ensure compliance. These steps are then constructed into a set of SOPs. And each SOP is assigned to a specific worker or team within the brewery.

Then, as records are generated from the operation of those SOPs, they are moved back to a central location. Periodically, the management/worker team performs a review of the GMP plan. Their review would necessarily include interviews with workers on the production floor and any other brewery personnel involved in the operation of the brewery. This review would also generate a report with recommendations that could include adjusting SOPs, modification of procedures, or even suggestions on how to modify production to avoid specific issues.

The final report should be disseminated back to the individual departments and teams within the brewery. This allows everyone to know what's going on within the brewery. And the specific recommendations to perform any modifications can be rapidly implemented.

13.3.2 Addressing Production Using 5Why and PDCA

Often during the production process, the brewer is posed with an issue that needs to be addressed. In some cases, the cause of the issue isn't easily noted. When this happens, the brewer needs a tool to perform the evaluation. Then, once the reason for the issue is found, adjustments can be made to solve a specific issue.

Two very useful tools include the *5 Why process* and the PDCA cycle. The 5 Why process is basically a tool to ensure that the reason for an issue is found. The process says that any investigator should ask at least 5 questions in order to find that reason.

For example, a brewery worker is walking through the brewhouse and notices another worker spreading sand on the floor. The conversation between the two might go like this:

Why are you spreading sand on the floor?
    I'm making the floor less slippery.
Why are you making the floor less slippery?
    Because I slipped when I walked by.
Why did you slip on the floor?
    Because it had oil on it.
Why did the floor have oil on it?
    Because that motor is leaking.
Why is the motor leaking?
    Because the oil fill cap is broken.

Granted this is a fairly simplistic evaluation of an issue, but it illustrates a couple of points. First, by asking "why" five times, the investigator arrived at the solution (broken oil fill cap) to a problem (oil on the floor). Now appropriate steps can be taken to address the problem. Second, the illustration reveals something else. Many workers are very well-intentioned but try to fix the problem at hand (oil on the floor) rather than seek out the real reason for the issue.

In more complex situations, it may be necessary for a brewery investigator to perform an analysis using the *PDCA cycle*. As the name implies, this is a cyclic process of improvement or implementation. The steps are:

- **Plan** – the problem is identified, and a plan constructed to address the problem.
- **Do** – the plan is implemented.
- **Check** – the validity of the plan is checked and how it works is evaluated.
- **Act** – the plan is integrated into SOPs and documentation.

When the final step in the cycle is followed, the process is not complete. Instead, the investigator moves back to the Plan stage and completes any missing specifications or any issues with implementation. A new plan to address those changes is put together. Then, the cycle is repeated.

The PDCA cycle can also be used for continuous improvement in the brewing process. For example, a brewery could implement a plan to increase wort yield. Then, after that plan was implemented and evaluated, the plan might be modified to further increase the yield. And this can continue indefinitely. The result is continuous improvement of the process.

13.4   Addressing Product Safety

Food safety and hygiene is paramount to any operation that makes and sells products that will be consumed by others. This is often referred to as wholesomeness – although the connotation of wholesomeness is that the product somehow is nutritious. This may be the case, especially when beer is the product, but that is not the intent of the word. Instead, wholesomeness is a measure of all of the practices that are in place to ensure the product being made is safe and fit for human consumption.

The biggest issue with determining the safety of a product and how fit it is for consumption is being able to verify that it has not been contaminated during production. Specifically, the product must be free of foreign objects, chemical contaminants, or pathogenic microbes.

Foreign Objects It is possible that a foreign object becomes introduced into the product during its manufacture. This could be a metal shard from a machine, a piece of glass from a bottle, a spider or other insect, or a rock or stone that somehow falls into the product. There are many locations in the process where it would be possible for a foreign object to get added into the stream. For example, a piece of glass could get introduced into a bottle if the capper accidentally hits the side of the bottle. A piece of the finish or a portion of the sealing surface could flake off and land in the beer. (Good practice when drinking beers is to not drink one where the finish, sealing surface, or lip has been damaged. Pouring the beer into a glass first is also a good idea.)

Foreign objects can also be introduced in the ingredients that come from other suppliers. For this reason alone, it is important that the brewer understands their ingredients and where they come from, works with suppliers to build a relationship of trust, and verifies the quality of the ingredients periodically. For example, it is possible that a metal shard is mixed into the malt used to make the beer. In this case, it's likely that the metal shard will be removed during the sparge step, but it is possible that it could make its way through to a consumer's glass at the end of the process.

Chemical Contaminants The job of the brewer involves about 90% cleaning, 15% paperwork, and 5% brewing. Yes, that does add up to over 100%, but the emphasis is on cleaning. Most brewers spend considerable time cleaning and sanitizing. The advent of CIP means that brewers don't have to babysit the cleaning process, but it is still a process that requires a significant amount of time. With the number of pipes, hoses, and tanks in the brewery, it is possible that one of these isn't rinsed as well as it should be. The result could be the addition of some cleaner in the finished beer. Mixproof valves, for instance, control the flow of multiple streams of liquids. With poor maintenance, it is possible that the valve leaks liquid from one stream into another.

Many of the possible chemical contaminants in the brewery are fairly toxic. After all, they are used to scrub residues off of stainless steel and kill microbes easily. They do have a flavor in large concentrations, but diluted into a large batch of beer, it might be impossible to taste that they are present. More importantly, many of the chemical contaminants can react with the flavor components in beer and end up changing the flavor of the product. Therefore, it is prudent to use the least toxic cleaners possible and to periodically test batches with laboratory analyses to ensure that no contamination exists. That, and a thorough and regular maintenance program will help ensure none of these compounds ends up in the finished product.

Pathogenic Microbes   The good news is that most pathogenic microbes that are harmful to human health tend not to survive in beer. The concentrations of alcohol and the presence of hop oil constituents (including the iso-α-acids) work as sterilizing agents for these microbes. While non-pathogenic microbes such as *lactobacillus* and *pediococcus* can be found in beer, their only damage is to sour the beer by producing production of acids and create other off-flavors that make the beer not taste good. Consumption of these microbes is not considered a human hazard (lactobacillus is one of the bacteria used to make yogurt – and exists as living bacteria in yogurt with "active cultures"). Other non-pathogenic bacteria can make their way into the finished beer, but each tends to produce an odor or flavor that would immediately signal the end of the consumer's enjoyment. For example, contamination of beer from *Megasphera cerevisiae* would be signaled with the aroma of rotten eggs and the flavor of vomit. Not many consumers would even swallow the first taste of a beer contaminated with this bacterium.

Some of the pathogenic bacteria, such as *Escherichia coli* (*E. coli*), also produce rancid odors that would make the beer very unpalatable. *Salmonella* and *E. coli* are potential issues in beer, but only when the alcohol concentration is very low (<1% ABV). Both cause health hazards if consumed, so care must be taken to avoid their entry into the product. Other potential pathogenic microbes include *cryptosporidium* and *Clostridium botulinum*. *Cryptosporidium*, which causes diarrhea, can enter the brewery through the use of untreated or un-boiled water in the brewery. For example, if a hose is cleaned and then rinsed with municipal water, traces of *cryptosporidium* could remain in the hose. If the hose is then used to transfer beer, it is likely that the microbe could end up in the finished beer. The same is true for *Clostridium botulinum*, a microbe responsible for the production of botulinum toxin. This toxin is extremely lethal. A dose of about 75 ng (1/4 the mass of a single grain of sand) gives the victim about a 50% chance of survival. This microbe is very common in soils across the globe and could enter the product stream if, for instance, unwashed and dirty strawberries were added to a beer just before packaging. Luckily, the acidity of the beer is the saving grace that should kill most microbes.

The preservative properties of beer have been a useful protection from the introduction of many of the more harmful microbes. In the United States, this is why the Food and Drug Administration (FDA) tends to allow other federal agencies to monitor the production of beer, wine, and other alcoholic beverages. However, any low- or non-alcoholic beverage is monitored by the FDA. For example, if a brewery also manufactures root beer to sell to those that are designated drivers, the product must adhere to the FDA standards for wholesomeness. These standards are much more stringent than the monitoring standards for beer production.

13.4.1 FSMA

Best practices in the production of beer should ensure the wholesomeness (or food safety) of the product. Having a stringent quality control program that ensures consistency in production will then evaluate and inform about the presence of things that may be classified as food safety hazards.

In 2011, the US Congress passed an Act to update the 1938 Food, Drug, and Cosmetic Act. This Act, known as the *Food Safety Modernization Act* or *FSMA*, gives the US Food and Drug Administration (FDA) a number of new powers in an effort to prevent and monitor the safety of the nation's food supply. The new regulations were focused primarily on the production of food and food products. And while beer is a consumable product, its inherent safety due to the presence of alcohol only delayed the application of the FSMA regulations into that industry.

There were laws that existed prior to the FSMA. However, the point to these regulations is that they work to protect the consumer from getting sick by eating or drinking something that was made. It wasn't too long ago that such protections were not in place and the consumer literally took a risk in eating a food that was produced by a company.

The Act requires that all breweries follow good manufacturing practices. This implies not only that each step in the brewing process is evaluated and followed with standard operating procedures, but that the entire process is monitored. Records should be maintained and evaluated periodically to ensure that the good manufacturing processes are being followed and that the product is safe for human consumption.

Based on the rules of the FSMA, the FDA can and should inspect every brewery periodically to verify that they are following good manufacturing processes. Their visit will include inspection of past records, current SOP evaluations, and other monitoring that would ensure food safety. The FSMA does provide that fines can be levied for those breweries that are not in compliance.

For these reasons, every brewer should be very familiar with the statutes found in the Act (available online through the FDA website). Every brewery should ensure that it is following good manufacturing practices (see Sect. 13.3.1) and that adequate records are being kept of the process. Those that aren't in compliance should seek to be online as quickly as possible.

The FSMA also requires that breweries include and follow a strict hazard analysis and risk-based preventative controls to ensure product safety. Good manufacturing practices (GMPs) do include many of the preventative controls already, such as pest control within the manufacturing facility. But, not all of the potential hazards are covered by GMPs. Therefore, a separate plan to address these is needed.

13.4.2 HACCP

One method to catch any hazards outside of those found via a thorough and complete GMP program is the application of the principles of *HACCP (hazard analysis and critical control points)* to the brewing process. This approach focuses on food safety, but the overall principles can be applied to the production of any material, product, or service by simply adjusting the definition of hazard. In general, a hazard is anything that can harm the safety of the finished product. In food safety, this could be microbial, chemical, or physical in nature. For example, a microbial hazard would be the accidental introduction of E. coli into the bottles during packaging. A chemical hazard might be the introduction of caustic cleaning solution into the product. And a physical hazard might be the presence of glass fragments in the finished beer.

HACCP evaluates the entire process stream from the inputs of ingredients coming into the brewery to the end of the stream where the consumer is sipping on the beer. Every point along that path is evaluated for a hazard to the quality of the final product. And if a hazard is identified, it is further evaluated to determine if a critical control point exists at that location. If one does exist, then monitoring the brewing process occurs at that point. If it doesn't exist, then there is no need to monitor the brewing process there.

A critical control point is a location, process, or step in the production process where some control could be made to prevent a hazard from impacting the safety of the product. Evaluation of each point in the process requires asking questions about the potential critical point to decide if any control can be had at that point. Typically, a flow chart can aid in making the decision. Questions such as "Does a hazard exist at this point?", "Could a hazard be introduced at this point?", "Will a future step reduce the hazard?", and "Does this step require control for safety?" can be helpful in determining if the point is a critical control point.

In some cases, determining a critical control point may involve determining the relative impact of the step or process on the final product quality versus the chance that the hazard will exist. In these cases, a risk assessment is performed. This bases the risk of the hazard happening versus the risk to product safety. Once performed, it may be determined that the risk of both things happening is high, and a critical control point should be considered. If the risk to both things happening is low, the step or process may not be a critical control point. The assessment comes about when the risk to one or both is intermediate. Then, a decision must be made whether to apply resources to the monitoring, recording, and reporting for the step or process.

Once the critical control points are identified, each of them is further evaluated and delineated into a SOP for monitoring and reporting. Specifically, the hazard is identified, the limits of the analysis are determined, corrective actions are identified if the limits are exceeded, and reporting and verification plan are outlined. An example of a critical control point is shown in Fig. 13.3. This example illustrates a critical control point for diacetyl in the conditioning tank.

Once the HACCP plan is in place, it must be followed. Each SOP outlined in the HACCP plan will generate data from monitoring the steps. The process and the data

Diacetyl HACCP

Process	Hazard	Standard	Corrective Action	Followup	Recording
Conditioning Tank	Diacetyl	< 0.10 ppm after 24 h in tank	add kreusen from next batch	repeat measurement 24 h after correction	Indicate value and corrective action if taken in brewer's log
			warm condition at +5°C for 24 h	repeat measurement 48 h after correction	

Record all diacetyl measurements in brewer's log and in conditioning tank log in the appropriate column. For all measurements that exceed the standard, highlight the value in the logs by drawing a box around the value. These values should be reported at the weekly review of the cellar.

Verification of the measurements by running duplicate analyses should be performed at least three times per week.

Fig. 13.3 Critical control point for diacetyl

collected need to be evaluated periodically in order to systematically assess the success of the plan. Verification that each of the corrective actions has been successful in mitigating the hazards is needed. This may be supported by further, more detailed, analyses at the critical control points to ensure that the process and its corrective actions are appropriately controlling the particular hazards. Finally, the HACCP plan itself needs to be assessed to ensure that the plan itself is working. In addition, if any changes to the process have been made since the plan was created, the periodic review allows those changes to be implemented into the plan.

CHECKPOINT 13.2

Use the Internet to look up and describe a quality assurance method that differs from the ones listed in this chapter.

Is a HACCP plan a form of quality control or quality assurance? Explain.

13.5   Sensory Analyses

While monitoring the hazards may seem like the entire process is complete, we must remember that the customer needs to be included in any quality management program. This means that the brewer needs to get information from the consumer so that it can be fed back into the system. In fact, in a HACCP plan, this information may be considered as verification of the data. (It cannot be a critical control point, because there is no corrective action that can be taken once the consumer is tasting the beer.)

One way to provide that consumer feedback is to use sensory analyses. Sensory analysis means exactly what it says: the beer is evaluated for its taste and appearance. Sensory analyses provide quality control information, product development for experimental batches, and troubleshooting for potential off-flavors. Some sensory analyses can provide basic information on consumer preferences, but they are not designed to give that information. For detailed information on customer preferences, specific testing geared to the desired outcomes must be undertaken. These tests include comparison tests using the brewer's and competitor's beers.

13.5.1 Types of Sensory Evaluations

There are many different sensory analyses that can be performed by the brewer. Each of these treats the analyzer as a variable measuring instrument. As discovered by Arthur Fox in the early 1930s, every individual is different. We all have different numbers of taste buds and the sensitivity of these buds differs. For example, 6-n-propylthiouracil is an intensely bitter compound that is only detected by some people. To others, it is tasteless. Because of their variability, the brewer must be well versed in statistical analysis, especially in the field of bias.

Standard sensory analyses include the following evaluative tests. Each is further described in some detail.

Paired Comparison Test This analysis is one of the easiest to perform. It provides information about two samples. This test can be used to determine if a preference for one or the other exists, or to determine if one has a different given flavor characteristic such as hoppiness. Each tester is only asked one question, such as "Which sample tastes more hoppy?" Statistical analysis to provide significance requires at least seven (7) evaluators. At the minimum number of tasters, however, all must agree for any significance to be found in the test. The level of agreement is less strict when more testers are used in the analysis.

Difference from Control Test This analysis is more advanced than the paired comparison test. It asks the tester to determine which of two samples has more of a flavor in it and then indicate the magnitude of the difference on a scale of 1–10. At least 20 testers are needed in order to obtain good information with statistical significance.

Triangular Test This is often one of the more common analyses employed in sensory studies. In this test, each evaluator is provided with three samples. Two of the samples are identical and the tester is asked to identify the one that is different. Comments can also be taken on why the selection was made. A rather simple test to perform, it provides usable information with only five testers. However, just as in the Paired Comparison Test, twenty or more testers will provide better statistical analysis of the results.

Duo-Trio Test A variation of the triangular test is to provide the evaluator with one sample denoted as the reference sample. Then two additional samples are provided and the evaluator must identify the one that is identical to the reference. This test allows the brewer to determine if there is a difference in flavor between the two samples. Only nine testers are needed to perform this analysis, but error is high until about 20 testers are included in the analysis.

Threshold Test This is a more advanced test that uses the triangle test method to determine the threshold at which a particular flavor can affect the flavor of the beer. It is performed by giving the tester an array of 18 samples in six columns. Each column of the array contains two identical control samples that have not been modified. The third sample in each column is the adulterated sample. The amount of the additive is increased from left to right. The tester is then asked to identify which of the three samples contains the one with the additive. The tester is also asked to give the identity of the additive flavor. This test requires at least 20 evaluators in order to give statistically useful information. Smaller numbers, such as 15 evaluators, have been shown to only provide useful information about themselves rather than about the flavor threshold of the general population.

Descriptive Test This is the most advanced test in the sensory analysis arsenal. In the test, the evaluators are provided a sample and asked to rate (typically using a Likert scale) the specific flavors or compounds in the sample. For example, the tester could be asked to analyze the diacetyl, bitterness, hoppiness, sweetness, aroma, and fruitiness of the sample. They would rate each of the categories. Because this test requires trained evaluators, it can provide statistically significant results with only three people; however, 20 or more are needed to reduce the error of the measurement.

Customer Preference Test A very simple customer preference test is to provide two kegs of equal weight to a bar-like setting. Evaluators should number at least 15–20 and should be provided with games and other social activities that are completely unstructured. The evaluators should be allowed to drink either beer and obtain additional samples as they wish. At the end of the event, the weight of the two kegs is determined. The one with the lighter weight was the beer that was more preferred.

These sensory analysis procedures were put together by an international coalition of organizations in the 1980s. Specific testing procedures and methods for the analysis of the results are found in the ASBC Methods of Analysis. Additional sensory evaluation methods can be found in other sources.

To perform any of the analyses, the brewer must take precautions to ensure that any bias is removed from the testing procedure. Bias in this type of analytical measurement occurs when the testing isn't completely random. For example, if the first beer sample in a triangle test is always the different beer and the second and third samples are always the identical beers, a bias will exist. The person who evaluates the data from the test would also have to consider whether the testers were just simply picking the first one as the different beer, or whether they actually could taste that the first one was different.

That bias can also exist in the facilities and way that the samples are presented. Thus, it is very important that the testing center be organized in such a way that the only things being evaluated are the samples. The facility itself must be:

- Easy to access and find in the brewery
- Free from noise, including background noise from the factory floor
- Free from unwanted odors such as those from nearby cafeterias
- Comfortable and well equipped
- Decorated in neutral soothing colors
- Well lit but not with overly bright lights, spotlights, or flood lamps
- Held at a constant comfortable temperature free from drafts and air conditioning noise
- Equipped with identical booths for each evaluator that provide privacy to evaluators

Bias can also exist in the evaluators themselves. Anyone with a cold, allergies, or other issues with their ability to smell should be excused from the testing. Similarly, anyone with dental issues should be asked to come back when those issues have been addressed. If visual examinations are required as part of the testing, it is important that each of the evaluators is not experiencing problems with their eyesight. Since odors can hinder any testing, it is imperative that none of the evaluators is wearing perfume, cologne, scented hand-creams, or has washed with heavily scented soaps. Finally, any evaluator that appears to be upset, sad, or otherwise emotionally compromised should be asked to reschedule their visit to complete the analysis.

So, who makes a good evaluator? Everyone can be a good evaluator of beer samples. Those that are untrained can provide valuable information, but care must be taken to use these people in sensory analyses that don't require identification of particular aromas or flavors. For example, untrained evaluators may not be able to discern a particular fusel alcohol in a beer. Trained evaluators can also be used in sensory analyses, and are required in some sensory tests. However, they should not be part of a sensory analysis that also contains untrained evaluators.

13.6   Safety in the Brewery

Quality control requires an adequate safety program for the brewery. Monitoring and measuring different parameters must include monitoring of the safety processes as well. Safety means not only the safety of the workers and the processes, but also the safety of the final product that the brewery produces.

13.6.1 Worker Safety

It is essential that every worker in the brewery is trained. The process of making beer involves, at a minimum, working with hot liquids (~100 °C, ~212 °F), electricity, natural gas, and caustic and acidic cleaning agents. While the brewing process can be fairly predictable, accidents do happen. The most common of these accidents typically results in someone getting burned from contact with a hot liquid or surface. Signage that reminds or warns people about a particularly hazardous area or process should also be plainly posted.

Any apprenticing or training in the brewery should include extensive safety training. Most people would recommend regular safety meetings and periodic safety re-training. Such practice can significantly reduce the chance of an accident. Yes, these programs require an investment of time and money, but the payback far exceeds the investment. For example, if a worker is accidentally electrocuted because the safety protocols were not known or practiced in the brewery, the cost to the brewery alone (not counting the harm done to the worker) in legal fees, judgments against the brewery if they lose a legal case, and medical expenses can be a catastrophic expense. In some cases, without proper insurance, the brewery may have to be sold to pay for the damages. With that in mind, the cost to run a safety program is a very wise investment.

Any safety training must provide the workers with standard operating procedures (SOPs) that describe in very specific steps what should be done at each stage of the brewing process. For example, when using a portable pump, the worker must understand and follow each step in the SOP in order to safely use the pump. These SOPs must be written down, posted, and distributed to each worker. It also is a good practice to remind and practice those working in the brewery about the SOPs to ensure that everyone is familiar with them. The lack of SOPs in a brewery setting is not an excusable omission from a successful operation.

In addition, every person allowed on the brewery floor must be trained in the SOP for every step in the brewing process. For larger breweries this training may be limited to the points of access for the workers. This mass training is required because it may be necessary for someone to step in and do a job that they don't typically perform. And if they aren't trained or know the SOP by heart, they probably shouldn't be stepping in to help in that area. Let's go back to our example of the portable pump. If one worker sets up and begins the transfer of hot wort to the boil kettle using a portable pump, but then leaves the area to accomplish another task elsewhere in the brewery, any worker that steps in to turn off and tear down the pump must be familiar with the SOP. It could be disastrous if an untrained worker incorrectly detached the hose to the pump without closing out the valves on either end.

Any safety training must also include the use of personal protective equipment (PPE). This category of equipment includes goggles, face shields, gloves, steel-toed boots, aprons, long pants, and other items designed to protect the wearer from hazards in the brewery. For example, under no circumstances should a worker be allowed to wear sandals while they work in the brewery. Yes, the wearer will argue that the goggles are uncomfortable, that it's too hot to wear jeans or coveralls, or that they'll only be doing the task for a couple of minutes. All of their statements may be true (likely they are just inconvenienced by the use of PPE), but the protective equipment must be worn or used in order to work in the brewery. There should never be an exception to anyone about the use of PPE.

Any safety training must include steps on what to do in case of an emergency. This includes the standard training on what to do in the event of a fire, but also should include tornado drills and active shooter drills. Moreover, and just as importantly, the training should provide every worker with the tools they need to assist others in the event that an accident occurs. For example, if a worker slips and falls down the stairs to the boil kettle, the nearby workers are likely the first on the scene. They should know how to provide at least minimal first aid, know whether it's safe to move the patient, and know how to seek assistance in the most efficient manner. They should be aware of how to crash (rapidly turn off) any process with which they are engaged in order to lend assistance (it is better to lose a batch of beer to provide assistance, comfort, and care for an injured worker, than to ignore the worker because it's time to add hops to the boil.) Some brewers may disagree with that practice, but they should not. Yes, the loss of a batch may impact the bottom line, but the safety, comfort, and health of the workforce are significantly more valuable. And every brewer would not want the bad publicity of the injured and disgruntled worker should they be ignored or placed second to the brewing process after an accident.

Trainings should address the proper use of safety equipment, such as fire blankets, fire extinguishers, eye wash stations, and safety showers. Each of these pieces of equipment should also be readily available in the brewery such that a worker would be able to obtain them for use immediately. In other words, storing all of the fire extinguishers in the main office when the fire is on the other end of the building is not good practice.

As we noted, regular trainings to go through each of the safety aspects of working in a brewery are also essential. Everyone tends to forget the little tricks and nuances of steps, and refreshing their memory is the best practice. The trainings, depending upon many different factors, should be no less than once a month. This gives time as well to providing feedback on accidents to the entire brewery. For example, an accident in a brewery can be used as a teaching tool to inform others about potential hazards. An accident can also inform everyone and potentially reveal a hazardous condition that was previously unseen.

Overall, the best practice is to promote a "culture of safety" in the operation of the brewery. Safety should be first in everyone's mind. Safety should be thought of at all times. General operations without thinking about safety should not be performed. And if every worker is thinking about safety first, the chance of an accident is greatly reduced.

Chapter Summary

Section 13.1
Quality is a subjective term that has different meanings to different people.

Because the determination of "high-quality" beer is ultimately made by the consumer, the brewer must include the customer's opinions and suggestions in the brewing process.

Section 13.2

Quality control evaluates and provides information to correct or reject a particular batch in the brewing process.

Food safety requires diligence in the identification of potential hazards in the brewery.

Charting allows brewers to visually identify measurements in the brewery and their comparison to a quality standard.

Section 13.3

Quality assurance is a system-wide approach to evaluation of the brewery and the consistent production of high-quality beer.

Good manufacturing practices help address product safety.

The PDCA cycle and the 5 Why process are very useful in identifying and addressing problems.

Section 13.4

FSMA is operated by the US Food and Drug Administration. It requires breweries to prevent the safety of its products.

HACCP analysis provides a useful quality management system for a brewery.

Section 13.5

Sensory analysis can provide useful information to the brewer about specific aspects of their beers.

Specific tests can either compare beer samples or can focus on the development of a flavor profile for the beer.

Section 13.6

Safety in the laboratory is paramount to the successful production of a consistent product and to protect workers from injury.

Questions to Consider

1. Define the term "quality" in your own words. Can you think of another word that has subjective meaning based upon who uses that word?
2. What is likely the main reason why startup microbreweries fail?
3. A brewery is considering whether to conduct a sensory analysis using only its employees. Describe the pros and cons of doing this.
4. Describe the differences between a quality control program and a quality assurance program.
5. Can a quality control program be sufficient to operate a brewery? Why or why not?
6. Outline a decision tree that would provide information on whether a particular process or step could be a critical control point.
7. Explain why an evaluator wearing heavy cologne could cause a bias in a sensory analysis.

8. Which of the sensory analyses would be best if the brewer wanted to verify that the current batch of beer was the same as a previously made batch? Explain.
9. Which of the sensory analyses would be best if the brewer wanted to identify why the current batch of beer was different than a previously made batch? Explain.
10. Why are pathogenic bacteria typically not found in beer?
11. Which sensory analysis requires the fewest number of evaluators to provide statistically useful information?
12. Describe the pros and cons for a CUSUM chart.
13. Does motivation to do a good job influence the outcome of an evaluator in a sensory analysis?
14. A particular off-flavor compound can be measured in a beer sample if its concentration is 1 ppm. This compound has a flavor threshold of 2 ppm. What should be the quality standard for this compound?
15. Would it be acceptable for a brewer to set the maximum limit for action to 0.5 ppm for the compound in question 14?
16. Assume that only 10% of all people can detect the compound in question 14. Describe the challenges in implementing a sensory analysis to check for this compound in beer.
17. A beer has an off-flavor. Describe how a brewer could use the PDCA cycle to address this issue.
18. If a future step in the brewing process will allow correction of a particular hazard, why is an earlier step that does the same thing not considered a critical control point?
19. In your own words, what is the purpose of good brewery practices?
20. How can stress testing provide information related to quality assurance?
21. What principles appear to be common among quality assurance programs?
22. Describe what is meant by a "culture of safety."
23. A particular measurement in the brewery appears to vary at each measurement equally. One day the measurement is higher than the quality standard and the next it is lower the same amount. Describe the appearance of the CUSUM chart.
24. A brewery hires a worker who is super sensitive to diacetyl. This person appears to find diacetyl in every beer sample that the brewery manufactures. Would this person be good to have on a sensory analysis? Why, or why not?
25. Use the Internet to look up Cryptosporidium. Describe where it comes from, what health hazards are associated with it and what the brewer needs to do if this organism is found in the brewery.
26. Use the Internet to determine background information on botulinum toxin. If a grain shipment was contaminated with this toxin, would it be safe to use in the brewery? Should it be used? Why or why not?

Laboratory Exercises

Turbidity in Beer

This laboratory-based experiment allows the measurement of haze in the finished beer. It may be part of a quality control analysis to determine if the beer needs to be treated with de-hazing compounds in the conditioning tank.

Equipment

Hydrazine sulfate solution – 0.24 g dissolved in 25 mL water.
Hexamethylenetetramine solution – 2.4 g dissolved in 25 mL water.
10,000 FTU solution – add both the hydrazine sulfate solution and the hexamethy-
    lenetetramine solutions together, swirl, and let sit overnight at room temperature
    in a stoppered flask.
Erlenmeyer flasks, 125 mL.
Graduated cylinders, 10 mL, 25 mL, 100 mL.
A selection of beers of varying haze.

Experiment

Create a 1000 FTU stock solution by diluting 10 mL of the 10,000 FTU solution to 100 mL using graduated cylinders. The solution should be stored in an Erlenmeyer flask and used to make standards.

Create at least four standards by dilution of the 1000 FTU stock solution. The best standards are made by diluting the amount of the stock solution indicated in the table below to 100 mL using graduated cylinders. The solution is then placed into an Erlenmeyer flask.

Amount of 1000 FTU diluted to 100 mL	FTU of resulting solution
5 mL	50
10 mL	100
20 mL	200
50 mL	500

The solutions are then placed into a cuvette and their absorbance measured at 580 nm. A plot of the absorbance (y-axis) and the FTU value (x-axis) is created and the equation of a straight line determined. This equation is then used to determine the FTU of a series of beer samples.

To measure the beer samples, a measurement of the turbidity is taken by adding the beer at room temperature to the cuvette and the absorbance at 580 nm recorded. This measurement is indicative of the permanent haze in the beer. A second measurement is taken at 0 °C by cooling the beer in an ice bath immediately prior to the measurement at 580 nm.

Do the beer samples contain permanent haze? ...chill haze?

Appendices

Appendix A – Math for the Brewer

Abstract Herein you'll find a handy reference to a variety of formulas you might need on brew day.

Keywords Formulae, Design

A.0   Introduction

This is a reference section for designing, and estimating parameters of a brew. Keep in mind that while there is considerable science in brewing, there is also room for the art of brewing. There are a lot of things we can control, and there are a lot of things we understand, through science, that the brewers of yesteryear did not understand. But there is still enough variability in the overall process that even the brewer's mood might influence the final product – the "art."

A.1   Designing Your Brew

A beer "style" will define things like the starting, original gravity (OG), and the relative percentage of each ingredient's contribution to the gravity. The "style" might also specify the overall hop character and bitterness in the product. So, in this section we will start with a simple "style" and use it as an example to illustrate how to calculate the needed materials. Our example style is defined as the following:

- Grain bill:
  - 85% pale 2-row
  - 10% crystal 20 L
  - 5% honey
- OG 1.085
- Hops:
  - Flowery, spicy, and mildly citrusy. e.g., Cascade or Centennial
- Bitterness 30IBU

M. Mosher, K. Trantham, *Brewing Science: A Multidisciplinary Approach*,
https://doi.org/10.1007/978-3-030-73419-0

A.1.1 Volume

Since all calculations depend on the volume, we must first decide how much beer we're making. A typical home-brew batch size might be 5 gallons, or about 19 l. When considering our batch volume, we also need to account for system losses after the boil, such as the volume of bitter wort left at the bottom of the boil which cannot be separated from the "muck" – the residual hops and hot-break proteins. So, we define the following variables

- V_{Target} = the target volume of beer we want to put into the fermenter. For the example, we'll use 19 liters.
- L_{HH} = the volume of post boil, bitter wort which cannot be easily separated from the spent hops and hot-break. With care, one should be able to keep this to less than 1 liter.

So, for the example at hand, we'll use 20 liters when calculating expected gravities and bitterness concentrations.

This does not mean that we'll only need 20 liters of water on brew day as there are other losses which much be accounted, such as evaporation of water during the boil. You never want to cover a wort boil to minimize this loss. (DMS!) Typical losses are about 2 liters per hour of boil. So,

- L_{evap}= losses to evaporation in the wort boil.

Again, there is room for art here. A faster, more vigorous boil will enhance Milliard reactions but will incur a higher loss rate. So, for a typical 1-hour boil, we might need a volume of 22 liters into the boil kettle.

Finally, there are losses to the grains after the mash and sparging. This last account of water will depend on the sparging method. Continuous sparging will require more water since there is always a full volume of water in the lauter tun. Batch sparging simply drains, refills, and drains (recall first and second runnings). This sparging method is easiest to account for water without knowing the lauter tun volume. The spent grains we toss out are not dry. Like a sponge, we do our best to drain the lauter tun but the grain will hold back some water. Then, we need to factor in system losses throughout the process. So these water losses in a home brewing setup using batch sparging will include:

- L_{grain} = loss to grain. This is the water that is left in the grain after sparging. We can start with a guess 0.5 of liter per kg of grain used. As we'll see in a moment, a round number for the grain bill might be 10 kg.
- L_{DV} = other losses in the mash/sparging due to dead volume. For example, the false bottom in a mash tun as well as hoses. Again, this depends on system, but for discussion purposes, we'll assume 1 liter.

Accounting for all of these losses, the starting volume of water one might need is determined from

$$V_{\text{Start}} = V_{\text{Target}} + L_{\text{grain}} + L_{\text{evap}} + L_{\text{HH}} + L_{\text{DV}} \tag{A.1}$$

Using the example volumes and assumptions above, we will need at least

$$V_{\text{Start}} = 19l + \left(0.5\frac{1}{\text{kg}} \cdot 10\text{kg}\right) + \left(2\frac{1}{\text{hr}} \cdot 1\text{hr}\right) + 1l + 1l = 28l$$

on brew day. Again, every system and brewer is different so this is just an estimate using reasonable assumptions. It is generally wise to add another 10–15% to this volume to be sure.

A.1.2 Designing the Grain Bill

Recall that we measure sugar concentration by measuring density. Density is defined as

$$\rho = \text{Mass} / \text{Volume} \tag{A.2}$$

In brewing, we measure specific gravity ("gravities" for short) which are simply a density measurement relative to the density of water. Since water has a specific gravity of 1.000, an OG of 1.085 means that there is 0.085 "gravity points" of fermentable sugar to be dissolved *per liter of water*. We approach the grain bill design through the use of "total gravity points" (Gpts), defined as

$$\text{Gpts} = (\text{OG} - 1) \cdot 1000 \cdot \text{Volume}, \tag{A.3}$$

or, using our example,

$$\text{Gpts} = 85\frac{\text{Gpts}}{\text{liter}} \cdot 20\text{liter} = 1700\text{Gpts}. \tag{A.4}$$

Note that this is essentially the total mass, in some undefined units, of fermentable sugars to be dissolved in water. We will determine each ingredient's contribution to the gravity, i.e. "extract potential," <u>relative to sucrose</u>. If we added pure sucrose to water, we would expect 384 gravity points per kilogram of sugar per liter of water (see Table 6.2).

So, to determine the gravity points that a given ingredient will contribute all we need to do is multiply by the extract potential of each grain. The extract potential of various malted grains is usually available from the malt vendor. However, most are in the vicinity of 70% and this will be a good number to use as an approximation.

The other thing we need to consider is brew house efficiency. This is something that must be determined empirically. Actual efficiency must be determined several times before one really knows how efficient a particular process or equipment arrangement is. That said, a good starting point is to estimate an efficiency of 70%. This is where some of the "art" of brewing comes in. There are a myriad of different equipment setup possibilities, compounded by the different approaches and methods that could be used by the brewer. Since we are making an estimate, 70% will be fine for the moment. So the gravity points contributed by each ingredient is determined from

$$\text{Gpts}_{\text{ingredient}} = \text{mass}(\text{kg}) \cdot \text{extractPotential}\% \cdot \text{efficiency}\% \cdot 384 \frac{\text{Gpts}}{\text{kg}}. \quad (A.5)$$

Rearranging,

$$\frac{\text{Gpts}_{\text{ingredient}}}{\text{extractPotential}\% \cdot \text{efficiency}\% \cdot 384 \dfrac{\text{Gpts}}{\text{kg}}} = \text{mass}(\text{kg}). \quad (A.6)$$

We will use the ingredient percentages determined by the style to determine how many gravity points each will contribute to the total. For example, since the pale ale malt is 85% of the recipe, we expect this ingredient to contribute

$$\text{Gpts}_{\text{PaleAle}} = 0.85 \cdot 1700 = 1445\text{Gpts}, \quad (A.7)$$

and from this we can find the needed mass of this recipe component. Table A.1 summarizes the rest of the calculations.

CHECKPOINT

Why was 100% used for both the extract potential and efficiency for the honey?

A.1.3 Hops

Beer bitterness is measured in International Bitterness units (IBUs). We can estimate this if we know the alpha-acid content of the hops, and something called "utilization." Utilization basically accounts for the rate at which the alpha-acid is isomerized during the boil, and thus solubilized into the wort. The formula for predicting the IBU is

$$\text{IBU} = \frac{10 \cdot m \cdot \%AA \cdot \text{Utl}}{\text{Volume}}. \quad (A.8)$$

with m the mass in grams of hops, $\%AA$ the percent of alpha acid in the hops, and Utl the (decimal) utilization. This number depends primarily on boil time, but some (reference) argue that it also depends on the original gravity as well. Utilization can be calculated (reference); however given that our IBU calculation is an estimation, it is far easier to look up the utilization from a table or a graph with very little loss of accuracy (Fig. A.1).

So, continuing with the above example, the question might be: how much Cascade hops (at 7% alpha acid) do we need to hit the target 30IBU? We are boiling for 1 h. Our OG = 1.085 is not on the graph, but we can extrapolate between the 1.07

Table A.1 Grain bill assumptions and calculated mass

Ingredient	Style%	Gpts	extPot%	Eff%	Required mass (kg)
Pale ale	85	1445	70	70	7.68
Crystal 10 L	10	170	70	70	0.90
Honey	5	85	100	100	0.22

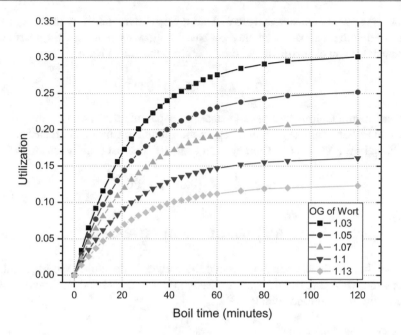

Fig. A.1 Hop utilization as a function of boil time and original gravity (OG). Data from (reference)

and 1.1 curve. Looking at the 60 minute mark, we can estimate a utilization of Utl = 0.17:

$$30\text{IBU} = \frac{10 \cdot m \cdot 7 \cdot 0.17}{20\text{liters}} \tag{A.9}$$
$$m = 50.4\text{grams}$$

A.1.4 Percent Alcohol by Volume (ABV)

Estimating the percent alcohol by volume (ABV) is as simple as measuring your original gravity (OG) (just before fermentation) and final gravity (FG), taken at the end of the fermentation process:

$$\text{ABV}(\%) = (\text{OG} - \text{FG}) \cdot 131 \tag{A.10}$$

One can estimate the FG based on the type of yeast used and the starting OG. Yeasts are characterized by their apparent attenuation. By definition, the apparent attenuation is

$$\text{AA} = \frac{(\text{OG} - \text{FG})}{(\text{OG} - 1)} \tag{A.11}$$

Actual attenuation is different from apparent attenuation. Attenuation basically refers to how much sugars the yeast ate. An (actual) attenuation of 75% means that one-quarter of the original sugars remain; 50%, half. Because the hydrometer

responds to *overall* density, and alcohol has a very different density than sugar solution, we define an apparent attenuation based on gravity readings. So, rearranging
the above we can estimate the FG (and therefore the ABV) from

$$OG - \left[AA \cdot (OG - 1) \right] = FG \tag{A.12}$$

Yeast manufacturers will usually state the apparent attenuation on the package.
Continuing our example, a typical apparent attenuation might be 65%, or in the
notation above, AA = 0.65. Then, we can estimate the final gravity and ABV:

$$1.085 - \left[0.65 \cdot (0.085) \right] = FG$$
$$1.030 = FG \tag{A.13}$$

$$ABV(\%) = (1.085 - 1.030) \cdot 131 = 7.24\% \tag{A.14}$$

Note that this is an estimate to guide the brewer in designing the beer and predicting the ABV. One should always measure the actual OG and FG gravities when
computing the final (and actual) ABV.

A.1.5 Color and SRM

Measuring color SRM by the home brewer is impractical. Measuring SRM color
with samples and color cards is extremely subjective and irreproducible.
Furthermore, this depends on the color of the ambient, illuminating light. Actual
determination of color must be done analytically using absorption spectroscopy.
The currently accepted method of reporting color in beer, as with any color-involved
measurement in the industry is using CIE L*a*b* coordinates. However, this is an
unfamiliar measurement system to most homebrewers and, as such, the older SRM
color measurement system is still common. Predicting SRM color is problematic
for the homebrewer since the final color is very dependent on process and malts used.

That said, the final color can be estimated based on a recipe. The malting industry will publish the SRM color to characterize their malts. In general, the larger the
SRM, the darker the malt. There does not exist an ab initio method of translating
this information into a SRM color of the beer since it depend on a) the quantity of
grain used, b) the volume of water, and c) the overall process.

We still try to estimate the final color by borrowing the gravity units idea from
above; each component of the recipe will contribute to the color. Deemed "malt
color units" (MCUs), each ingredient will contribute its share to the total and will
depend its "color potential," amount used, and volume of water. An MCU has been
traditionally defined using

$$MCU = SRM_{Grain} \cdot \frac{Weight_{(lbs)}}{Volume_{(gallon)}}. \tag{A.15}$$

Adjusting the formula to metric units,

Table A.2 Example recipe for calculating color

Ingredient	Weight(lbs)	Mass (kg)	SRM (rating)	MCU
Pale 2-row	20.7	9.39	2	6.369
Cyrstal 30 L	3.4	1.54	30	15.692
Rye	1.9	0.86	3	0.877
Corn grits	0.79	0.36	3	0.365

$$MCU = SRM_{Grain} \cdot 8.35 \cdot \frac{Mass_{(kg)}}{Volume_{(liters)}}. \tag{A.16}$$

The unfortunate thing about MCUs is that their contribution to the final color isn't always a linear, one-to-one relationship as the gravity calculation was. This part of the calculation relies on the empirical work of others to make the final estimation. Before we illustrate each method, let's start with a recipe, Table A.2, as an example to use as we work through the math for each approach. For the recipe, we assume a volume of 6.5 gallons, or 24.6 liters as the final volume.

Equivalence Model (SRM < 10)

In this approach, we simply add the MCUs and say that the final SRM is the MCU total:

$$SRM = MCU_{Total} = \sum_{ingredient,i} MCU_i \tag{A.17}$$

This approach is really only valid for lighter colored beers, say, SRM less than 10. Adding the MCUs from the example in Table A.2 we get SRM = 23.3, so we conclude that this approach may not be the best for our example.

Linear Model (R. Mosher) (9 < SRM < 20)

This method of calculation scales and shifts the final SRM, but the model is still linear with MCU. The formula suggested for home brewers is

$$SRM = 0.2 \cdot MCU_{Total} + 8 \tag{A.18}$$

and inserting the above recipe example, an SRM = 11.69 is predicted. It cannot be stressed enough that the final color of the beer is very much process dependent. So, these calculations are, at best, good to two digits. Thus, the example gives SRM = 12. The issue with this approach is that it gives unrealistic SRM color for small MCU totals. Thus, the limited range of validity with this method.

Power Law (Morley & Daniels)

This approach tries to account for the non-linearity in the observed data. The formula:

$$SRM = 1.49 \cdot MCU_{total}^{0.69} \tag{A.19}$$

and using the example recipe, this formula predicts an SRM = 13. There is another way in which this approach could be interpreted. Let's assume that the SRM contribution component of each ingredient is linear, but the MCU conversion to SRM for each follows the above power law. So, then the final contributes

$$\text{SRM} = \sum_{\text{ingredient},i} 1.49 \cdot \text{MCU}_i^{0.69} \tag{A.20}$$

Following this approach, the color prediction for the example recipe is SRM = 17.

So which of the above methods are "correct"? Without an absorption spectrometer it is impossible to verify any of the above calculations. These only serve as estimations to guide the brewer.

A.2.1 Misc-Strike Water Temperature

Chapter 8 introduced us to the science behind the energy balance equations to determine strike water temperature. In summary, we just set the thermal energy gained by the grains and any water that is already there equal to the energy loss by the hot, strike water. In this example, we assume a single infusion mash in which only one addition of hot strike water added to dry grains. For the example, we'll also include the thermal mass of a Styrofoam mash tun.

The change in heat energy for the system looks like

$$\Delta Q_{\text{water}} + \Delta Q_{\text{grain}} + \Delta Q_{\text{tun}} = 0 \tag{A.21}$$

And inserting specific heats and temperatures, this is

$$m_{\text{water}} \cdot c_{\text{water}} \cdot \left(T_f - T_i\right)_{\text{water}} + m_{\text{grain}} \cdot c_{\text{grain}} \cdot \left(T_f - T_i\right)_{\text{grain}} + m_{\text{tun}} \cdot c_{\text{tun}} \cdot \left(T_f - T_i\right)_{\text{tun}} = 0 \tag{A.22}$$

The density of water is $\rho = 1000 \text{ kg/m}^3$, or 1 kg/liter. Inserting this, and the specific heats for all three components (see Table 8.3):

$$V_{\text{water}} \cdot 1\frac{\text{kg}}{\text{l}} \cdot 4184\frac{\text{J}}{\text{kg}^\circ\text{C}} \cdot \left(T_f - T_i\right)_{\text{water}}$$

$$+ m_{\text{grain}} \cdot 1674\frac{\text{J}}{\text{kg}^\circ\text{C}} \cdot \left(T_f - T_i\right)_{\text{grain}}$$

$$+ m_{\text{tun}} \cdot 1300\frac{\text{J}}{\text{kg}^\circ\text{C}} \cdot \left(T_f - T_i\right)_{\text{tun}} \tag{A.23}$$

$$= 0$$

The variable T_f is the final equilibrium temperature of the system, and is the desired mash temperature; $T_f = T_{\text{mash}}$. The variable T_i for the grain and the mash tun is the beginning temperature, say room temperature $T_i = 20°C$. The variable T_i for the water is the desired strike temperature; we'll call it T_{strike}:

$$V_{water} \cdot 1\frac{kg}{l} \cdot 4184\frac{J}{kg\,^\circ C} \cdot \left(T_{mash} - T_{strike}\right)_{water}$$

$$+m_{grain} \cdot 1674\frac{J}{kg\,^\circ C} \cdot \left(T_{mash} - 20\,^\circ C\right)_{grain} \tag{A.24}$$

$$+m_{tun} \cdot 1300\frac{J}{kg\,^\circ C} \cdot \left(T_{mash} - 20\,^\circ C\right)_{tun}$$

$$= 0$$

So now let's apply some specific numbers as an example. This example will also take us through some units conversions. A brewer desires a mash temperature of 155 °F (68 °C) with 8 lbs (3.6 kg) of grain using a water to grist ratio of 1.4 qts/lb. The mash tun is a 3 lb. (1.4 kg) container. The volume of water expected in the mash tun is

$$8 lbs \cdot 1.4\frac{qts}{lb} = 11.2 qts \cdot \frac{1 liter}{1.06 qts} = 10.6 liter \tag{A.25}$$

Putting this information in,

$$10.6 l \cdot 1\frac{kg}{l} \cdot 4184\frac{J}{kg\,^\circ C} \cdot \left(68\,^\circ C - T_{strike}\right)_{water}$$

$$+3.6 kg \cdot 1674\frac{J}{kg\,^\circ C} \cdot \left(68\,^\circ C - 20\,^\circ C\right)_{grain} \tag{A.26}$$

$$+1.4 kg \cdot 1300\frac{J}{kg\,^\circ C} \cdot \left(68\,^\circ C - 20\,^\circ C\right)_{tun}$$

$$= 0$$

$$44350 \cdot \left(68\,^\circ C - T_{strike}\right)$$

$$+289267$$

$$+87360 \tag{A.27}$$

$$= 0$$

or

$$3015800 \cdot -44350 T_{strike} = -376627$$

$$T_{strike} = 76.5\,^\circ C \tag{A.28}$$

References

Daniels, Ray. Designing Great Beers – TheUltimate Guide to Brewing Classic Beer Styles, Brewers Publications, 1996.

Tinseth, Glenn. The Hop Page, http://realbeer.com/hops/, as of June 1997.

Appendix B – R134a Refrigerant Data

Abstract R134a - Tetrafluoroethane data including specific volume (density), specific energy, specific enthalpy, and specific entropy.

Keywords R134a data

B.0　Introduction

Refrigerant R134a – Tetrafluoroethane data including specific volume (density), specific energy, specific enthalpy, and specific entropy. Source of data: NIST Chemistry WebBook - http://webbook.nist.gov/chemistry/fluid/

B.1　Saturated, Organized by Temperature

Pressure	Temp	volume (m³/kg)		Energy (kJ/kg)		Enthalpy (kJ/kg)		Entropy (kJ/kg)	
kPa	°C	v_f	v_g	u_f	u_g	h_f	h_g	s_f	s_g
−40	51.2	7.05E-04	0.3611	-0.04	207.37	0.00	225.86	0.0000	0.9687
−36	62.9	7.11E-04	0.2977	4.99	209.66	5.04	228.39	0.0214	0.9632
−32	76.7	7.17E-04	0.2473	10.05	211.96	10.10	230.92	0.0425	0.9582
−28	92.7	7.23E-04.	0.2068	15.13	214.26	15.20	233.43	0.0634	0.9536
−26	101.7	7.27E-04	0.1896	17.69	215.41	17.76	234.68	0.0738	0.9515
−24	111.3	7.30E-04	0.1741	20.25	216.55	20.33	235.93	0.0841	0.9495
−22	121.7	7.33E-04	0.1601	22.82	217.70	22.91	237.17	0.0944	0.9476
−20	132.7	7.36E-04	0.1474	25.39	218.85	25.49	238.41	0.1046	0.9457
−18	144.6	7.40E-04	0.1359	27.98	219.99	28.09	239.64	0.v	0.9440
−16	157.3	7.43E-04	0.1255	30.57	221.13	30.69	240.87	0.1249	0.9423
−14	170.8	7.46E-04	0.1161	33.17	222.27	33.30	242.09	0.1350	0.9407
−12	185.2	7.50E-04	0.1074	35.78	223.41	35.92	243.31	0.1451	0.9392
−10	200.6	7.54E-04	0.0996	38.40	224.54	38.55	244.52	0.1550	0.9377
−8	216.5	7.57E-04	0.0924	41.03	225.67	41.19	245.72	0.1650	0.9364
−6	234.3	7.6E-04	0.0859	43.67	226.80	43.84	246.92	0.1749	0.9351
−4	252.7	7.65E-04	0.0799	46.31	227.93	46.50	248.11	0.1848	0.9338
−2	272.2	7.68E-04	0.0744	48.97	229.05	49.17	249.29	0.1946	0.9326
0	292.8	7.72E-04	0.0693	51.63	230.17	51.86	250.46	0.2044	0.9315
2	314.6	7.76E-04	0.0647	54.30	231.28	54.55	251.62	0.2142	0.9304
4	337.7	7.80E-04	0.0604	56.99	232.39	57.25	252.78	0.2239	0.9294
6	362.0	7.85E-04	0.0564	59.68	233.49	59.97	253.92	0.2336	0.9284
8	387.6	7.89E-04	0.0528	62.39	234.58	62.69	255.05	0.2432	0.9274
12	443.0	7.97E-04	0.0463	67.83	236.76	68.19	257.29	0.2625	0.9256
16	504.3	8.07E-04	0.0408	73.32	238.91	73.73	259.47	0.2816	0.9240
20	571.7	8.16E-04	0.0360	78.86	241.02	79.32	261.60	0.3006	0.9224
24	645.8	8.26E-04	0.0319	84.45	243.11	84.98	263.68	0.3196	0.9210
26	685.4	8.31E-04	0.0300	87.26	244.13	87.83	264.70	0.3290	0.9203

Pressure	Temp	volume (m³/kg)		Energy (kJ/kg)		Enthalpy (kJ/kg)		Entropy (kJ/kg)	
kPa	°C	v_f	v_g	u_f	u_g	h_f	h_g	s_f	s_g
28	726.9	8.37E-04	0.0283	90.09	245.15	90.70	265.69	0.3385	0.9196
30	770.2	8.42E-04	0.0266	92.93	246.16	93.58	266.67	0.3479	0.9189
32	815.4	8.48E-04	0.0251	95.79	247.15	96.48	267.64	0.3573	0.9182
34	862.6	8.54E-04	0.0237	98.66	248.13	99.40	268.58	0.3667	0.9175
36	911.9	8.60E-04	0.0224	101.55	249.10	102.33	269.50	0.3761	0.9168
38	963.2	8.66E-04	0.0211	104.46	250.05	105.29	270.41	0.3855	0.9162
40	1016.6	8.72E-04	0.0200	107.38	250.99	108.27	271.38	0.3949	0.9155
42	1072.2	8.79E-04	0.0189	110.32	251.91	111.26	272.14	0.4043	0.9147
44	1130.1	8.85E-04	0.0178	113.28	252.81	114.28	272.97	0.4136	0.9140
48	1252.9	9.00E-04	0.0160	119.26	254.56	120.39	274.55	0.4324	0.9125
52	1385.4	9.15E-04	0.0143	125.33	256.23	126.60	276.01	0.4513	0.9108
56	1528.2	9.32E-04	0.0128	131.49	257.19	132.92	277.32	0.4702	0.9089
60	1681.8	9.50E-04	0.0114	137.76	259.24	139.36	278.49	0.4892	0.9068
70	2116.8	1.00E-03	0.0087	154.01	262.19	156.14	280.51	0.5376	0.9000
80	2633.2	1.08E-03	0.0064	171.41	263.69	174.25	280.67	0.5880	0.8894
90	3244.2	1.19E-03	0.0046	190.91	262.30	194.78	277.27	0.6434	0.8706
100	3972.4	1.54E-03	0.0027	219.05	248.89	225.15	259.54	0.7232	0.8153
101.06	4059.1	1.95E-03	0.0020	233.57	233.57	241.49	241.49	0.7665	0.7665

B.2   Saturated, Organized by Pressure

Pressure	Temp	volume (m³/kg)		energy (kJ/kg)		enthalpy (kJ/kg)		entropy (kJ/kg·K)	
kPa	°C	v_f	v_g	u_f	u_g	h_f	h_g	s_f	s_g
60	−36.9	7.10E-04	0.3112	3.8	209.1	3.9	227.8	0.0164	0.9645
80	−31.1	7.19E-04	0.2376	11.2	212.5	11.2	231.5	0.0472	0.9572
100	−26.4	7.26E-04	0.1926	17.2	215.2	17.3	234.5	0.0720	0.9519
120	−22.3	7.32E-04	0.1621	22.4	217.5	22.5	237.0	0.0928	0.9478
140	−18.8	7.38E-04	0.1402	27.0	219.6	27.1	239.2	0.1110	0.9446
160	−15.6	7.44E-04	0.1235	31.1	221.4	31.2	241.1	0.1270	0.9420
180	−12.7	7.49E-04	0.1104	34.9	223.0	35.0	242.9	0.1415	0.9397
200	−10.1	7.53E-04	0.0999	38.3	224.5	38.5	244.5	0.1547	0.9378
220	−7.6	7.58E-04	0.0912	41.5	225.9	41.7	245.9	0.1668	0.9361
240	−5.4	7.62E-04	0.0839	44.5	227.2	44.7	247.3	0.1780	0.9347
260	−3.2	7.66E-04	0.0777	47.3	228.4	47.5	248.6	0.1885	0.9333
280	−1.2	7.70E-04	0.0724	50.0	229.5	50.2	249.7	0.1984	0.9322
300	0.7	7.74E-04	0.0677	52.5	230.5	52.8	250.9	0.2077	0.9311
320	2.5	7.77E-04	0.0636	54.9	231.5	55.2	251.9	0.2165	0.9301
340	4.2	7.81E-04	0.0600	57.3	232.5	57.5	252.9	0.2248	0.9293
360	5.8	7.84E-04	0.0567	59.5	233.4	59.8	253.8	0.2328	0.9284
400	8.9	7.91E-04	0.0512	63.7	235.1	64.0	255.6	0.2477	0.9270
500	15.7	8.06E-04	0.0411	73.0	238.8	73.4	259.3	0.2803	0.9241
600	21.6	8.20E-04	0.0343	81.0	241.9	81.5	262.4	0.3081	0.9219

Pressure	Temp	volume (m³/kg)		energy (kJ/kg)		enthalpy (kJ/kg)		entropy (kJ/kg·K)	
kPa	°C	v_f	v_g	u_f	u_g	h_f	h_g	s_f	s_g
700	26.7	8.33E-04	0.0294	88.3	244.5	88.8	265.1	0.3324	0.9200
800	31.3	8.46E-04	0.0256	94.8	246.8	95.5	267.3	0.3541	0.9184
900	35.5	8.58E-04	0.0227	100.9	248.9	101.6	269.3	0.3739	0.9170
1000	39.4	8.70E-04	0.0203	106.5	250.7	107.4	271.0	0.3920	0.9157
1200	46.3	8.94E-04	0.0167	116.7	253.8	117.8	273.9	0.4245	0.9131
1400	52.4	9.17E-04	0.0141	126.0	256.4	127.3	276.2	0.4533	0.9106
1600	57.9	9.40E-04	0.0121	134.5	258.5	136.0	277.9	0.4792	0.9080
1800	62.9	9.64E-04	0.0106	142.4	260.2	144.1	279.2	0.5031	0.9051
2000	67.5	9.89E-04	0.0093	149.8	261.6	151.8	280.1	0.5252	0.9020
2500	77.6	1.06E-03	0.0069	167.1	263.5	169.7	280.9	0.5755	0.8925
3000	86.2	1.14E-03	0.0053	183.1	263.4	186.6	279.2	0.6215	0.8792

B.3 Superheated Vapor

Temperature in parenthesis is the saturation temperature at the indicated pressure.

	P = 0.06 MPa (-36.9 C)					P = 0.10 MPa (-26.4 C)			
Temp	volume	energy	enthalpy	entropy	Temp	volume	energy	enthalpy	entropy
°C	v (m³/ kg)	u (kJ/ kg)	h (kJ/ kg)	s (kJ/ kg.K)	°C	v (m³/ kg)	u (kJ/ kg)	h (kJ/ kg)	s (kJ/ kg.K)
−20	0.3361	220.6	240.8	1.018	−20	0.1984	219.7	239.5	0.972
−10	0.3505	227.6	248.6	1.048	−10	0.2074	226.8	247.5	1.003
0	0.3648	234.7	256.5	1.077	0	0.2163	234.0	255.6	1.033
10	0.3789	241.9	264.7	1.107	10	0.2251	241.3	263.8	1.063
20	0.3930	249.4	272.9	1.135	20	0.2337	248.8	272.2	1.092
30	0.4071	257.0	281.4	1.164	30	0.2423	256.5	280.7	1.120
40	0.4210	264.7	290.0	1.192	40	0.2509	264.3	289.3	1.149
50	0.4350	272.6	298.7	1.219	50	0.2594	272.2	298.2	1.176
60	0.4488	280.7	307.7	1.246	60	0.2678	280.4	307.1	1.204
70	0.4627	289.0	316.8	1.273	70	0.2763	288.6	316.3	1.231
80	0.4765	297.4	326.0	1.300	80	0.2847	297.1	325.6	1.257
90	0.4903	306.0	335.4	1.326	90	0.2930	305.7	335.0	1.284
100	0.5041	314.8	345.0	1.352	100	0.3014	314.5	344.6	1.310
	P = 0.14 MPa (−18.8 C)					P = 0.18 MPa (−12.7 C)			
Temp	volume	energy	enthalpy	entropy	Temp	volume	energy	enthalpy	entropy
°C	v (m³/ kg)	u (kJ/ kg)	h (kJ/ kg)	s (kJ/ kg.K)	°C	v (m³/ kg)	u (kJ/ kg)	h (kJ/ kg)	s (kJ/ kg.K)
−20	0.1402	219.6	239.2	0.945	−10	0.1119	225.0	245.2	0.948
−10	0.1461	225.9	246.4	0.972	0	0.1172	232.5	253.6	0.980
0	0.1526	233.2	254.6	1.003	10	0.1224	240.0	262.0	1.010
10	0.1591	240.7	262.9	1.033	20	0.1275	247.6	270.6	1.040
20	0.1654	248.2	271.4	1.062	30	0.1325	255.4	279.3	1.069
30	0.1717	255.9	280.0	1.091	40	0.1374	263.3	288.1	1.098
40	0.1780	263.8	288.7	1.120	50	0.1423	271.4	297.0	1.126
50	0.1841	271.8	297.6	1.147	60	0.1472	279.6	306.1	1.153

60	0.1903	280.0	306.6	1.175	70	0.1520	287.9	315.3	1.181
70	0.1964	288.3	315.8	1.202	80	0.1567	296.4	324.6	1.207
80	0.2024	296.8	325.1	1.229	90	0.1615	305.1	334.1	1.234
90	0.2085	305.4	334.6	1.255	100	0.1662	313.9	343.8	1.260
100	0.2145	314.2	344.2	1.282					

P = 0.20 MPa (−10.1 C)					**P = 0.24 Mpa (−5.4 C)**				
Temp	volume	energy	enthalpy	entropy	Temp	volume	energy	enthalpy	entropy
°C	v (m³/kg)	u (kJ/kg)	h (kJ/gq)	s (kJ/kg.K)	°C	v (m³/kg)	u (kJ/kg)	h (kJ/kg)	s (kJ/kg.K)
−10					−10				
0	0.1048	232.1	253.1	0.970	0	0.0862	231.3	252.0	0.952
10	0.1096	239.7	261.6	1.001	10	0.0903	239.0	260.7	0.983
20	0.1142	247.4	270.2	1.030	20	0.0942	246.8	269.4	1.013
30	0.1187	255.2	278.9	1.060	30	0.0981	254.6	278.2	1.043
40	0.1232	263.1	287.7	1.088	40	0.1019	262.6	287.1	1.072
50	0.1277	271.2	296.7	1.116	50	0.1057	270.7	296.1	1.100
60	0.1321	279.4	305.8	1.144	60	0.1094	279.0	305.2	1.128
70	0.1364	287.7	315.0	1.171	70	0.1131	287.4	314.5	1.156
80	0.1407	296.3	324.4	1.198	80	0.1168	295.9	323.9	1.183
90	0.1451	304.9	333.9	1.225	90	0.1204	304.6	333.5	1.209
100	0.1493	313.7	343.6	1.251	100	0.1240	313.5	343.2	1.236

P = 0.28 MPa (−1.2 C)					**P = 0.32 MPa (2.5 C)**				
Temp	volume	energy	enthalpy	entropy	Temp	volume	energy	enthalpy	entropy
°C	v (m³/kg)	u (kJ/kg)	h (kJ/kg)	s (kJ/kg.K)	°C	v (m³/kg)	u (kJ/kg)	h (kJ/kg) s	(kJ/kg.K)
10	0.0765	238.3	259.7	0.968	10	0.0661	237.5	258.7	0.954
20	0.0800	246.1	268.5	0.999	20	0.0693	245.5	267.7	0.986
30	0.0834	254.1	277.4	1.029	30	0.0723	253.5	276.7	1.016
40	0.0867	262.1	286.4	1.058	40	0.0753	261.6	285.7	1.045
50	0.0900	270.3	295.5	1.086	50	0.0782	269.8	294.9	1.074
60	0.0932	278.6	304.7	1.114	60	0.0811	278.2	304.1	1.102
70	0.0964	287.0	314.0	1.142	70	0.0839	286.6	313.5	1.130
80	0.0996	295.6	323.5	1.169	80	0.0868	295.2	323.0	1.157
90	0.1028	304.3	333.1	1.196	90	0.0895	304.0	332.6	1.184
100	0.1059	313.2	342.8	1.222	100	0.0923	312.9	342.4	1.211
110	0.1090	322.2	352.7	1.248	110	0.0950	321.9	352.3	1.237
120	0.1121	331.3	362.7	1.274	120	0.0977	331.1	362.4	1.263

P = 0.40 MPa (8.9 C)					**P = 0.32 MPa (2.5 C)**				
Temp	volume	energy	enthalpy	entropy	Temp	volume	energy	enthalpy	entropy
°C	v (m³/kg)	u (kJ/kg)	h (kJ/kg)	s (kJ/kg.K)	°C	v (m³/kg)	u (kJ/kg)	h (kJ/kg)	s (kJ/kg.K)
10	0.0515	236.0	256.6	0.931	20	0.0421	242.4	263.5	0.938
20	0.0542	244.2	265.9	0.963	30	0.0443	250.8	273.0	0.970
30	0.0568	252.4	275.1	0.994	40	0.0465	259.3	282.5	1.001
40	0.0593	260.6	284.3	1.024	50	0.0485	267.7	292.0	1.031
50	0.0617	268.9	293.6	1.053	60	0.0505	276.3	301.5	1.060
60	0.0641	277.3	303.0	1.081	70	0.0524	284.9	311.1	1.088
70	0.0664	285.9	312.4	1.109	80	0.0543	293.6	320.8	1.116

80	0.0687	294.5	322.0	1.137	90	0.0562	302.5	330.6	1.144
90	0.0710	303.3	331.7	1.164	100	0.0581	311.5	340.5	1.171
100	0.0733	312.3	341.6	1.191	110	0.0599	320.6	350.6	1.197
110	0.0755	321.3	351.5	1.217	120	0.0617	329.9	360.7	1.223
120	0.0777	330.6	361.6	1.243	130	0.0635	339.3	371.0	1.249
					140	0.0653	348.8	381.5	1.275

P=0.60 MPa (21.6 C)				P=0.70 MPa (26.7 C)				
Temp °C	volume v (m^3/kg)	energy u (kJ/kg)	enthalpy h (kJ/kg)	entropy s (kJ/kg.K)	Temp °C	volume v (m^3/kg)	energy u (kJ/kg)	enthalpy h (kJ/kg)	entropy s (kJ/kg.K)
20					20				
30	0.0360	249.2	270.8	0.950	30	0.0300	247.5	268.5	0.931
40	0.0379	257.9	280.6	0.982	40	0.0317	256.4	278.6	0.964
50	0.0397	266.5	290.3	1.012	50	0.0333	265.2	288.5	0.995
60	0.0414	275.2	300.0	1.042	60	0.0349	274.0	298.4	1.026
70	0.0431	283.9	309.7	1.071	70	0.0364	282.9	308.3	1.055
80	0.0447	292.7	319.6	1.099	80	0.0378	291.8	318.3	1.084
90	0.0463	301.7	329.5	1.126	90	0.0393	300.8	328.3	1.111
100	0.0479	310.7	339.5	1.154	100	0.0406	310.0	338.4	1.139
110	0.0495	319.9	349.6	1.180	110	0.0420	319.2	348.6	1.166
120	0.0510	329.2	359.8	1.207	120	0.0434	328.6	358.9	1.192
130	0.0525	338.7	370.2	1.233	130	0.0447	338.0	369.3	1.219
140	0.0540	348.3	380.7	1.258	140	0.0460	347.7	379.9	1.244

P = 0.80 MPa (31.3 C)				P = 0.90 MPa (35.5 C)				
Temp °C	volume v (m³/kg)	energy u (kJ/kg)	enthalpy h (kJ/kg)	entropy s (kJ/kg.K)	Temp °C	volume v (m³/kg)	energy u (kJ/kg)	enthalpy h (kJ/kg)	entropy s (kJ/kg.K)
40	0.0270	254.8	276.5	0.948	40	0.0234	253.1	274.2	0.933
50	0.0285	263.9	286.7	0.980	50	0.0248	262.4	284.8	0.966
60	0.0300	272.8	296.8	1.011	60	0.0261	271.6	295.1	0.998
70	0.0313	281.8	306.9	1.041	70	0.0274	280.7	305.4	1.028
80	0.0327	290.8	317.0	1.070	80	0.0286	289.9	315.6	1.057
90	0.0339	300.0	327.1	1.098	90	0.0298	299.1	325.9	1.086
100	0.0352	309.2	337.3	1.126	100	0.0310	308.3	336.2	1.114
110	0.0364	318.5	347.6	1.153	110	0.0321	317.7	346.6	1.141
120	0.0376	327.9	358.0	1.180	120	0.0332	327.2	357.0	1.168
130	0.0388	337.4	368.5	1.206	130	0.0342	336.8	367.6	1.195
140	0.0400	347.1	379.1	1.232	140	0.0353	346.5	378.2	1.221
150	0.0411	356.9	389.8	1.258	150	0.0363	356.3	389.0	1.247
160	0.0423	366.8	400.6	1.283	160	0.0374	366.2	399.9	1.272

P = 1.00 MPa (39.4 C)				P = 1.2 MPa (46.3 C)				
Temp °C	volume v (m³/kg)	energy u (kJ/kg)	enthalpy h (kJ/kg)	entropy s (kJ/kg.K)	Temp °C	volume v (m³/kg)	energy u (kJ/kg)	enthalpy h (kJ/kg)	entropy s (kJ/kg.K)
40	0.0204	251.3	271.7	0.918	50	0.0172	257.6	278.3	0.927
50	0.0218	260.9	282.7	0.953	60	0.0184	267.6	289.6	0.961
60	0.0231	270.3	293.4	0.985	70	0.0195	277.2	300.6	0.994

Temp	v	u	h	s	Temp	v	u	h	s
70	0.0243	279.6	303.9	1.016	80	0.0205	286.8	311.4	1.025
80	0.0254	288.9	314.3	1.046	90	0.0215	296.3	322.1	1.055
					100	0.0224	305.8	332.7	1.084
90	0.0265	298.2	324.7	1.075	110	0.0233	315.4	343.4	1.112
100	0.0276	307.5	335.1	1.103	120	0.0242	325.0	354.1	1.139
110	0.0286	316.9	345.5	1.131	130	0.0251	334.8	364.9	1.166
120	0.0296	326.5	356.1	1.158	140	0.0259	344.6	375.7	1.193
130	0.0306	336.1	366.7	1.185	150	0.0268	354.6	386.7	1.219
140	0.0316	345.9	377.4	1.211	160	0.0276	364.6	397.7	1.245
150	0.0325	355.7	388.2	1.237	170	0.0284	374.8	408.8	1.270
160	0.0335	365.7	399.2	1.262	180	0.0292	385.1	420.1	1.295

P = 1.4 MPa (52.4 C)					**P = 1.6 MPa (57.9 C)**				
Temp	volume	energy	enthalpy	entropy	Temp	volume	energy	enthalpy	entropy
°C	v (m^3/kg)	u (kJ/kg)	h (kJ/kg)	s (kJ/kg.K)	°C	v (m^3/kg)	u (kJ/kg)	h (kJ/kg)	s (kJ/kg.K)
50					50				
60	0.0150	264.5	285.5	0.939	60	0.0124	260.9	280.7	0.916
70	0.0161	274.6	297.1	0.973	70	0.0134	271.8	293.3	0.954
80	0.0170	284.5	308.3	1.006	80	0.0144	282.1	305.1	0.987
90	0.0179	294.3	319.4	1.036	90	0.0152	292.2	316.5	1.019
100	0.0188	304.0	330.3	1.066	100	0.0160	302.1	327.8	1.050
110	0.0196	313.8	341.2	1.095	110	0.0168	312.1	338.9	1.080
120	0.0204	323.6	352.1	1.123	120	0.0175	322.0	350.0	1.108
130	0.0212	333.4	363.0	1.150	130	0.0182	332.0	361.1	1.136
140	0.0219	343.4	374.0	1.177	140	0.0189	342.1	372.3	1.163
150	0.0226	353.4	385.1	1.204	150	0.0195	352.2	383.5	1.190
160	0.0234	363.5	396.2	1.230	160	0.0202	362.4	394.7	1.216
170	0.0241	373.8	407.4	1.255	170	0.0208	372.7	406.0	1.242
180	0.0248	384.1	418.8	1.281	180	0.0215	383.1	417.4	1.268

P = 1.8 MPa (62.9 C)					**P = 2.0 MPa (67.5 C)**				
Temp	volume	energy	enthalpy	entropy	Temp	volume	energy	enthalpy	entropy
°C	v (m³/kg)	u (kJ/kg)	h (kJ/kg) s	(kJ/kg.K)	°C	v (m³/kg)	u (kJ/kg)	h (kJ/kg)	s (kJ/kg.K)
70	0.0113	416.7	437.1	1.729	70	0.0096	412.9	432.1	1.709
80	0.0123	427.6	449.7	1.766	80	0.0105	424.7	445.8	1.748
90	0.0131	438.1	461.6	1.799	90	0.0114	435.7	458.4	1.783
100	0.0139	448.3	473.3	1.831	100	0.0121	446.2	470.5	1.816
110	0.0146	458.5	484.7	1.861	110	0.0128	456.6	482.2	1.847
120	0.0152	468.6	496.0	1.890	120	0.0134	466.9	493.8	1.877
130	0.0159	478.7	507.3	1.918	130	0.0141	477.2	505.3	1.906
140	0.0165	488.9	518.6	1.946	140	0.0146	487.5	516.8	1.934
150	0.0171	499.1	529.9	1.973	150	0.0152	497.8	528.2	1.961
160	0.0177	509.4	541.3	2.000	160	0.0158	508.2	539.7	1.988
170	0.0183	519.8	552.7	2.026	170	0.0163	518.7	551.3	2.015
180	0.0189	530.3	564.3	2.051	180	0.0168	529.2	562.9	2.041

Index